D0083107

Statistical Models

This lively and engaging textbook explains the things you have to know in order to read empirical papers in the social and health sciences, as well as techniques you need to build statistical models of your own. The author, David A. Freedman, explains the basic ideas of association and regression, and takes you through the current models that link these ideas to causality.

The focus is on applications of linear models, including generalized least squares and two-stage least squares, with probits and logits for binary variables. The bootstrap is developed as a technique for estimating bias and computing standard errors. Careful attention is paid to the principles of statistical inference. There is background material on study design, bivariate regression, and matrix algebra. To develop technique, there are computer labs, with sample computer programs. The book is rich in exercises, most with answers.

Target audiences include undergraduates and beginning graduate students in statistics, as well as students and professionals in the social and health sciences. The discussion in the book is organized around published studies, as are many of the exercises. Relevant journal articles are reprinted at the back of the book.

Freedman makes a thorough appraisal of the statistical methods in these papers, and in a variety of other examples. He illustrates the principles of modeling, and the pitfalls. The book shows you how to think about the critical issues—including the connection (or lack of it) between the statistical models and the real phenomena.

Features of the book

- authoritative guide by a well-known author with wide experience in teaching, research, and consulting
- will be of interest to anyone who deals in applied statistics
- no-nonsense, direct style will appeal to both new and experienced users of statistics
- careful analysis of statistical issues that come up in substantive applications, mainly in the social and health sciences
- can be used as a text in a course, or read on its own
- developed over many years at Berkeley, thoroughly class-tested
- background material on regression and matrix algebra
- plenty of exercises
- extra material for instructors, including data sets and MATLAB code for lab projects (email to solutions@cambridge.org)

The author

David A. Freedman is Professor of Statistics at the University of California, Berkeley. He has also taught in Athens, Caracas, Jerusalem, Kuwait, London, Mexico City, and Stanford. He has written several previous books, including a widely used elementary text. He is one of the leading researchers in probability and statistics, with 150 papers in the professional literature.

He is a member of the American Academy of Arts and Sciences. In 2003, he received the John J. Carty Award for the Advancement of Science from the National Academy of Sciences, recognizing his "profound contributions to the theory and practice of statistics."

Freedman has consulted for the Carnegie Commission, the City of San Francisco, and the Federal Reserve, as well as several departments of the U.S. government. He has testified as an expert witness on statistics in law cases that involve employment discrimination, fair loan practices, duplicate signatures on petitions, railroad taxation, ecological inference, flight patterns of golf balls, price scanner errors, sampling techniques, and census adjustment.

Cover illustration

The ellipse on the cover shows the region in the plane where a bivariate normal probability density exceeds a threshold level. The correlation coefficient is 0.50. The means of x and y are equal. So are the standard deviations. The dashed line is both the major axis of the ellipse and the SD. The solid line gives the regression of y on x. The normal density (with suitable means and standard deviations) serves as a mathematical idealization of the Pearson-Lee data on heights, discussed in chapter 2. Normal densities are reviewed in chapter 3.

Statistical Models: Theory and Practice

David A. Freedman

University of California, Berkeley

CAMBRIDGE
UNIVERSITY PRESS

CAMBRIDGE UNIVERSITY PRESS
Cambridge, New York, Melbourne, Madrid, Cape Town, Singapore, São Paulo

Cambridge University Press
40 West 20th Street, New York, NY 10011-4211, USA

www.cambridge.org
Information on this title: www.cambridge.org/9780521854832

© David A. Freedman 2005

This publication is in copyright. Subject to statutory exception
and to the provisions of relevant collective licensing agreements,
no reproduction of any part may take place without
the written permission of Cambridge University Press.

First published 2005

Printed in the United States of America

A catalog record for this publication is available from the British Library.

Library of Congress Cataloging in Publication Data
Freedman, David, 1938–
Statistical models : theory and practice / David A. Freedman.
p. cm.
Includes bibliographical references and index.
ISBN-13: 978-0-521-85483-2
ISBN-10: 0-521-85483-0
1. Social sciences – Statistics – Methodology. 2. Medical
statistics – Methodology. 3. Regression analysis.
4. Statistics – Methodology. I. Title.
HA29.F678 2005
300′.1′519536 – dc22 2005047097

ISBN-13 978-0-521-85483-2 hardback
ISBN-10 0-521-85483-0 hardback

ISBN-13 978-0-521-67105-7 paperback
ISBN-10 0-521-67105-1 paperback

Cambridge University Press has no responsibility for
the persistence or accuracy of URLs for external or
third-party Internet Web sites referred to in this publication
and does not guarantee that any content on such
Web sites is, or will remain, accurate or appropriate.

Table of Contents

Preface

This book is primarily intended for advanced undergraduates or beginning graduate students in statistics. It should also be of interest to many students and professionals in the social and health sciences. Although written as a textbook, it can be read on its own. The focus is on applications of linear models, including generalized least squares, two-stage least squares, probits and logits. The bootstrap is explained as a technique for estimating bias and computing standard errors.

The contents of the book can fairly be described as what you have to know in order to start reading empirical papers that use statistical models. The emphasis throughout is on the connection—or lack of connection—between the models and the real phenomena. Much of the discussion is organized around published studies; key papers are reprinted here for ease of reference. Some may find the tone of the discussion too skeptical. If you are among them, I would make an unusual request: suspend belief until you finish reading the book. (Suspension of disbelief is all too easily obtained, but that is a topic for another day.)

The first chapter contrasts observational studies with experiments, and introduces regression as a technique that may help to adjust for confounding in observational studies. There is a chapter that explains the regression line, and another chapter with a quick review of matrix algebra. (At Berkeley, half the statistics majors need these chapters.) The going would be much easier with students who knew such material. Another big plus would be a solid upper-division course introducing the basics of probability and statistics.

Technique is developed by practice. At Berkeley, we have lab sessions where students use the computer to analyze data. There is a baker's dozen of these labs at the back of the book, with outlines for several more, and there are sample computer programs. Data are available to instructors from the publisher, along with source files for the labs and computer code: send email to solutions@cambridge.org.

A textbook is only as good as its exercises, and there are plenty of exercises in the pages that follow. Some are mathematical and some are hypothetical, but many of them are based on actual studies. That kind of exercise says, here is a summary of the data and the analysis; here is a specific issue: where do you come down? Answers to most of the exercises are at the back of the book. Beyond exercises and labs, students at Berkeley write papers during the semester. (The best are presented in class, with discussion.) Instructions for projects are also available from the publisher.

A text is defined in part by what it chooses to discuss, and in part by what it chooses to ignore; the topics of interest are not all to be covered in one book, no matter how thick. ANOVA would be natural to discuss, but ANOVA can be viewed—with only some distortion—as a special case of regression. (The ANOVA table for regression is covered in chapter 4, along with the F-test.)

Some discussion of proportional hazards would also be natural. However, logistic regression (chapter 6) is a more common technique in the biomedical literature. Furthermore, proportional-hazard models require a substantial investment in time on risk, survival curves, and hazard rates. All tradeoffs are debatable; otherwise, they wouldn't be tradeoffs. I can only plead the finitude of semesters—never mind quarters—and the necessity of examining the logic of the enterprise as well as the mechanics.

There is enough material in the book for 15–20 weeks of lectures and discussion at the undergraduate level, or 10–15 weeks at the graduate level. With undergraduates on the semester system, I cover chapters 1–6, and introduce simultaneity (sections 8.1–4). This usually takes 13 weeks. If things go quickly, I do the examples in chapter 8 and the bootstrap. During the last two weeks of the term, students present their projects. I often have a review period on the last day of class. On a quarter system with ten-week terms, I would skip the student presentations and chapters 7–8; the bivariate probit model in chapter 6 could also be dispensed with. For a graduate course, I supplement the material with additional case studies and discussion of technique.

Acknowledgements

I've taught graduate and undergraduate courses based on this material for many years at Berkeley, and on occasion at Stanford and Athens. I would like to thank the students in those courses for their help and support. I would also like to thank Dick Berk, Máire Ní Bhrolcháin, Taylor Boas, Derek Briggs, David Collier, Persi Diaconis, Thad Dunning, Mike Finkelstein, Paul Humphreys, Jon McAuliffe, Doug Rivers, Mike Roberts, David Tranah, Don Ylvisaker, and Peng Zhao, along with several anonymous reviewers, for many useful comments. Russ Lyons was incredibly helpful, and Roger Purves was a virtual coauthor.

David A. Freedman
Berkeley, California
June 2005

1

Observational Studies and Experiments

1.1 Introduction

This book is about regression models and variants like path models, simultaneous-equation models, logits and probits. Regression models can be used for different purposes:

> (i) to summarize data,
> (ii) to predict the future,
> (iii) to predict the results of interventions.

The third—causal inference—is the most interesting and the most slippery. It will be our focus. For background, this section covers some basic principles of study design.

Causal inferences are made from *observational studies*, *natural experiments*, and *randomized controlled experiments*. When using observational (non-experimental) data to make causal inferences, the key problem is *confounding*. Sometimes this problem is handled by subdividing the study population (*stratification*, also called *cross-tabulation*), and sometimes by modeling. These strategies have various strengths and weaknesses, which need to be explored.

In medicine and social science, causal inferences are most solid when based on randomized controlled experiments, where investigators assign subjects at random—by the toss of a coin—to a *treatment group* or to a *control group*. Up to random error, the coin balances the two groups with respect to all relevant factors other than treatment. Differences between the treatment group and the control group are therefore due to treatment. That is why causation is relatively easy to infer from experimental data. However, experiments tend to be expensive, and may be impossible for ethical or practical reasons. Then statisticians turn to observational studies.

In an observational study, it is the subjects who assign themselves to the different groups. The investigators just watch what happens. Studies on the effects of smoking, for instance, are necessarily observational. However, the treatment-control terminology is still used. The investigators compare smokers—the treatment group, also called the *exposed group*—with non-smokers (the control group) to determine the effect of smoking. The jargon is a little confusing, because the word "control" has two senses:

  (i) a control is a subject who did not get the treatment;
  (ii) a controlled experiment is a study where the investigators decide who will be in the treatment group.

Smokers come off badly in comparison with nonsmokers. Heart attacks, lung cancer, and many other diseases are more common among smokers. There is a strong *association* between smoking and disease. If cigarettes cause disease, that explains the association: e.g., death rates are higher for smokers because cigarettes kill. Generally, association is circumstantial evidence for causation. However, the proof is incomplete. There may be some hidden confounding factor which makes people smoke and also makes them sick. If so, there is no point in quitting: that will not change the hidden factor. Association is not the same as causation.

> Confounding means a difference between the treatment and control groups—other than the treatment—which affects the response being studied.

Typically, a confounder is a third variable, which is associated with exposure and influences the risk of disease.

Statisticians like Joseph Berkson and R. A. Fisher did not believe the evidence against cigarettes, and suggested possible confounding variables. Epidemiologists (including Richard Doll and Bradford Hill in England, as well as Wynder, Graham, Hammond, Horn, and Kahn in the United States) ran careful observational studies to show these alternative explanations were not plausible. Taken together, the studies make a powerful case that smoking causes heart attacks, lung cancer, and other diseases. If you give up smoking, you will live longer.

Epidemiological studies often make comparisons separately for smaller and more homogeneous groups, assuming that within these groups, subjects have been assigned to treatment or control as if by randomization. For example, a crude comparison of death rates among smokers and nonsmokers could be misleading if smokers are disproportionately male, because men are more likely than women to have heart disease and cancer. Gender is therefore a confounder. To control for this confounder—a third use of the word "control"—epidemiologists compared male smokers to male nonsmokers, and females to females.

Age is another confounder. Older people have different smoking habits, and are more at risk for heart disease and cancer. So the comparison between smokers and nonsmokers was made separately by gender and age: for example, male smokers age 55–59 were compared to male nonsmokers in the same age group. This controls for gender and age. Air pollution would be a confounder, if air pollution causes lung cancer and smokers live in more polluted environments. To control for this confounder, epidemiologists made comparisons separately in urban, suburban, and rural areas. In the end, explanations for health effects of smoking in terms of confounders became very, very implausible.

Of course, as we control for more and more variables this way, study groups get smaller and smaller, leaving more and more room for chance effects. This is a problem with cross-tabulation as a method for dealing with confounders, and a reason for using statistical models. Furthermore, most observational studies will be less compelling than the ones on smoking. The following (slightly artificial) example illustrates the problem.

Example 1. In cross-national comparisons, there is a striking correlation between the number of telephone lines per capita in a country and the death rate from breast cancer in that country. This is not because talking on the telephone causes cancer. Richer countries have more phones and higher cancer rates. The probable explanation for the excess cancer risk is that women in richer countries have fewer children. Pregnancy—especially early first pregnancy—is protective. Differences in diet and other lifestyle factors across countries may also play some role.

Randomized controlled experiments minimize the problem of confounding. That is why causal inferences from randomized controlled experiments are stronger than those from observational stud-

ies. With observational studies of causation, you always have to
worry about confounding. What were the treatment and control
groups? How were they different, apart from treatment? What
adjustments were made to take care of the differences? Are these
adjustments sensible?

The rest of this chapter will discuss examples: the HIP trial of mammography,
Snow on cholera, and the causes of poverty.

1.2 The HIP trial

Breast cancer is one of the most common malignancies among women in
Canada and the United States. If the cancer is detected early enough—before
it spreads—chances of successful treatment are better. "Mammography"
means screening women for breast cancer by X-rays. Does mammography
speed up detection by enough to matter? The first large-scale randomized
controlled experiment was HIP (Health Insurance Plan) in New York, followed
by the Two-County study in Sweden. There were about half a dozen other
trials as well. Some were negative (screening doesn't help) but most were
positive. By the late 1980s, mammography had gained general acceptance.

The HIP study was done in the early 1960s. HIP was a group medical
practice which had at the time some 700,000 members. Subjects in the experiment were 62,000 women age 40–64, members of HIP, who were randomized
to treatment or control. "Treatment" consisted of invitation to 4 rounds of
annual screening—a clinical exam and mammography. The control group
continued to receive usual health care. Results from the first 5 years of followup are shown in table 1. In the treatment group, about 2/3 of the women
accepted the invitation to be screened, and 1/3 refused. Death rates (per 1000
women) are shown, so groups of different sizes can be compared.

Table 1. HIP data. Group sizes (rounded), deaths in 5 years of
followup, and death rates per 1000 women randomized.

	Group size	Breast cancer No.	Breast cancer Rate	All other No.	All other Rate
Treatment					
Screened	20,200	23	1.1	428	21
Refused	10,800	16	1.5	409	38
Total	31,000	39	1.3	837	27
Control	31,000	63	2.0	879	28

Which rates show the efficacy of treatment? It seems natural to compare those who accepted screening to those who refused. However, this is an observational comparison, even though it occurs in the middle of an experiment. The investigators decided which subjects would be invited to screening, but it is the subjects themselves who decided whether or not to accept the invitation. Richer and better-educated subjects were more likely to participate than those who were poorer and less well educated. Furthermore, breast cancer (unlike most other diseases) hits the rich harder than the poor. Social status is therefore a confounder—a factor associated with the outcome and with the decision to accept screening.

The tip-off is the death rate from other causes (not breast cancer) in the last column of table 1. There is a big difference between those who accept screening and those who refuse. The refusers have almost double the risk of those who accept. There must be other differences between those who accept screening and those who refuse, in order to account for the doubling in the risk of death from other causes—because screening has no effect on this risk. One major difference is social status. It is the richer women who come in for screening. Richer women are less vulnerable to other diseases, but more vulnerable to breast cancer. So the comparison of those who accept screening with those who refuse is biased, and the bias is against screening.

Comparing the death rate from breast cancer for those who accept screening and those who refuse is *analysis by treatment received*. This analysis is seriously biased, as we have just seen. The experimental comparison is between the whole treatment group—all those invited to be screened, whether or not they accepted screening—and the whole control group. This is the *intention-to-treat analysis*.

Intention-to-treat is the recommended analysis.

HIP, which was a very well-run study, made the intention-to-treat analysis. The investigators compared the breast cancer death rate in the total treatment group to the rate in the control group, and showed that screening works.

The effect of the invitation is small in absolute terms: $63 - 39 = 24$ lives saved (table 1). Since the absolute risk from breast cancer is small, no intervention can have a large effect in absolute terms. On the other hand, in relative terms, the 5-year death rates from breast cancer are in the ratio $39/63 = 62\%$. Followup continued for 18 years, and the savings in lives persisted over that period. The Two-County study—a huge randomized controlled experiment in Sweden—confirmed the results of HIP. So did other studies in Finland, Scotland, and Sweden. That is why mammography became so widely accepted.

1.3 Snow on cholera

A *natural experiment* is an observational study where assignment to treatment or control is as if randomized by nature. In this section, we look at one of the first natural experiments. In 1855, some twenty years before Koch and Pasteur laid the foundations of modern microbiology, John Snow discovered that cholera is a waterborne infectious disease. At the time, the germ theory of disease was only one of many theories. Miasma (bad air) was often said to cause epidemics. Imbalance in the humors of the body (black bile, yellow bile, blood, phlegm) was an older explanation for disease. Poison in the ground was an explanation that came into vogue slightly later.

Snow was a physician in London. By observing the course of the disease, he concluded that cholera was caused by a living organism, which entered the body with water or food, multiplied in the body, and made the body expel water containing copies of the organism. The dejecta then contaminated food or reentered the water supply, and the organism proceeded to infect other victims. Snow explained the lag between infection and disease—a matter of hours or days—as the time needed for the infectious agent to multiply in the body of the victim. This multiplication is characteristic of life: inanimate poisons do not reproduce themselves. (Of course, they may take some time to do their evil: the lag is not compelling evidence.)

Snow developed a series of arguments in support of the germ theory. For instance, cholera spread along the tracks of human commerce. Furthermore, when a ship entered a port where cholera was prevalent, sailors contracted the disease only when they came into contact with residents of the port. These facts were easily explained if cholera was an infectious disease, but were hard to explain by the miasma theory.

There was a cholera epidemic in London in 1848. Snow identified the first or "index" case in this epidemic:

> "a seaman named John Harnold, who had newly arrived by the *Elbe* steamer from Hamburgh, where the disease was prevailing." [p. 3]

He also identified the second case: a man named Blenkinsopp who took Harnold's room after the latter died, and became infected by contact with the bedding. Next, Snow was able to find adjacent apartment buildings, one hard hit by cholera and one not. In each case, the affected building had a water supply contaminated by sewage, the other had relatively pure water. Again, these facts are easy to understand if cholera is an infectious disease—but not if miasmas are the cause.

There was an outbreak of the disease in August and September of 1854. Snow made a "spot map," showing the locations of the victims. These clustered near the Broad Street pump. (Broad Street is in Soho, London; at the time, public pumps were used as a source of drinking water.) By contrast, there were a number of institutions in the area with few or no fatalities. One was a brewery. The workers seemed to have preferred ale to water; if any wanted water, there was a private pump on the premises. Another institution almost free of cholera was a poor-house, which too had its own private pump. (Poor-houses will be discussed again, in section 4.)

People in other areas of London did contract the disease. In most cases, Snow was able to show they drank water from the Broad Street pump. For instance, one lady in Hampstead so much liked the taste that she had water from the Broad Street pump delivered to her house by carter.

So far, we have persuasive anecdotal evidence that cholera is an infectious disease, spread by contact or through the water supply. Snow also used statistical ideas. There were a number of water companies in the London of his time. Some took their water from heavily contaminated stretches of the Thames river. For others, the intake was relatively uncontaminated.

Snow made "ecological" studies, correlating death rates from cholera in various areas of London with the quality of the water. Generally speaking, areas with contaminated water had higher death rates. The Chelsea water company was exceptional. This company started with contaminated water, but had quite modern methods of purification—with settling ponds and careful filtration. Its service area had a low death rate from cholera.

In 1852, the Lambeth water company moved its intake pipe upstream to get purer water. The Southwark and Vauxhall company left its intake pipe where it was, in a heavily contaminated stretch of the Thames. Snow made an ecological analysis comparing the areas serviced by the two companies in the epidemics of 1853–54 and in earlier years. Let him now continue in his own words.

> "Although the facts shown in the above table [the ecological analysis] afford very strong evidence of the powerful influence which the drinking of water containing the sewage of a town exerts over the spread of cholera, when that disease is present, yet the question does not end here; for the intermixing of the water supply of the Southwark and Vauxhall Company with that of the Lambeth Company, over an extensive part of London, admitted of the subject being sifted in such a way as to yield the most incontrovertible proof on one side or the other. In the subdistricts enumerated in the above table as being supplied by both Companies, the mixing of the supply is of the most intimate kind. The pipes of each Company go down all the streets, and into nearly all the courts and alleys. A few houses are supplied by one Company and a few by the other, according to the decision of the owner or occupier at that time when the Water Companies were in active competition.

In many cases a single house has a supply different from that on either side. Each company supplies both rich and poor, both large houses and small; there is no difference either in the condition or occupation of the persons receiving the water of the different Companies. Now it must be evident that, if the diminution of cholera, in the districts partly supplied with improved water, depended on this supply, the houses receiving it would be the houses enjoying the whole benefit of the diminution of the malady, whilst the houses supplied with the [contaminated] water from Battersea Fields would suffer the same mortality as they would if the improved supply did not exist at all. As there is no difference whatever in the houses or the people receiving the supply of the two Water Companies, or in any of the physical conditions with which they are surrounded, it is obvious that no experiment could have been devised which would more thoroughly test the effect of water supply on the progress of cholera than this, which circumstances placed ready made before the observer.

"The experiment, too, was on the grandest scale. No fewer than three hundred thousand people of both sexes, of every age and occupation, and of every rank and station, from gentlefolks down to the very poor, were divided into groups without their choice, and in most cases, without their knowledge; one group being supplied with water containing the sewage of London, and amongst it, whatever might have come from the cholera patients; the other group having water quite free from such impurity.

"To turn this grand experiment to account, all that was required was to learn the supply of water to each individual house where a fatal attack of cholera might occur." [pp. 74–75]

Snow's data are shown in table 2. The denominator data—the number of houses served by each water company—were available from parliamentary records. For the numerator data, however, a house-to-house canvass was needed to determine the source of the water supply at the address of each cholera fatality. (The "bills of mortality," as death certificates were called at the time, showed the address but not the water source for each victim.) The death rate from the Southwark and Vauxhall water is about 9 times the death rate for the Lambeth water. Snow explains that the data could be analyzed as

Table 2. Death rate from cholera by source of water. Rate per 10,000 houses. London. Epidemic of 1853–54. Snow's table IX.

	No. of Houses	Cholera Deaths	Rate per 10,000
Southwark & Vauxhall	40,046	1,263	315
Lambeth	26,107	98	37
Rest of London	256,423	1,422	59

if they had resulted from a randomized controlled experiment: there was no difference between the customers of the two water companies, except for the water. The data analysis is simple—a comparison of rates. It is the design of the study and the size of the effect that compel conviction.

1.4 Yule on the causes of poverty

Legendre (1805) and Gauss (1809) developed regression techniques to fit data on orbits of astronomical objects. The relevant variables were known from Newtonian mechanics, and so were the functional forms of the equations connecting them. Measurement could be done with high precision. Much was known about the nature of the errors in the measurements and equations. Furthermore, there was ample opportunity for comparing predictions to reality. A century later, investigators were using regression on social science data where these conditions did not hold, even to a rough approximation—with consequences that need to be explored (chapters 4–8).

Yule (1899) was studying the causes of poverty. At the time, paupers in England were supported either inside grim Victorian institutions called "poor-houses" or outside, depending on the policy of local authorities. Did policy choices affect the number of paupers? To study this question, Yule proposed a regression equation,

$$(1) \qquad \Delta\text{Paup} = a + b \times \Delta\text{Out} + c \times \Delta\text{Old} + d \times \Delta\text{Pop} + \text{error}.$$

In this equation,

Δ is percentage change over time,
Paup is the number of paupers,
Out is the out-relief ratio N/D,
$\quad N =$ number on welfare outside the poor-house,
$\quad D =$ number inside,
Old is the population over 65,
Pop is the population.

Data are from the English Censuses of 1871, 1881, 1891. There are two Δ's, one for 1871–81 and one for 1881–91. (Error terms will be discussed later.)

Relief policy was determined separately in each "union" (an administrative district comprising several parishes). At the time, there were about 600 unions, and Yule divided them into four kinds: rural, mixed, urban, metropolitan. There are 4×2 = 8 equations, one for each type of union and time period. Yule fitted his equations to the data by least squares. That is, he determined a, b, c, and d by minimizing the sum of squared errors,

$$\sum \left(\Delta\text{Paup} - a - b \times \Delta\text{Out} - c \times \Delta\text{Old} - d \times \Delta\text{Pop}\right)^2.$$

The sum is taken over all unions of a given type in a given time period, which assumes (in effect) that coefficients are constant for those combinations of geography and time.

Table 3. Pauperism, Out-relief ratio, Proportion of Old, Population. Ratio of 1881 data to 1871 data, times 100. Metropolitan Unions, England. Yule (1899, table XIX).

	Paup	Out	Old	Pop
Kensington	27	5	104	136
Paddington	47	12	115	111
Fulham	31	21	85	174
Chelsea	64	21	81	124
St. George's	46	18	113	96
Westminster	52	27	105	91
Marylebone	81	36	100	97
St. John, Hampstead	61	39	103	141
St. Pancras	61	35	101	107
Islington	59	35	101	132
Hackney	33	22	91	150
St. Giles'	76	30	103	85
Strand	64	27	97	81
Holborn	79	33	95	93
City	79	64	113	68
Shoreditch	52	21	108	100
Bethnal Green	46	19	102	106
Whitechapel	35	6	93	93
St. George's East	37	6	98	98
Stepney	34	10	87	101
Mile End	43	15	102	113
Poplar	37	20	102	135
St. Saviour's	52	22	100	111
St. Olave's	57	32	102	110
Lambeth	57	38	99	122
Wandsworth	23	18	91	168
Camberwell	30	14	83	168
Greenwich	55	37	94	131
Lewisham	41	24	100	142
Woolwich	76	20	119	110
Croydon	38	29	101	142
West Ham	38	49	86	203

For example, consider the metropolitan unions. Fitting the equation to the data for 1871–81, Yule got

(2) ΔPaup $= 13.19 + 0.755\Delta$Out $- 0.022\Delta$Old $- 0.322\Delta$Pop $+$ error.

For 1881–91, his equation was

(3) ΔPaup $= 1.36 + 0.324\Delta$Out $+ 1.37\Delta$Old $- 0.369\Delta$Pop $+$ error.

The coefficient of ΔOut being relatively large and positive, Yule concludes that out-relief causes poverty.

Let's take a look at some of the details. Table 3 has the ratio of 1881 data to 1871 data for Pauperism, Out-relief ratio, Proportion of Old, and Population. If we subtract 100 from each entry, column 1 gives ΔPaup in the regression equation (2). Columns 2, 3, 4 give the other variables. For Kensington (the first union in the table),

$$\Delta\text{Out} = 5 - 100 = -95, \quad \Delta\text{Old} = 104 - 100 = 4, \quad \Delta\text{Pop} = 136 - 100 = 36.$$

The predicted value for ΔPaup from (2) is therefore

$$13.19 + 0.755 \times (-95) - 0.022 \times 4 - 0.322 \times 36 = -70.$$

The actual value for ΔPaup is -73. So the error is -3. As noted before, the coefficients were chosen by Yule to minimize the sum of squared errors. (In chapter 4, we will see how to do this.)

Look back at equation (2). The causal interpretation of the coefficient 0.755 is this. Other things being equal, if ΔOut is increased by 1 percentage point—the administrative district supports more people outside the poorhouse—then ΔPaup will go up by 0.755 percentage points. This is a *quantitative inference*. Out-relief causes an increase in pauperism—a *qualitative inference*. The point of introducing ΔPop and ΔOld into the equation is to control for possible confounders, implementing the idea of "other things being equal." For Yule's argument, it is important that the coefficient of ΔOut be significantly positive. With regression, the qualitative and quantitative aspects are woven together.

Quetelet (1835) wanted to uncover "social physics"—the laws of human behavior—by using statistical technique. Yule was using regression to infer the social physics of poverty. But this is not so easily to be done. Confounding is one problem. According to Pigou, a leading welfare economist of Yule's era, districts with more efficient administrations were building poor-houses and reducing poverty. Efficiency of administration is then a confounder, influencing both the presumed cause and its effect. Economics may be another confounder. Yule occasionally tries to control for this, using the rate of population change as a proxy for economic growth. Generally, however, he pays little attention to economics. The explanation:

"A good deal of time and labour was spent in making trial of this idea, but the results proved unsatisfactory, and finally the measure was abandoned altogether." [p. 253]

The form of Yule's equation is somewhat arbitrary, and the coefficients are not consistent across time and geography: compare equations (2) and (3) to see differences across time. (Differences across geography are reported in table C of Yule's paper.) The inconsistencies may not be fatal. However, unless the coefficients have some existence of their own—apart from the data—how can they predict the results of interventions that would change the data? The distinction between parameters and estimates is a basic one, and we will return to this issue several times in chapters 4–8.

There are other problems too. At best, Yule has established association. Conditional on the covariates, there is a positive association between ΔPaup and ΔOut. Is this association causal? If so, which way do the causal arrows point? For instance, a parish may choose not to build poor-houses in response to a short-term increase in the number of paupers, in which case pauperism causes out-relief. Likewise, the number of paupers in one area may well be affected by relief policy in neighboring areas. Such issues are not resolved by the data analysis. Instead, answers are assumed a priori. Yule's enterprise is substantially more problematic than Snow on cholera, or the HIP trial, or the epidemiology of smoking.

Yule was aware of the problems. Although he was busily parceling out changes in pauperism—so much is due to changes in the out-relief ratio, so much to changes in other variables, and so much to random effects—there is one deft footnote (number 25) that withdraws all causal claims: "Strictly speaking, for 'due to' read 'associated with.'"

Yule's approach is strikingly modern, except there is no causal diagram with stars to indicate statistical significance. Figure 1 brings him up to date. The arrow from ΔOut to ΔPaup indicates that ΔOut is included in the regression equation explaining ΔPaup. "Statistical significance" is indicated by an asterisk, and three asterisks signal a high degree of significance. The idea is that a statistically significant coefficient differs from zero, so that ΔOut has a causal influence on ΔPaup. By contrast, an insignificant coefficient is thought to be zero: e.g., ΔOld does not have a causal influence on ΔPaup. We return to these diagrams in chapter 5.

Figure 1. Yule's model. Metropolitan unions, 1871–81.

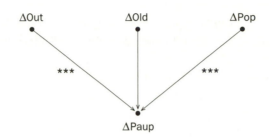

Yule could have used regression to summarize his data: for a given time period and unions of a specific type, with certain values of the explanatory variables, the change in pauperism was about so much and so much. In other words, he could have used his equations to approximate the average value of ΔPaup, given the values of ΔOut, ΔOld, ΔPop. This assumes linearity. If we turn to prediction, there is another assumption: the system will remain stable over time. Prediction is already more complicated than description. On the other hand, if we make a series of predictions and test them against data, it may be possible to show that the system is stable enough for regression to be helpful.

Causal inference is different, because a change in the system is contemplated—an intervention. Descriptive statistics tell you about the data that you happen to have. Causal models claim to tell you what will happen to some of the numbers if you intervene to change other numbers. This is a claim worth examining. Something has to remain constant amidst the changes. What is this, and why is it constant? Chapter 4 will explain how to fit regression equations like (2) and (3). Chapter 5 discusses some examples from contemporary social science, and examines the constancy-in-the-midst-of-changes assumptions that justify causal inference by statistical models. *Response schedules* will be used to formalize the constancy assumptions.

Exercise set A

1. In the HIP trial (table 1), what is the evidence confirming that treatment has no effect on death from other causes?

2. Someone wants to analyze the HIP data by comparing the women who accept screening to the controls. Is this a good idea?

3. Was Snow's study of the epidemic of 1853–54 (table 2) a randomized controlled experiment or a natural experiment? Why does it matter that the Lambeth company moved its intake point in 1852? Explain briefly.

4. Was Yule's study a randomized controlled experiment or an observational study?

5. In Yule's equation (2), suppose the coefficient of ΔOut had been -0.755. What would he have had to conclude? If the coefficient had been $+.005$?

Exercises 6–8 prepare for the next chapter. If the material is unfamiliar, you might want to read chapters 16–18 in Freedman-Pisani-Purves (1998), or similar material in another text. Keep in mind that

$$\text{variance} = (\text{standard error})^2.$$

6. Suppose $X_1, X_2, \ldots, X_n$ are independent random variables, with common expectation μ and variance σ^2. Let $S_n = X_1 + X_2 + \cdots + X_n$. Find the expectation and variance of S_n. Repeat for S_n/n.

7. Suppose $X_1, X_2, \ldots, X_n$ are independent random variables, with a common distribution: $P(X_i = 1) = p$ and $P(X_i = 0) = 1 - p$, where $0 < p < 1$. Let $S_n = X_1 + X_2 + \cdots + X_n$. Find the expectation and variance of S_n. Repeat for S_n/n.

8. What is the law of large numbers?

9. Keefe et al (2001) summarize their data as follows:

   "Thirty-five patients with rheumatoid arthritis kept a diary for 30 days. The participants reported having spiritual experiences, such as a desire to be in union with God, on a frequent basis. On days that participants rated their ability to control pain using religious coping methods as high, they were much less likely to have joint pain."

   Does the study show that religious coping methods are effective at controlling joint pain? If not, how would you explain the data?

10. According to many textbooks, association is not causation. To what extent do you agree? Discuss briefly.

1.5 End notes for chapter 1

Experimental design is a topic in itself. For instance, many experiments *block* subjects into relatively homogeneous groups. Within each group, some are chosen at random for treatment, and the rest serve as controls. *Blinding* is another important topic. Of course, experiments can go off the rails. For one example, see EC/IC Bypass Study Group (1985), with commentary by Sundt (1987) and others. The commentary makes the case that management and reporting of this large multi-center surgery trial broke down, with the result that many patients likely to benefit from surgery were operated on outside the trial and excluded from tables in the published report.

Epidemiology is the study of medical statistics. More formally, epidemiology is "the study of the distribution and determinants of health-related states or events in specified populations and the application of this study to control of health problems." See Last (2001, p. 62) and Gordis (2004, p. 3).

Health effects of smoking. See Cornfield et al (1959), International Agency for Research on Cancer (1986). For a brief summary, see Freedman (1999). There have been some experiments on smoking cessation, but these are inconclusive at best. Likewise, animal experiments can be done, but there are difficulties in extrapolating from one species to another. Critical commentary on the smoking hypothesis includes Berkson (1955) and Fisher (1959). The latter makes arguments that are almost perverse. (Nobody's perfect.)

Telephones and breast cancer. The correlation is 0.74 with 165 countries. Breast cancer death rates (age standardized) are from

http://www-dep.iarc.fr/globocan/globocan.html

Population figures, counts of telephone lines (and much else) are available at

http://www.cia.gov/cia/publications/factbook

HIP. The best source is Shapiro et al (1988). The actual randomization mechanism involved list sampling. The differentials in table 1 persist throughout the 18-year followup period, and are more marked if we take cases incident during the first 7 years of followup, rather than 5. Screening ended after 4 or 5 years and it takes a year or two for the effect to be seen, so 7 years is probably the better time period to use.

Intention-to-treat measures the effect of assignment, not the effect of screening. The effect of screening is diluted by crossover—only 2/3 of the women came in for screening. When there is crossover from the treatment arm to the control arm, but not the reverse, it is straightforward to correct for dilution. The effect of screening is to reduce the death rate from breast cancer by a factor of 2. This estimate is confirmed by results from the Two-County study. See Freedman et al (2004) for a review; correcting for dilution is discussed there, on p. 72.

Subjects in the treatment group who accepted screening had a much lower death rate from all causes other than breast cancer (table 1). Why? For one thing, the compliers were richer and better educated; mortality rates decline as income and education go up. Furthermore, the compliers probably took better care of themselves in general. See section 2.2 in Freedman-Pisani-Purves (1998); also see Petitti (1994).

Recently, questions about the value of mammography have again been raised, but the evidence from the screening trials is quite solid. For reviews, see Smith (2003) and Freedman et al (2004).

Snow on cholera. At the end of the 19th century, there was a burst of activity in microbiology. In 1878, Pasteur published *La théorie des germes et ses applications à la médecine et à la chirurgie.* Around that time, Pasteur and Koch isolated the anthrax bacillus and developed techniques for vaccination. The tuberculosis bacillus was next. In 1883, there was a cholera epidemic in Egypt, and Koch isolated the vibrio; he was perhaps anticipated by Filipo Pacini. There was another epidemic in Hamburg in 1892. The city fathers turned to Max von Pettenkofer, a leading figure in the German hygiene movement of the time. He did not believe Snow's theory, holding instead that cholera was caused by poison in the ground.

Hamburg was a center of the slaughterhouse industry: von Pettenkofer had the carcasses of dead animals dug up and hauled away, in order to reduce pollution of the ground. The epidemic continued until the city lost faith in von Pettenkofer, and turned in desperation to Koch. References on the history of cholera include Rosenberg (1962), Howard-Jones (1975), Evans (1987), Winkelstein (1995). Today, the molecular biology of the cholera vibrio is reasonably well understood. There are recent surveys by Colwell (1996) and Raufman (1998). For a synopsis, see Alberts et al (1994, pp. 484, 738).

In the history of epidemiology, there are many examples like Snow's work on cholera. For instance, Semmelweis (1860) discovered the cause of puerperal fever. There is a lovely book by Loudon (2000) that tells the history, although Semmelweiss could perhaps have been treated a little more gently. Around 1914, to mention another example, Goldberger showed that pellagra was the result of a diet deficiency. Terris (1964) reprints many of Goldberger's articles; also see Carpenter (1981).

Quetelet. A few sentences will indicate the flavor of his enterprise.

> "In giving my work the title of Social Physics, I have had no other aim than to collect, in a uniform order, the phenomena affecting man, nearly as physical science brings together the phenomena appertaining to the material world.... in a given state of society, resting under the influence of certain causes, regular effects are produced, which oscillate, as it were, around a fixed mean point, without undergoing any sensible alterations....

> "This study ... has too many attractions—it is connected on too many sides with every branch of science, and all the most interesting questions in philosophy—to be long without zealous observers, who will endeavour to carry it farther and farther, and bring it more and more to the appearance of a science." (Quetelet 1842, pp. vii, 103)

Yule. The "errors" in (1) and (2) play different roles in the theory. In (1), we have random errors which are unobservable parts of a statistical model. In (2), we have residuals which can be computed as part of model fitting; (3) is like (2). Details are in chapter 4. For sympathetic accounts of the history, see Stigler (1986) and Desrosières (1993). Meehl (1954) provides some well-known examples of success in prediction by regression. Predictive validity is best demonstrated by making real "ex ante"—before the fact—forecasts in several different contexts: predicting the future is a lot harder than fitting regression equations to the past (Ehrenberg and Bound 1993). Chapters 5 and 8 continue the discussion.

John Stuart Mill. The contrast between experiment and observation goes back to Mill (1843), as does the idea of confounding. (In the seventh edition, see Book III, Chapters VII and X, esp. pp. 423 and 503.)

Experiments vs observational studies. Fruits-and-vegetables epidemiology is a well-known case where experiments contradict observational data. In brief, the observational data say that people who eat a vitamin-rich diet get cancer at lower rates, "so" vitamins prevent cancer. The experiments say that vitamin supplements either don't help or actually increase the risk. The problem with the observational studies is that people who eat (say) five servings of fruit and vegetables every day are different from the rest of us in many other ways. It is hard to adjust for all these differences by purely statistical methods (Freedman-Pisani-Purves, 1998, p. 26 and note 23 on p. A6). More recent references include Clarke and Armitage (2002), Virtamo et al (2003). Hercberg et al (2004) get a positive effect for men not women.

Hormone replacement therapy (HRT) is another example (Petitti 1998, 2002). The observational studies say that HRT prevents heart disease in women, after menopause. The experiments show that HRT has no benefit. The women who chose HRT were different from other women, in ways that the observational studies missed. We will discuss HRT again in chapter 6.

Anecdotal evidence—based on individual cases, without a systematic comparison of different groups—is a weak basis for causal inference. If there is no control group in a study, considerable skepticism is justified, especially if the effect is small or hard to measure. When the effect is dramatic, as with penicillin for wound infection, these statistical caveats can be set aside. On penicillin, see Fleming (1947), Goldsmith (1946), Hare (1970), Waxman and Strominger (1983). Smith and Pell (2004) have a good—and brutally funny—discussion of causal inference when effects are large.

2

The Regression Line

2.1 Introduction

This chapter is about the *regression line*. The regression line is important on its own (to statisticians), and it will help us with multiple regression in chapter 4. The first example is a scatter diagram showing the heights of 1078 fathers and their sons (figure 1). Each pair of fathers and sons becomes a dot on the diagram. The height of the father is plotted on the x-axis; the height of his son, on the y-axis. The left hand vertical strip (inside the chimney) shows the families where the father is 64 inches tall to the nearest inch; the right hand vertical strip, families where the father is 72 inches tall. Many other strips could be drawn too. The regression line (solid) approximates the average height of the sons, given the heights of their fathers. This line goes through the centers of all the vertical strips. The regression line is flatter than the SD line, which is dashed. "SD" is shorthand for "standard deviation"; definitions come next.

2.2 The regression line

We have n subjects indexed by $i = 1, \ldots, n$, and two *data variables* x and y. A data variable stores a value for each subject in a study. Thus, x_i is the value of x for subject i, and y_i is the value of y. In figure 1, a "subject" is a family: x_i is the height of the father in family i, and y_i is the height of

Figure 1. Heights of fathers and sons. Pearson and Lee (1903).

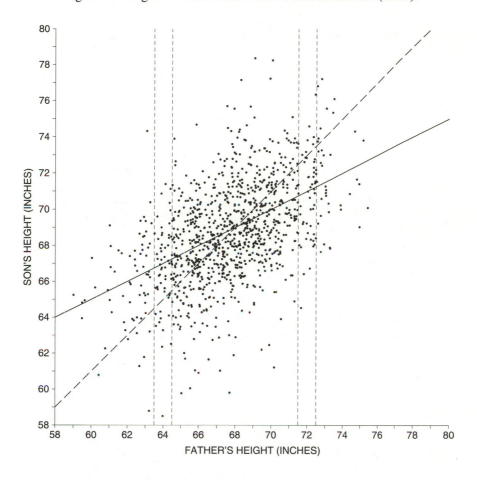

the son. For Yule (section 1.4), a "subject" might be a metropolitan union, with $x_i = \Delta$Out for union i, and $y_i = \Delta$Paup.

The regression line is computed from five summary statistics: (i) the average of x, (ii) the average of y, (iii) the SD of x, (iv) the SD of y, and (v) the correlation between x and y. The calculations can be organized as follows, with "variance" abbreviated to "var:"

(1) $$\bar{x} = \frac{1}{n} \sum_{i=1}^{n} x_i, \quad \text{var } x = \frac{1}{n} \sum_{i=1}^{n} (x_i - \bar{x})^2,$$

(2) $$\text{the SD of } x \text{ is } s_x = \sqrt{\text{var } x},$$

(3) x_i in standard units is $z_i = \dfrac{x_i - \overline{x}}{s_x},$

(4)   and the correlation coefficient is

$$r = \frac{1}{n} \sum_{i=1}^{n} \left(\frac{x_i - \overline{x}}{s_x} \cdot \frac{y_i - \overline{y}}{s_y} \right).$$

We're tacitly assuming $s_x \neq 0$ and $s_y \neq 0$. Necessarily, $-1 \leq r \leq 1$: see exercise B16 below. The correlation between x and y is often written as $r(x, y)$. Let $\text{sign}(r) = +1$ when $r > 0$ and $\text{sign}(r) = -1$ when $r < 0$. Equations (5) and (6) show that the regression line is flatter than the SD line.

(5)   The regression line of y on x goes through the point of averages
           $(\overline{x}, \overline{y})$. The slope is $r s_y / s_x$. The intercept is $\overline{y} - \text{slope} \cdot \overline{x}$.

(6)   The SD line also goes through the point of averages. The slope
           is $\text{sign}(r) s_y / s_x$. The intercept is $\overline{y} - \text{slope} \cdot \overline{x}$.

Figure 2. Graph of averages. The dots show the average height
of the sons, for each value of father's height. The regression line
(solid) follows the dots: it is flatter than the SD line (dashed).

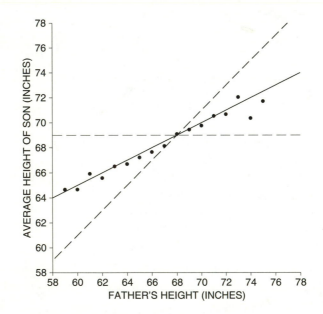

The regression of y on x, also called the regression line for predicting y from x, is a linear approximation to the *graph of averages*, which shows the average value of y for each x (figure 2).

Correlation is a key concept. Figure 3 shows the correlation coefficient for three scatter diagrams. All the diagrams have the same number of points ($n = 50$), the same means ($\overline{x} = \overline{y} = 50$), and the same SDs ($s_x = s_y = 15$). The shapes are very different. The correlation coefficient r tells you about the shapes. (If the variables aren't paired—two numbers for each subject—you won't be able to compute the correlation coefficient or regression line.)

Figure 3. Three scatter diagrams. The correlation measures the extent to which the scatter diagram is packed in around a line. If the sign is positive, the line slopes up. If sign is negative, the line slopes down (not shown here).

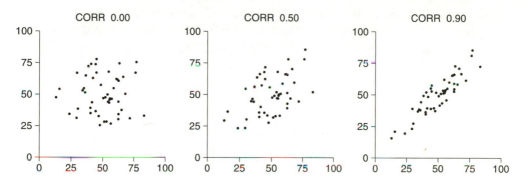

If you use the line $y = a + bx$ to predict y from x, the error or *residual* for subject i is $e_i = y_i - a - bx_i$, and the MSE is

$$\frac{1}{n} \sum_{i=1}^{n} e_i^2.$$

The RMS error is the square root of the MSE. For the regression line, as will be seen later, the MSE equals $(1 - r^2)$ var y. The abbreviations: MSE stands for mean square error; RMS, for root mean square.

A THEOREM DUE TO C.-F. GAUSS. Among all lines, the regression line has the smallest MSE.

A more general theorem will be proved in chapter 3. If the material in sections 1–2 is unfamiliar, you might want to read chapters 8–12 in Freedman-Pisani-Purves (1998).

2.3 Hooke's law

A weight is hung on the end of a spring whose length under no load is a. The spring stretches to a new length. According to Hooke's law, the amount of stretch is proportional to the weight. If you hang weight x_i on the spring, the length is

$$(7) \qquad\qquad Y_i = a + bx_i + \epsilon_i, \text{ for } i = 1, \ldots, n.$$

Equation (7) is a *regression model*. In this equation, a and b are constants that depend on the spring. The values are unknown, and have to be estimated from data. These are *parameters*. The ϵ_i are independent, identically distributed, mean 0, variance σ^2. These are *random errors*, or *disturbances*. The variance σ^2 is another parameter. You choose x_i, the weight on occasion i. The response Y_i is the length of the spring under the load. You do not see a, b, or the ϵ_i.

Table 1 shows the results of an experiment on Hooke's law, done in a physics class at U.C. Berkeley. The first column shows the load. The second column shows the measured length. (The "spring" was a long piece of piano wire hung from the ceiling of a big lecture hall.)

Table 1. An experiment on Hooke's law

Weight (kg)	Length (cm)
0	439.00
2	439.12
4	439.21
6	439.31
8	439.40
10	439.50

We use the method of least squares to estimate the parameters a and b. In other words, we fit the regression line. The intercept is

$$\hat{a} \doteq 439.01 \text{ cm}.$$

A hat over a parameter denotes an estimate: we estimate a as 439.01 cm. The slope is

$$\hat{b} \doteq 0.05 \text{ cm per kg}.$$

We estimate b as 0.05 cm per kg. (The dotted equals sign "$\doteq$" means nearly equal; numerical results are rounded off.)

There are two conclusions. (i) Putting a weight on the spring makes it longer. (ii) Each extra kilogram of weight makes the spring about 0.05 centimeters longer. The first is a (pretty obvious) qualitative inference; the second is quantitative. The distinction between qualitative and quantitative inference will come up again in chapter 5.

Exercise set A

1. In the Pearson-Lee data, the average height of the fathers was 67.7 inches; the SD was 2.74 inches. The average height of the sons was 68.7 inches; the SD was 2.81 inches. The correlation was 0.501.

    (a) True or false and explain: if the father is 72 inches tall, it's 50–50 whether the son is taller than 73 inches—because the sons average an inch taller than the fathers.

    (b) Find the regression line of son's height on father's height, and its RMS error.

2. Can you determine a in equation (7) by measuring the length of spring with no load? With one measurement? Ten measurements? Explain briefly.

3. Use the data in table 1 to find the MSE and the RMS error for the regression line predicting length from weight. Which statistic gives a better sense of how far the data are from the regression line? Hint: keep track of the units, or plot the data, or both.

4. The correlation coefficient is a good descriptive statistic for one of the three diagrams below. Which one, and why?

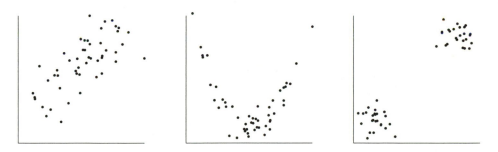

2.4 Complexities

Compare equation (7) with equation (8):

(7)
$$Y_i = a + bx_i + \epsilon_i,$$

(8)
$$Y_i = \hat{a} + \hat{b}x_i + e_i.$$

Looks the same? Take another look. In the regression model (7), we can't see the parameters a, b or the disturbances ϵ_i. In the fitted model (8), the estimates $\hat{a}$, $\hat{b}$ are observable, and so are the residuals e_i. With a large sample, $\hat{a} \doteq a$ and $\hat{b} \doteq b$, so $e_i \doteq \epsilon_i$. But

$$\doteq \neq =$$

The e_i in (8) is called a residual rather than a disturbance term or random error term. Often, e_i is called an "error," although this can be confusing. "Residual" is clearer.

Estimates aren't parameters, and residuals aren't random errors.

The Y_i in (7) are random variables, because the ϵ_i are random. How are random variables connected to data? The answer, which involves *observed values*, will be developed by example. The examples will also show how ideas of mean and variance can be extended from data to random variables—with some pointers on going back and forth between the two realms. We begin with the mean. Consider the list $\{1, 2, 3, 4, 5, 6\}$. This has mean 3.5 and variance 35/12, by formula (1). So far, we have a tiny data set. Random variables are coming next.

Throw a die n times. (A die has six faces, all equally likely; one face has 1 spot, another face has 2 spots, and so forth, up to 6.) Let U_i be the number of spots on the ith roll, for $i = 1, \ldots, n$. The U_i are independent, identically distributed random variables—like choosing numbers at random from the list $\{1, 2, 3, 4, 5, 6\}$. Each random variable has mean (expectation, aka expected value) equal to 3.5, and variance equal to 35/12. Here, mean and variance have been applied to a random variable—the number of spots when you throw a die.

The sample mean and the sample variance are

$$(9) \qquad \overline{U} = \frac{1}{n}\sum_{i=1}^{n} U_i \quad \text{and} \quad \text{var}\{U_1, \ldots, U_n\} = \frac{1}{n}\sum_{i=1}^{n}(U_i - \overline{U})^2.$$

The sample mean and variance in (9) are themselves random variables. In principle, they differ from $E(U_i)$ and $\text{var}(U_i)$, which are fixed numbers—the expectation and variance, respectively, of U_i. When n is large,

$$(10) \qquad \overline{U} \doteq E(U_i) = 3.5, \qquad \text{var}\{U_1, \ldots, U_n\} \doteq \text{var}(U_i) = 35/12.$$

That is how the expectation and variance of a random variable are estimated from repeated observations.

- Random variables have means; so do data sets.
- Random variables have variances; so do data sets.

The discussion has been a little abstract. Now someone actually throws the die $n = 100$ times. That generates some data. The total number of spots is 371. The average number of spots per roll is $371/100 = 3.71$. This is not $\overline{U}$, but the *observed value* of $\overline{U}$. After all, $\overline{U}$ has a probability distribution: 3.71 just sits there. Similarly, the measurements on the spring in Hooke's law (table 1) aren't random variables. According to the regression model, the lengths are observed values of the random variables Y_i defined by (7).

> In a regression model, as a rule, the data are observed values of random variables.

Sometimes, observed values are called *realizations*.

There is one more issue to take up. Variance is often used to measure spread. However, as the next example shows, variance usually has the wrong units and the wrong size: take the square root to get the SD.

Example 1. American men age 18–24 have an average weight of 170 lbs. The typical person in this group weighs around 170 lbs, but will not weigh exactly 170 lbs. The typical deviation from average is _____. Do not put variance into the blank. The variance of weight is 900 square pounds: wrong units, wrong size. The SD is $\sqrt{\text{variance}} = 30$ lbs. The typical deviation from average weight is something like 30 lbs.

Example 2. Roll a die 100 times. Let $S = X_1 + \cdots + X_{100}$ be the total number of spots. This is a random variable, with $E(S) = 100 \times 3.5 = 350$. You will get around 350 spots, give or take _____ or so. The variance is var $S = 100 \times 35/12 \doteq 292$. (The 35/12 is the variance of the list $\{1, 2, 3, 4, 5, 6\}$, as mentioned earlier.) Do not put 292 into the blank. To use the variance, take the square root. The SE—*standard error*—is $\sqrt{292} \doteq 17$. Put 17 into the blank.

The number of spots will be around 350, but will be off 350 by something like 17. The number of spots is unlikely to be more than two or three SEs away from its expected value. For random variables, the standard error is the square root of the variance. (The standard error of a random variable is often called its standard deviation, which can be confusing.)

2.5 Simple vs multiple regression

A *simple* regression equation has on the right hand side an intercept and an explanatory variable with a slope coefficient. A *multiple* regression equation has several explanatory variables on the right hand side, each with its own slope coefficient. To study multiple regression, we will need matrix algebra. That is covered in chapter 3.

Exercise set B

1. In equation (1), variance applies to data, or random variables? What about correlation in (4)?

2. On page 22, below table 1, you will find the number 439.01. Is this a parameter or an estimate? What about the 0.05?

3. Suppose we didn't have the last line in table 1. Find the regression of length on weight, based on the data in the first 5 lines of the table.

4. In example 1, is 900 square pounds the variance of a random variable? or of data? Discuss briefly.

5. In example 2, is 35/12 the variance of a random variable? of data? maybe both? Discuss briefly.

6. A die is rolled 180 times. Find the expected number of aces, and the variance for the number of aces. The number of aces will be around _____, give or take _____ or so. (A die has six faces, all equally likely; the face with one spot is the "ace.")

7. A die is rolled 250 times. The fraction of times it lands ace will be around _____, give or take _____ or so.

8. One hundred draws are made at random with replacement from the box $\boxed{\;\boxed{1}\;\boxed{2}\;\boxed{2}\;\boxed{5}\;}$. The draws come out as follows: 17 $\boxed{1}$'s, 54 $\boxed{2}$'s, and 29 $\boxed{5}$'s. Fill in the blanks.

   (a) For the _____, the observed value is 0.8 SEs above the expected value. (Reminder: SE = standard error.)
   (b) For the _____, the observed value is 1.33 SEs above the expected value.

   Options (two will be left over):

       number of 1's   number of 2's   number of 5's   sum of the draws

If exercises 6–8 cover unfamiliar material, you might want to read chapters 16–18 in Freedman-Pisani-Purves (1998), or similar material in another text.

9. Equation (7) is _____. Options:

             a model   a parameter   a random variable

10. In equation (7), a is _____. Options (more than one may be right):

observable   unobservable   a parameter   a random variable

Repeat for b. For ϵ_i. For Y_i.

11. According to equation (7), the 439.00 in table 1 is _____. Options:

      a parameter

      a random variable

      the observed value of a random variable

12. Suppose $x_1, \ldots, x_n$ are real numbers. Let $\bar{x} = (x_1 + \cdots + x_n)/n$. Let c be a real number.

(a) Show that $\sum_{i=1}^{n} (x_i - \bar{x}) = 0$.

(b) Show that $\sum_{i=1}^{n} (x_i - c)^2 = \left[\sum_{i=1}^{n} (x_i - \bar{x})^2 \right] + n(\bar{x} - c)^2$.

Hint: $(x_i - c) = (x_i - \bar{x}) + (\bar{x} - c)$.

(c) Show that $\sum_{i=1}^{n} (x_i - c)^2$, as a function of c, has a unique minimum at $c = \bar{x}$.

(d) Show that $\sum_{i=1}^{n} x_i^2 = \left[\sum_{i=1}^{n} (x_i - \bar{x})^2 \right] + n\bar{x}^2$.

13. A statistician has a sample, and is computing the sum of the squared deviations of the sample numbers from a number q. The sum of the squared deviations will be smallest when q is the _____. Fill in the blank (25 words or less) and explain.

14. Suppose $x_1, \ldots, x_n$ and $y_1, \ldots, y_n$ have means $\bar{x}, \bar{y}$; the standard deviations are $s_x > 0$, $s_y > 0$; and the correlation is r. Let

$$\text{cov}(x, y) = \frac{1}{n} \sum_{i=1}^{n} (x_i - \bar{x})(y_i - \bar{y}).$$

("cov" is shorthand for covariance.) Show that—

(a) $\text{cov}(x, y) = r s_x s_y$.

(b) The slope of the regression line for predicting y from x is

$$\text{cov}(x, y)/\text{var}(x).$$

(c) $\text{var}(x) = \text{cov}(x, x)$.

(d) $\text{cov}(x, y) = \overline{xy} - \bar{x}\,\bar{y}$.

(e) $\text{var}(x) = \overline{x^2} - \bar{x}^2$.

15. Suppose $x_1, \ldots, x_n$ and $y_1, \ldots, y_n$ are real numbers, with $s_x > 0$ and $s_y > 0$. Let x^* be x in standard units; similarly for y. Show that $r(x, y) = r(x^*, y^*)$.

16. Suppose $x_1, \ldots, x_n$ and $y_1, \ldots, y_n$ are real numbers, with $\bar{x} = \bar{y} = 0$ and $s_x = s_y = 1$. Show that $\frac{1}{n} \sum_{i=1}^{n} (x_i + y_i)^2 = 2(1 + r)$ and $\frac{1}{n} \sum_{i=1}^{n} (x_i - y_i)^2 = 2(1 - r)$, where $r = r(x, y)$. Show that

$$-1 \le r \le 1.$$

17. A die is rolled twice. Let X_i be the number of spots on the ith roll for $i = 1, 2$.

   (a) Find $P(X_1 = 3 \mid X_1 + X_2 = 8)$, the *conditional probability* of a 3 on the first roll given a total of 8 spots.

   (b) Find $P(X_1 + X_2 = 7 \mid X_1 = 3)$.

   (c) Find $E(X_1 \mid X_1 + X_2 = 6)$, the *conditional expectation* of X_1 given that $X_1 + X_2 = 6$.

18. (Hard.) Suppose $x_1, \ldots, x_n$ are real numbers. Suppose n is odd and the x_i are all distinct. There is a unique *median* μ: the middle number when the x's are arranged in increasing order. Let c be a real number. Show that $f(c) = \sum_{i=1}^{n} |x_i - c|$, as a function of c, is minimized when $c = \mu$. Hints. You can't do this by calculus, because f isn't differentiable. Instead, show that $f(c)$ is (i) continuous, (ii) strictly increasing as c increases for $c > \mu$, i.e., $\mu < c_1 < c_2$ implies $f(c_1) < f(c_2)$, and (iii) strictly decreasing as c increases for $c < \mu$. It's easier to think about points (ii) and (iii) when c differs from all the x's. You may as well assume that the x_i are increasing with i. If you pursue this line of reasoning far enough, you will find that f is linear between the x's, with corners at the x's. Moreover, f is convex, i.e., $f[(x + y)/2] \leq [f(x) + f(y)]/2$.

2.6 End notes for chapter 2

In (6), if $r = 0$, you can take the slope of the SD line to be either s_y/s_x or $-s_y/s_x$. In other applications, however, sign(0) is usually defined as 0.

Hooke's law (7) is a good approximation when the weights are relatively small. When the weights are larger, a quadratic term may be needed. Close to the "elastic limit" of the spring, things get more complicated.

Many statisticians find it surprising that regression is commonly used in the social and life sciences to infer causation from observational data, with qualitative inference perhaps more common than quantitative: X causes (or doesn't cause) Y, the magnitude of the effect being of lesser interest. Eyebrows are sometimes raised about the whole idea of causation:

> "Beyond such discarded fundamentals as 'matter' and 'force' lies still another fetish amidst the inscrutable arcana of even modern science, namely, the category of cause and effect. Is this category anything but a conceptual limit to experience, and without any basis in perception beyond a statistical approximation?" (Pearson 1911, p. vi)

3

Matrix Algebra

3.1 Introduction

Matrix algebra is the key to multiple regression (chapter 4), so we review the basics here. Section 4 covers *positive definite matrices*, with a quick introduction to the normal distribution in section 5. A matrix is a rectangular array of numbers. (In this book, we only consider matrices of real numbers.) For example, M is a 3×2 matrix—3 rows, 2 columns—and b is a 2×1 *column vector*:

$$
M = \begin{pmatrix} 3 & -1 \\ 2 & -1 \\ -1 & 4 \end{pmatrix}, \qquad b = \begin{pmatrix} 3 \\ -3 \end{pmatrix}.
$$

The ijth element of M is written M_{ij}, e.g., $M_{32} = 4$; similarly, $b_2 = -3$. Matrices can be multiplied (element-wise) by a scalar. Matrices of the same size can be added (again, element-wise). For instance,

$$
2 \times \begin{pmatrix} 3 & -1 \\ 2 & -1 \\ -1 & 4 \end{pmatrix} = \begin{pmatrix} 6 & -2 \\ 4 & -2 \\ -2 & 8 \end{pmatrix}, \qquad \begin{pmatrix} 3 & -1 \\ 2 & -1 \\ -1 & 4 \end{pmatrix} + \begin{pmatrix} 3 & 2 \\ 4 & -1 \\ -1 & 1 \end{pmatrix} = \begin{pmatrix} 6 & 1 \\ 6 & -2 \\ -2 & 5 \end{pmatrix}.
$$

An $m \times n$ matrix A can be multiplied by a matrix B of size $n \times p$. The product is an $m \times p$ matrix, whose ikth element is $\sum_j A_{ij} B_{jk}$. For example,

$$Mb = \begin{pmatrix} 3 \times 3 + (-1) \times (-3) \\ 2 \times 3 + (-1) \times (-3) \\ (-1) \times 3 + 4 \times (-3) \end{pmatrix} = \begin{pmatrix} 12 \\ 9 \\ -15 \end{pmatrix}.$$

Matrix multiplication is *not* commutative. This may seem tricky at first, but you get used to it. Exercises 1–2 (below) provide a little more explanation.

The matrix $0_{m \times n}$ is an $m \times n$ matrix all of whose entries are 0. For instance,

$$0_{2 \times 3} = \begin{pmatrix} 0 & 0 & 0 \\ 0 & 0 & 0 \end{pmatrix}.$$

The $m \times m$ identity matrix is written I_m or $I_{m \times m}$. This matrix has 1's on the diagonal and 0's off the diagonal:

$$I_{3 \times 3} = \begin{pmatrix} 1 & 0 & 0 \\ 0 & 1 & 0 \\ 0 & 0 & 1 \end{pmatrix}.$$

If A is $m \times n$, then $I_{m \times m} \times A = A = A \times I_{n \times n}$.

An $m \times n$ matrix A can be "transposed." The result is an $n \times m$ matrix denoted A' or A^T. E.g.,

$$\begin{pmatrix} 3 & -1 \\ 2 & -1 \\ -1 & 4 \end{pmatrix}^T = \begin{pmatrix} 3 & 2 & -1 \\ -1 & -1 & 4 \end{pmatrix}.$$

If $A' = A$, then A is *symmetric*.

If u and v are $n \times 1$ column vectors, the *inner product* or *dot product* is $u \cdot v = u' \times v$. If this is 0, then u and v are *orthogonal*: we write $u \perp v$. The *norm* or *length* of u is $\|u\|$, where $\|u\|^2 = u \cdot u$. People often write $|u|$ instead of $\|u\|$. The inner product $u \cdot v$ equals the length of u, times the length of v, times the cosine of the angle between the two vectors. If $u \perp v$, the angle is $90°$, and $\cos(90°) = 0$.

For square matrices, the *trace* is the sum of the diagonal elements:

$$\text{trace} \begin{pmatrix} 1 & 2 \\ 5 & 3 \end{pmatrix} = 4.$$

Exercise set A

1. Suppose A is $m \times n$ and B is $n \times p$. For i and j with $1 \leq i \leq m$ and $1 \leq j \leq p$, let r_i be the ith row of A and let c_j be the jth column of B.

What is the size of r_i? of c_j? How is $r_i \times c_j$ related to the ijth element of $A \times B$?

2. Suppose A is $m \times n$, while u, v are $n \times 1$ and α is scalar. Show that $Au \in R^m$, $A(\alpha u) = \alpha Au$, and $A(u + v) = Au + Av$. As they say, A is a linear map from R^n to R^m, where R^n is n-dimensional Euclidean space.

3. If A is $m \times n$, check that

$$A + 0_{m \times n} = 0_{m \times n} + A = A.$$

For exercises 4 and 5, let

$$M = \begin{pmatrix} 3 & -1 \\ 2 & -1 \\ 1 & -4 \end{pmatrix}.$$

4. Show that $I_{3 \times 3} M = M = M I_{2 \times 2}$.

5. Compute $M'M$ and MM'. Find the trace of $M'M$ and the trace of MM'.

6. Find the lengths of u and v, defined below. Are these vectors orthogonal? Compute the *outer product* $u \times v'$. What is the trace of the outer product?

$$u = \begin{pmatrix} 1 \\ 2 \\ -1 \end{pmatrix}, \quad v = \begin{pmatrix} 1 \\ 2 \\ 4 \end{pmatrix}.$$

3.2 Determinants and inverses

Matrix inversion will be needed to get regression estimates and their standard errors. One way to find inverses begins with *determinants*. The determinant of a square matrix is computed by an inductive procedure:

$$\det(4) = 4, \qquad \det \begin{pmatrix} 1 & 2 \\ 5 & 3 \end{pmatrix} = (1 \times 3) - (2 \times 5) = -7,$$

$$\det \begin{pmatrix} 1 & 2 & 3 \\ 2 & 3 & 1 \\ 0 & 1 & 1 \end{pmatrix} = 1 \times \det \begin{pmatrix} 3 & 1 \\ 1 & 1 \end{pmatrix} - 2 \times \det \begin{pmatrix} 2 & 3 \\ 1 & 1 \end{pmatrix} + 0 \times \det \begin{pmatrix} 2 & 3 \\ 3 & 1 \end{pmatrix}$$

$$= 1 \times (3 - 1) - 2 \times (2 - 3) + 0 \times (2 - 9) = 4.$$

Here, we work our way down the first column, getting the determinants of the smaller matrices obtained by striking out the row and column through each current position. The determinants pick up extra signs, which alternate $+$ and $-$. With a 4×4 matrix, for instance, the extra signs are

$$\begin{pmatrix} + & - & + & - \\ - & + & - & + \\ + & - & + & - \\ - & + & - & + \end{pmatrix}.$$

The determinants with the extra signs tacked on are called *cofactors*. The determinant of a matrix is $\sum_{i=1}^{n} a_{i1} c_{i1}$, where a_{ij} is the ijth element in the matrix, and c_{ij} is the cofactor. (Watch it: the determinants have signs of their own, as well as the extra signs shown above.) It turns out that you can use any row or column, not just column 1, for computing the determinant. As a matter of notation, people often write $|A|$ instead of $\det(A)$.

The *rank* of a matrix is the number of linearly independent columns (or rows—has to be the same). If $n > p$, an $n \times p$ matrix X has *full rank* if the rank is p; otherwise, X is *rank deficient*. An $n \times n$ matrix A has full rank if and only if $\det(A) \neq 0$. Then the matrix has an inverse A^{-1}:

$$A \times A^{-1} = A^{-1} \times A = I_{n \times n}.$$

Such matrices are *invertible* or *non-singular*. The inverse is unique; this follows from existence. Conversely, if A is invertible, then $\det(A) \neq 0$ and the rank of A is n.

The inverse can be computed as follows:

$$A^{-1} = \operatorname{adj}(A) / \det(A),$$

where $\operatorname{adj}(A)$ is the transpose of the matrix of cofactors. (This is the *classical adjoint*.) For example,

$$\operatorname{adj} \begin{pmatrix} 1 & 2 \\ 5 & 3 \end{pmatrix} = \begin{pmatrix} 3 & -2 \\ -5 & 1 \end{pmatrix},$$

$$\operatorname{adj} \begin{pmatrix} 1 & 2 & 3 \\ 2 & 3 & 1 \\ 0 & 1 & 1 \end{pmatrix} = \begin{pmatrix} a & b & c \\ d & e & f \\ g & h & i \end{pmatrix},$$

where

$$a = \det \begin{pmatrix} 3 & 1 \\ 1 & 1 \end{pmatrix}, \quad b = -\det \begin{pmatrix} 2 & 3 \\ 1 & 1 \end{pmatrix}, \quad c = \det \begin{pmatrix} 2 & 3 \\ 3 & 1 \end{pmatrix} \dots.$$

Exercise set B

For exercises 1–7 below, let

$$A = \begin{pmatrix} 1 & 2 \\ 5 & 3 \end{pmatrix}, \quad B = \begin{pmatrix} 1 & 2 & 3 \\ 2 & 3 & 1 \\ 0 & 1 & 1 \end{pmatrix}, \quad C = \begin{pmatrix} 1 & 2 \\ 2 & 4 \\ 3 & 6 \end{pmatrix}.$$

1. Find $\text{adj}(B)$. This is just to get on top of the definitions; later, we do all this sort of thing on the computer.

2. Show that $A \times \text{adj}A = \text{adj}A \times A = \det(A) \times I_n$. Repeat, for B. What is n in each case?

3. Find the rank and the trace of A. Repeat, for B.

4. Find the rank of C.

5. If possible, find the trace and determinant of C. If not, why not?

6. If possible, find A^2. If not, why not? (Hint: $A^2 = A \times A$.)

7. If possible, find C^2. If not, why not?

8. If M is $m \times n$ and N is $m \times n$, show that $(M + N)' = M' + N'$.

9. Suppose M is $m \times n$ and N is $n \times p$.

    (a) Show that $(MN)' = N'M'$.

    (b) Suppose $m = n = p$, and M, N are both invertible. Show that $(MN)^{-1} = N^{-1}M^{-1}$ and $(M')^{-1} = (M^{-1})'$.

10. Suppose X is $n \times p$ with $p \le n$. If X has rank p, show that $X'X$ has rank p, and conversely. Hints. Suppose X has rank p and c is $p \times 1$. Then $X'Xc = 0_{p \times 1} \Rightarrow c'X'Xc = 0 \Rightarrow \|Xc\|^2 = 0 \Rightarrow Xc = 0_{n \times 1}$.

    Notes. The matrix $X'X$ is $p \times p$. The rank is p if and only if $X'X$ is invertible. The $\Rightarrow$ is shorthand for "implies."

11. If A is $m \times n$ and B is $n \times m$, show that $\text{trace}(AB) = \text{trace}(BA)$. Hint: the iith element of AB is $\sum_j A_{ij}B_{ji}$, while the jjth element of BA is $\sum_i B_{ji}A_{ij}$.

12. If u and v are $n \times 1$, show that $\|u + v\|^2 = \|u\|^2 + \|v\|^2 + 2u \cdot v$.

13. If u and v are $n \times 1$, show that $\|u + v\|^2 = \|u\|^2 + \|v\|^2$ if and only if $u \perp v$. (This is Pythagoras' theorem in n dimensions.)

14. Suppose X is $n \times p$ with rank $p < n$. Suppose Y is $n \times 1$. Let $\hat{\beta} = (X'X)^{-1}X'Y$ and $e = Y - X\hat{\beta}$.

    (a) Show that $X'X$ is $p \times p$, while $X'Y$ is $p \times 1$.

    (b) Show that $X'X$ is symmetric. Hint: look at exercise 9(a).

    (c) Show that $X'X$ is invertible. Hint: look at exercise 10.

(d) Show that $(X'X)^{-1}$ is $p \times p$, so $\hat{\beta} = (X'X)^{-1}X'Y$ is $p \times 1$.

(e) Show that $(X'X)^{-1}$ is symmetric. Hint: look at exercise 9(b).

(f) Show that $X\hat{\beta}$ and $e = Y - X\hat{\beta}$ are $n \times 1$.

(g) Show that $X'X\hat{\beta} = X'Y$, and hence $X'e = 0_{p \times 1}$.

(h) Show that $e \perp X\hat{\beta}$, so $\|Y\|^2 = \|X\hat{\beta}\|^2 + \|e\|^2$.

(i) If γ is $p \times 1$, show that $\|Y - X\gamma\|^2 = \|Y - X\hat{\beta}\|^2 + \|X(\hat{\beta} - \gamma)\|^2$. Hint: $Y - X\gamma = Y - X\hat{\beta} + X(\hat{\beta} - \gamma)$.

(j) Show that $\|Y - X\gamma\|^2$ is minimized when $\gamma = \hat{\beta}$.

(k) If $\tilde{\beta}$ is $p \times 1$ with $Y - X\tilde{\beta} \perp X$, show that $\tilde{\beta} = \hat{\beta}$. Notation: $v \perp X$ if v is orthogonal to each column of X. Hint: what is $X'(Y - X\tilde{\beta})$?

(l) Is XX' invertible? Hints. By assumption, $p < n$. Can you find an $n \times 1$-vector $c \neq 0_{n \times 1}$ with $c'X = 0_{1 \times p}$?

(m) Is $(X'X)^{-1} = X^{-1}(X')^{-1}$?

Notes. $\hat{\beta}$ is called "the OLS estimator," where OLS is shorthand for "ordinary least squares." This exercise develops a lot of the theory for OLS estimators. The geometry in brief: $X'e = 0_{p \times 1}$ means that e is orthogonal—perpendicular—to each column of X. Hence $\hat{Y} = X\hat{\beta}$ is the projection of Y onto the columns of X, and the closest point in column space to Y. Part (j) is Gauss' theorem for multiple regression.

15. In exercise 14, suppose $p = 1$, so X is a column vector. Show that $\hat{\beta} = X \cdot Y / \|X\|^2$.

16. In exercise 14, suppose $p = 1$ and X is a column of 1's. Show that $\hat{\beta}$ is the mean of the Y's. How is this related to exercise 2B12(c), i.e., part (c), exercise 12, set B, chapter 2?

17. This exercise explains a stepwise procedure for computing $\hat{\beta}$ in exercise 14. There are hints, but there is also some work to do. Let M be the first $p - 1$ columns of X, so M is $n \times (p - 1)$. Let N be the last column of X, so N is $n \times 1$.

(i) Let $\hat{\gamma}_1 = (M'M)^{-1}M'Y$ and $f = Y - M\hat{\gamma}_1$.

(ii) Let $\hat{\gamma}_2 = (M'M)^{-1}M'N$ and $g = N - M\hat{\gamma}_2$.

(iii) Let $\hat{\gamma}_3 = f \cdot g / \|g\|^2$ and $e = f - g\hat{\gamma}_3$.

Show that $e \perp X$. (Hint: begin by checking $f \perp M$ and $g \perp M$.) Finally, show that

$$\hat{\beta} = \begin{pmatrix} \hat{\gamma}_1 - \hat{\gamma}_2\hat{\gamma}_3 \\ \hat{\gamma}_3 \end{pmatrix}.$$

Note. The procedure amounts to (i) regressing Y on M, (ii) regressing N on M, then (iii) regressing the first set of residuals on the second.

18. Suppose u, v are $n \times 1$; neither is identically 0. What is the rank of $u \times v'$?

3.3 Random vectors

Let $U = \begin{pmatrix} U_1 \\ U_2 \\ U_3 \end{pmatrix}$, a 3×1 column vector of random variables. Then

$E(U) = \begin{pmatrix} E(U_1) \\ E(U_2) \\ E(U_3) \end{pmatrix}$, a 3×1 column vector of numbers. On the other hand,

$\mathrm{cov}(U)$ is 3×3 matrix of real numbers:

$$\mathrm{cov}(U) = E\left\{ \begin{pmatrix} U_1 - E(U_1) \\ U_2 - E(U_2) \\ U_3 - E(U_3) \end{pmatrix} \begin{pmatrix} U_1 - E(U_1) & U_2 - E(U_2) & U_3 - E(U_3) \end{pmatrix} \right\}.$$

Here, cov applies to random vectors, not to data ("cov" is shorthand for covariance). The same definitions can be used for vectors of any size.

People sometimes use correlations for random variables: the correlation between U_1 and U_2, for instance, is $\mathrm{cov}(U_1, U_2)/\sqrt{\mathrm{var}(U_1)\mathrm{var}(U_2)}$.

Exercise set C

1. Show that the 1,1 element of $\mathrm{cov}(U)$ equals $\mathrm{var}(U_1)$, and the 2,3 element equals $\mathrm{cov}(U_2, U_3)$.

2. Show that $\mathrm{cov}(U)$ is symmetric.

3. If A is a fixed (i.e., non-random) matrix of size $n \times 3$ and B is a fixed matrix of size $1 \times m$, show that $E(AUB) = AE(U)B$.

4. Show that $\mathrm{cov}(AU) = A\mathrm{cov}(U)A'$.

5. If c is a fixed vector of size 3×1, show that $\mathrm{var}(c'U) = c'\mathrm{cov}(U)c$ and $\mathrm{cov}(U + c) = \mathrm{cov}(U)$.

6. What's the difference between $\overline{U} = (U_1 + U_2 + U_3)/3$ and $E(U)$?

7. Suppose ξ and ζ are two random vectors of size 7×1. If $\xi'\zeta = 0$, are ξ and ζ independent? What about the converse: if ξ and ζ are independent, is $\xi'\zeta = 0$?

8. Suppose ξ and ζ are two random variables with $E(\xi) = E(\zeta) = 0$. Show that $\mathrm{var}(\xi) = E(\xi^2)$ and $\mathrm{cov}(\xi, \zeta) = E(\xi\zeta)$.

Notes. More generally, $\mathrm{var}(\xi) = E(\xi^2) - [E(\xi)]^2$ and $\mathrm{cov}(\xi, \zeta) = E(\xi\zeta) - E(\xi)E(\zeta)$.

9. Suppose ξ is an $n \times 1$ random vector with $E(\xi) = 0$. Show that $\mathrm{cov}(\xi) = E(\xi\xi')$.

   Notes. Generally, $\mathrm{cov}(\xi) = E(\xi\xi') - E(\xi)E(\xi')$ and $E(\xi') = [E(\xi)]'$.

10. Suppose ξ_i, ζ_i are random variables for $i = 1, \ldots, n$. As pairs, they are independent and identically distributed in i. Let $\bar{\xi} = \frac{1}{n}\sum_{i=1}^n \xi_i$, and likewise for ζ. True or false, and explain:

   (a) $\mathrm{cov}(\xi_i, \zeta_i)$ is the same for every i.
   (b) $\mathrm{cov}(\xi_i, \zeta_i) = \frac{1}{n}\sum_{i=1}^n (\xi_i - \bar{\xi})(\zeta_i - \bar{\zeta})$.

11. The random variable X has density f on the line; σ and μ are real numbers. What is the density of $\sigma X + \mu$? of X^2? Reminder: if X has density f, then $P(X < x) = \int_{-\infty}^x f(u)du$.

3.4 Positive definite matrices

Material in this section will be used when we discuss generalized least squares (section 4.4). Detailed proofs are beyond our scope. An $n \times n$ *orthogonal* matrix R has $R'R = I_{n \times n}$. (These matrices are also said to be "unitary.") Necessarily, $RR' = I_{n \times n}$. Geometrically, R is a rotation, which preserves angles and distances; R can also reverse certain directions. A *diagonal* matrix D is square and vanishes off the main diagonal: e.g., D_{11} and D_{22} may be non-zero but $D_{12} = D_{21} = 0$. An $n \times n$ matrix G is *non-negative definite* if

   (i) G is symmetric, and
   (ii) $x'Gx \geq 0$ for any n-vector x.

The matrix G is *positive definite* if $x'Gx > 0$ for any n-vector x except $x = 0_{n \times 1}$. (Non-negative definite matrices are also called "positive semi-definite.")

THEOREM 1. The matrix G is non-negative definite if and only if there is a diagonal matrix D whose elements are non-negative, and an orthogonal matrix R such that $G = RDR'$. The matrix G is positive definite if and only if the diagonal entries of D are all positive.

The columns of R are the *eigenvectors* of G, and the diagonal elements of D are the *eigenvalues*. For instance, if c is the first column of R and $\lambda = D_{11}$, then $Gc = c\lambda$. (This is because $GR = RD$.) It follows from theorem 1 that a non-negative definite G has a non-negative definite square root, $G^{1/2} = RD^{1/2}R'$, where the square root of D is taken element by element. A positive definite G has a positive definite inverse, $G^{-1} = RD^{-1}R'$. (See exercises below.) If G is non-negative definite rather than positive definite, that is, $x'Gx = 0$ for some $x \neq 0$, then G is not invertible. Theorem 1 is an elementary version of the "spectral theorem."

Exercise set D

1. Which of the following matrices are positive definite? non-negative definite?

$$\begin{pmatrix} 2 & 0 \\ 0 & 1 \end{pmatrix} \quad \begin{pmatrix} 2 & 0 \\ 0 & 0 \end{pmatrix} \quad \begin{pmatrix} 0 & 1 \\ 1 & 0 \end{pmatrix} \quad \begin{pmatrix} 0 & 0 \\ 1 & 0 \end{pmatrix}$$

   Hint: work out $(u \ v) \begin{pmatrix} a & b \\ c & d \end{pmatrix} \begin{pmatrix} u \\ v \end{pmatrix} = (u \ v) \left[\begin{pmatrix} a & b \\ c & d \end{pmatrix} \begin{pmatrix} u \\ v \end{pmatrix} \right]$.

2. Suppose X is an $n \times p$ matrix with rank $p \leq n$.
   (a) Show that $X'X$ is $p \times p$ positive definite. Hint: if c is $p \times 1$, what is $c'X'Xc$?
   (b) Show that XX' is $n \times n$ non-negative definite.

For exercises 3–6, suppose R is an $n \times n$ orthogonal matrix and D is an $n \times n$ diagonal matrix, with $D_{ii} > 0$ for all i. Let $G = RDR'$. Work the exercises directly, without appealing to theorem 1.

3. Show that $\|Rx\| = \|x\|$ for any $n \times 1$ vector x.

4. Show that D and G are positive definite.

5. Let $\sqrt{D}$ be the $n \times n$ matrix whose ijth element is $\sqrt{D_{ij}}$. Show that $\sqrt{D} \times \sqrt{D} = D$. Show also that $R\sqrt{D}R' \times R\sqrt{D}R' = G$.

6. Let D^{-1} be the matrix whose ijth element is 0 for $i \neq j$, while the iith element is $1/D_{ii}$. Show that $D^{-1} \times D = I_{n \times n}$ and $RD^{-1}R' \times G = I_{n \times n}$.

7. Suppose G is positive definite. Show that—

   (a) G is invertible and G^{-1} is positive definite.
   (b) G has a positive definite square root $G^{1/2}$.
   (c) G^{-1} has a positive definite square root $G^{-1/2}$.

8. Let U be a random 3×1 vector. Show that $\text{cov}(U)$ is non-negative definite, and positive definite unless there is a 3×1 fixed (i.e., non-random) vector such that $c'U = c'E(U)$ with probability 1. Hints. Can you compute $\text{var}(c'U)$ from $\text{cov}(U)$? If that hint isn't enough, try the case $E(U) = 0_{3 \times 1}$. Comment: if $c'U = c'E(U)$ with probability 1, then $U - E(U)$ concentrates in a fixed hyperplane.

3.5 The normal distribution

This is a quick review; proofs will not be given. A random variable X is $N(\mu, \sigma^2)$ if it is normally distributed with mean μ and variance σ^2. Then the density of X is

$$\frac{1}{\sigma\sqrt{2\pi}} \exp\left[-\frac{1}{2}\frac{(x-\mu)^2}{\sigma^2}\right], \quad \text{where } \exp(t) = e^t.$$

If X is $N(\mu, \sigma^2)$, then $(X - \mu)/\sigma$ is $N(0, 1)$, i.e., $(X - \mu)/\sigma$ is *standard normal*. The standard normal density is

$$\phi(x) = \frac{1}{\sqrt{2\pi}} \exp\left(-\frac{1}{2}x^2\right).$$

Random variables $X_1, \ldots, X_n$ are *jointly normal* if all their linear combinations are normally distributed. If X_1, X_2 are independent normal variables, they are jointly normal, because $a_1 X_1 + a_2 X_2$ is normally distributed for any pair a_1, a_2 of real numbers. Later on, a couple of examples will involve jointly normal variables, and the following theorem will be helpful. (If you want to construct normal variables, see exercise 1 below for the method.)

THEOREM 2. The distribution of jointly normal random variables is determined by the mean vector α and covariance matrix G; the latter must be non-negative definite. If G is positive definite, the density of the random variables at x is

$$\left(\frac{1}{\sqrt{2\pi}}\right)^n \frac{1}{\sqrt{\det G}} \exp\left[-\frac{1}{2}(x-\alpha)'G^{-1}(x-\alpha)\right].$$

For any pair X_1, X_2 of random variables, normal or otherwise, if X_1 and X_2 are independent then $\text{cov}(X_1, X_2) = 0$. The converse is generally false, although counter-examples may seem contrived. For normal random variables, the converse is true: if X_1, X_2 are jointly normal and $\text{cov}(X_1, X_2) = 0$, then X_1 and X_2 are independent.

The central limit theorem. With a big sample, the probability distribution of the sum (or average) will be close to normal. More formally, suppose $X_1, X_2, \ldots$ are independent and identically distributed with $E(X_i) = \mu$ and $\text{var}(X_i) = \sigma^2$. Then $S_n = X_1 + X_2 + \cdots + X_n$ has expected value $n\mu$ and variance $n\sigma^2$. To standardize, subtract the expected value and divide by the standard error (the square root of the variance):

$$Z_n = \frac{S_n - n\mu}{\sigma\sqrt{n}}.$$

The central limit theorem says that if n is large, the distribution of Z_n is close to standard normal. For example,

$$P\{|S_n - n\mu| < \sigma\sqrt{n}\} = P\{|Z_n| < 1\} \to \frac{1}{\sqrt{2\pi}} \int_{-1}^{1} \exp\left(-\frac{1}{2}x^2\right) dx \doteq 0.6827.$$

There are many extensions of the theorem. Thus, the sum of independent random variables with different distributions is asymptotically normal, provided each term in the sum is only a small part of the total. There are also versions of the central limit theorem for random vectors. Feller (1971) has careful statements and proofs, as do other texts on probability.

Terminology. (i) Symmetry is built into the definition of positive definite matrices. (ii) Orthogonal matrices have orthogonal rows, and the length of each row is 1. The rows are said to be "orthonormal." Similar comments apply to the columns. (iii) "Multivariate normal" is a synonym for jointly normal. (iv) Sometimes, the phrase "jointly normal" is contracted to "normal," although this can be confusing. (v) "Asymptotically" means, as the sample size—the number of terms in the sum—gets large.

Exercise set E

1. Suppose G is $n \times n$ non-negative definite, and α is $n \times 1$.
    (a) Find an $n \times 1$ vector of normal random variables with mean 0 and $\mathrm{cov}(U) = G$. Hint: let V be an $n \times 1$ vector of independent $N(0, 1)$ variables, and let $U = G^{1/2}V$.
    (b) How would you modify the construction to get $E(U) = \alpha$?

2. Suppose R is an orthogonal $n \times n$ matrix. If U is an $n \times 1$ vector of IID $N(0, \sigma^2)$ variables, show that RU is an $n \times 1$ vector of IID $N(0, \sigma^2)$ variables. Hint: what is $E(RU)$? $\mathrm{cov}(RU)$? ("IID" is shorthand for "independent and identically distributed.")

3. Suppose ξ and ζ are two random variables. If $E(\xi\zeta) = E(\xi)E(\zeta)$, are ξ and ζ independent? What about the converse: if ξ and ζ are independent, is $E(\xi\zeta) = E(\xi)E(\zeta)$?

4. If U and V are random variables, show that $\mathrm{cov}(U, V) = \mathrm{cov}(V, U)$ and $\mathrm{var}(U + V) = \mathrm{var}(U) + \mathrm{var}(V) + 2\mathrm{cov}(U, V)$. Hint: what is $[(U - \alpha) + (V - \beta)]^2$?

5. Suppose ξ and ζ are jointly normal variables, with $E(\xi) = \alpha$, $\mathrm{var}(\xi) = \sigma^2$, $E(\zeta) = \beta$, $\mathrm{var}(\zeta) = \tau^2$, and $\mathrm{cov}(\xi, \zeta) = \rho\sigma\tau$. Find the mean and variance of $\xi + \zeta$. Is $\xi + \zeta$ normal?

Comments. Exercises 6–8 prepare for the next chapter. Exercise 6 is covered, for instance, by Freedman-Pisani-Purves (1998) in chapter 18. Exercises 7 and 8 are covered in chapters 20–21.

6. A coin is tossed 1000 times. Use the central limit theorem to approximate the chance of getting 475–525 heads (inclusive).

7. A box has red marbles and blue marbles. The fraction p of reds is unknown. 250 marbles are drawn at random with replacement, and 102 turn out to be red. Estimate p. Attach a standard error to your estimate.

8. Let $\hat{p}$ be the estimator in exercise 7.
    (a) About how big is the difference between $\hat{p}$ and p?
    (b) Can you find an approximate 95% confidence interval for p?

9. The "error function" Ψ is defined as follows:

$$\Psi(x) = \frac{2}{\sqrt{\pi}} \int_0^x \exp(-u^2)\, du.$$

   Show that Ψ is the distribution function of $|W|$, where W is $N(0, \sigma^2)$. Find σ^2. If Z is $N(0, 1)$, how would you compute $P(Z < x)$ from Ψ?

10. If U, V are IID $N(0, 1)$, show that $(U + V)/\sqrt{2}, (U - V)/\sqrt{2}$ are IID $N(0, 1)$.

3.6 If you want a book on matrix algebra

Blyth TS, Robertson EF (2002). *Basic Linear Algebra.* 2nd ed. Springer. Clear, mathematical.

Lang S (1997). *Introduction to Linear Algebra.* 3rd ed. Springer. More advanced. Also clear and mathematical.

Lang S (1996). *Linear Algebra.* 3rd ed. Springer. Still more advanced.

Meyer CD (2001). *Matrix Analysis and Applied Linear Algebra.* SIAM. More of a conventional textbook.

Strang G (1988). *Linear Algebra and Its Applications.* 3rd ed. Harcourt, Brace, Jovanovich. Love it or hate it.

4

Multiple Regression

4.1 Introduction

In this chapter, we set up the regression model and derive the main results about least squares estimators. The model is

$$(1) \qquad\qquad Y = X\beta + \epsilon.$$

On the left, Y is an $n \times 1$ vector of observable random variables. The Y vector is the *dependent* or *response* variable; Y is being "explained" or "modeled." As usual, Y_i is the ith component of Y.

On the right hand side, X is an $n \times p$ matrix of observable random variables, called the *design matrix*. We assume that $n > p$, and the design matrix has *full rank*, i.e., the rank of X is p. (In other words, the columns of X are linearly independent.) Next, β is a $p \times 1$ vector of parameters. Usually, these are unknown, to be estimated from data. The final term on the right is ϵ, an $n \times 1$ random vector. This is the *random error* or *disturbance* term. Generally, ϵ is not observed. We write ϵ_i for the ith component of ϵ.

In applications, there is a Y_i for each unit of observation i. Similarly, there is one row in X for each unit of observation, and one column for each data variable. These are the *explanatory* or *independent* variables, although seldom will any column of X be statistically independent of any other column. Orthogonality is rare too, except in designed experiments.

Columns of X are often called *covariates* or *control variables*, especially if they are put into the equation to control for confounding; "covariate" can have a more specific meaning, discussed in chapter 8. Sometimes, Y is called the "left hand side" variable. The columns in X are then (surprise) the "right hand side" variables. If the equation—like (1.1) or (2.7)—has an intercept, the corresponding column in the matrix is a "variable" only by courtesy: this column is all 1's.

We'll write X_i for the ith row of X. The matrix equation (1) unpacks into n ordinary equations, one for each unit of observation. For the ith unit, the equation is

$$(2) \qquad\qquad Y_i = X_i \beta + \epsilon_i.$$

To estimate β, we need some data—and some assumptions connecting the data to the model. A basic assumption is that

$$(3) \qquad\qquad \text{the data on } Y \text{ are observed values of } X\beta + \epsilon.$$

We have observed values for X and Y, not the random variables themselves. We do not know β and do not observe ϵ. These remain at the level of concepts.

The next assumption is:

(4)   The ϵ_i are independent and identically distributed, with mean 0 and variance σ^2.

Here, mean and variance apply to random variables not data; $E(\epsilon_i) = 0$, and $\text{var}(\epsilon_i) = \sigma^2$ is a parameter. Now comes another assumption:

(5) If X is random, we assume ϵ is independent of X. In symbols, $\epsilon \perp\!\!\!\perp X$.

(Note: $\perp\!\!\!\perp \neq \perp$.) Assumptions (3)-(4)-(5) are not easy to check, because ϵ is not observable. By contrast, the rank of X is easy to determine.

A matrix X is "random" if some of the entries X_{ij} are random variables rather than constants. This is an additional complication. People often prefer to condition on X. Then X is fixed; expectations, variances, and covariances are conditional on X.

We will estimate β using the OLS (ordinary least squares) estimator:

$$(6) \qquad\qquad \hat{\beta} = (X'X)^{-1}X'Y,$$

as in exercise 3B14 (shorthand for exercise 14, set B, chapter 3). This $\hat{\beta}$ is a $p \times 1$ vector. Let

$$(7) \qquad\qquad e = Y - X\hat{\beta}.$$

This is an $n \times 1$ vector of "residuals" or "errors." Exercise 3B14 suggests the origin of the name "least squares:" a sum of squares is being minimized. The exercise contains enough hints to prove the following theorem.

THEOREM 1.
(i) $e \perp X$.
(ii) As a function of the $p \times 1$ vector γ, $\|Y - X\gamma\|^2$ is minimized when $\gamma = \hat{\beta}$.

THEOREM 2. OLS is conditionally unbiased, that is, $E(\hat{\beta}|X) = \beta$.

Proof. $\hat{\beta} = (X'X)^{-1}X'Y$ by (6). The model (1) says that $Y = X\beta + \epsilon$, so

$$\hat{\beta} = (X'X)^{-1}X'(X\beta + \epsilon)$$
$$= (X'X)^{-1}X'X\beta + (X'X)^{-1}X'\epsilon$$
$$= \beta + (X'X)^{-1}X'\epsilon.$$

For the last step, $(X'X)^{-1}X'X = (X'X)^{-1}(X'X) = I_{p\times p}$ and $I_{p\times p}\beta = \beta$. Thus,

(8) $$\hat{\beta} = \beta + \eta \quad \text{where } \eta = (X'X)^{-1}X'\epsilon.$$

Now $E(\eta|X) = E((X'X)^{-1}X'\epsilon|X) = (X'X)^{-1}X'E(\epsilon|X)$. We've conditioned on X, so X is fixed (not random). Ditto for matrices that only depend on X. They factor out of the expectation (exercise 3C3). What we've shown so far is

(9) $$E(\hat{\beta}|X) = \beta + (X'X)^{-1}X'E(\epsilon|X).$$

Next, $X \perp\!\!\!\perp \epsilon$ by assumption (5): conditioning on X does not change the distribution of ϵ. But $E(\epsilon) = 0_{n\times 1}$ by assumption (4). Thus, $E(\hat{\beta}|X) = \beta$, completing the proof.

Example 1. Hooke's law (section 2.3, i.e., section 3 in chapter 2). Look at equation (2.7). The parameter vector β is 2×1:

$$\beta = \begin{pmatrix} a \\ b \end{pmatrix}.$$

The design matrix X is 6×2. The first column is all 1's, to accommodate the intercept a. The second column is the column of weights in table 2.1. In matrix form, then, the model is $Y = X\beta + \epsilon$, where

$$Y = \begin{pmatrix} Y_1 \\ Y_2 \\ Y_3 \\ Y_4 \\ Y_5 \\ Y_6 \end{pmatrix}, \quad X = \begin{pmatrix} 1 & 0 \\ 1 & 2 \\ 1 & 4 \\ 1 & 6 \\ 1 & 8 \\ 1 & 10 \end{pmatrix}, \quad \epsilon = \begin{pmatrix} \epsilon_1 \\ \epsilon_2 \\ \epsilon_3 \\ \epsilon_4 \\ \epsilon_5 \\ \epsilon_6 \end{pmatrix}.$$

Let's check the first row. Since $X_1 = (1 \ 0)$, the first row in the matrix equation says that $Y_1 = X_1\beta + \epsilon_1 = a + 0b + \epsilon_1 = a + \epsilon_1$. This is equation (2.7) for $i = 1$. Similarly for the other rows.

We want to compute $\hat{\beta}$ from (6), so data on Y are needed. That is where the "length" column in table 2.1 comes into the picture. The model says that the lengths of the spring under the various loads are the observed values of $Y = X\beta + \epsilon$. These observed values are

$$\begin{pmatrix} 439.00 \\ 439.12 \\ 439.21 \\ 439.31 \\ 439.40 \\ 439.50 \end{pmatrix}.$$

Now we can compute the OLS estimates from (6).

$\hat{\beta} = (X'X)^{-1}X'Y$

$$= \left[\begin{pmatrix} 1 & 1 & 1 & 1 & 1 & 1 \\ 0 & 2 & 4 & 6 & 8 & 10 \end{pmatrix} \begin{pmatrix} 1 & 0 \\ 1 & 2 \\ 1 & 4 \\ 1 & 6 \\ 1 & 8 \\ 1 & 10 \end{pmatrix} \right]^{-1} \begin{pmatrix} 1 & 1 & 1 & 1 & 1 & 1 \\ 0 & 2 & 4 & 6 & 8 & 10 \end{pmatrix} \begin{pmatrix} 439.00 \\ 439.12 \\ 439.21 \\ 439.31 \\ 439.40 \\ 439.50 \end{pmatrix}$$

$$= \begin{pmatrix} 439.01 \text{ cm} \\ .05 \text{ cm/kg} \end{pmatrix}.$$

Exercise set A

1. One of the following is true and the other is false. Which is which, and why?

$$\text{(i) } \epsilon \perp X \qquad \text{(ii) } \epsilon \perp\!\!\!\perp X$$

2. One of the following is true and the other is false. Which is which, and why?

$$\text{(i) } e \perp X \qquad \text{(ii) } e \perp\!\!\!\perp X$$

3. Does $e \perp X$ help validate assumption (5)?

4. Suppose the first column of X is all 1's, so the regression equation has an intercept—

   (a) Show that $\sum_i e_i = 0$.

   (b) Does $\sum_i e_i = 0$ help validate assumption (4)?

   (c) Is $\sum_i \epsilon_i = 0$? Or is $\sum_i \epsilon_i$ around $\sigma\sqrt{n}$ in size?

5. Show that $E(\epsilon'\epsilon|X) = n\sigma^2$ and $E(\epsilon\epsilon'|X) = \sigma^2 I_{n\times n}$.

6. How is column 2 in table 2.1 related to the regression model for Hooke's law? (Cross-references: table 2.1 is table 1 in chapter 2.)

7. Yule's regression model (1.1) for pauperism can be translated into matrix notation: $Y = X\beta + \epsilon$. We assume (3)-(4)-(5). For the metropolitan unions and the period 1871–81:

   (a) What are X and Y? (Hint: look at table 1.3.)

   (b) What are the observed values of X_{41}? X_{42}? Y_4?

   (c) Where do we look in $(X'X)^{-1}X'Y$ to find the estimated coefficient of ΔOut?

Note. These days, we use the computer to work out $(X'X)^{-1}X'Y$. Yule did it by hand, with a slide rule for multiplication and division. The slide rule got plenty of exercise.

4.2 Standard errors

Once we've computed the regression estimates, we need to see how accurate they are. If the model is right, this is pretty easy. Standard errors do the job. The first step is getting the covariance matrix of $\hat{\beta}$.

THEOREM 3. $\text{cov}(\hat{\beta}|X) = \sigma^2(X'X)^{-1}$.

Proof. Recall from (8) that $\eta = \hat{\beta} - \beta = (X'X)^{-1}X'\epsilon$. As theorem 2 shows, $E(\hat{\beta}|X) = \beta$. So

$$\begin{aligned}
\text{cov}(\hat{\beta}|X) &= E\big((\hat{\beta} - \beta)(\hat{\beta} - \beta)'|X\big) \\
&= E(\eta\eta'|X) \\
&= E\big((X'X)^{-1}X'\epsilon\epsilon'X(X'X)^{-1}|X\big) \\
&\qquad\qquad \text{because } X'X \text{ is symmetric and } (AB)' = B'A'
\end{aligned}$$

$$= (X'X)^{-1}X'E(\epsilon\epsilon'|X)X(X'X)^{-1}$$
$$= (X'X)^{-1}X'\sigma^2 I_{n\times n}X(X'X)^{-1}$$
$$= \sigma^2(X'X)^{-1}X'X(X'X)^{-1}$$
$$= \sigma^2(X'X)^{-1}.$$

The argument is like the one in theorem 2. Conditionally, X is fixed. So are matrices that only involve X. They factor out of the expectation. Since $\epsilon \perp\!\!\!\perp X$, conditioning on X does not change the distribution of ϵ. And $E(\epsilon\epsilon'|X) = \sigma^2 I_{n\times n}$ by exercise A5, which completes the proof of theorem 3.

Usually, σ^2 is unknown and has to be estimated from the data. If we knew the ϵ_i, we could estimate σ^2 as

$$\frac{1}{n}\sum_{i=1}^{n}\epsilon_i^2.$$

But we don't know the ϵ's. The next thing to try might be

$$\frac{1}{n}\sum_{i=1}^{n}e_i^2.$$

This is a little too small. The e_i are generally smaller than the ϵ_i, because $\hat{\beta}$ was chosen to make the sum of the e_i^2 as small as possible. The usual fix is to divide by the *degrees of freedom* $n - p$ rather than n:

$$(10) \qquad \hat{\sigma}^2 = \frac{1}{n-p}\sum_{i=1}^{n}e_i^2.$$

This is conditionally unbiased (theorem 4 below). Equation (10) is the reason we need $n > p$ not just $n \geq p$. If $n = p$, the estimator $\hat{\sigma}^2$ is undefined: you would get $0/0$. See exercise B12 below.

The proof that $\hat{\sigma}^2$ is unbiased is a little complicated, so let's postpone it for a minute and look at the bigger picture. We can estimate the parameter vector β in the model (1) by OLS: $\hat{\beta} = (X'X)^{-1}X'Y$. Conditionally on X, this estimator is unbiased, and the covariance matrix is $\sigma^2(X'X)^{-1}$. All is well, except that σ^2 is unknown. We just plug in $\hat{\sigma}^2$, which is (almost) the mean square of the residuals—the sum of squares is divided by the *degrees of freedom* $n - p$ not by n. To sum up,

$$(11) \qquad \widehat{\text{cov}}(\hat{\beta}|X) = \hat{\sigma}^2(X'X)^{-1}.$$

The variances are on the diagonal. Variances are the wrong size and have the wrong units: take the square root of the variances to get the standard errors. (What are the off-diagonal elements good for? See theorem 5 below, and the discussion that follows: you will need the off-diagonal elements to compute the standard error of, e.g., $\hat{\beta}_2 - \hat{\beta}_3$.)

Back to the mathematics. Before tackling theorem 4, we discuss the "hat matrix,"

$$H = X(X'X)^{-1}X',$$

and the "predicted" or "fitted" values,

$$\hat{Y} = X\hat{\beta}.$$

The terminology of "predicted values" can be misleading, since these are computed from the actual values. Nothing is being predicted. "Fitted values" is better.

The hat matrix is $n \times n$, because X is $n \times p$, $X'X$ is $p \times p$, $(X'X)^{-1}$ is $p \times p$, and X' is $p \times n$. On the other hand, $\hat{Y}$ is $n \times 1$. The fitted values are connected to the hat matrix by the equation

(12) $$\hat{Y} = X(X'X)^{-1}X'Y = HY.$$

(The equation, and the hat on Y, might explain the name "hat matrix.") Check these facts, with $I_{n \times n}$ abbreviated to I:

    (i) $e = (I - H)Y$.

    (ii) H is symmetric, and so is $I - H$.

    (iii) H is idempotent ($H^2 = H$), and so is $I - H$.

    (iv) X is invariant under H, that is, $HX = X$.

    (v) $e = Y - HY \perp X$.

Thus, H projects Y into the column space of X: in more detail, $HY = \hat{Y} = X\hat{\beta} \in \text{cols } X$, and (v) finishes the argument. Next,

    (vi) $(I - H)X = 0$.

    (vii) $(I - H)H = H(I - H) = 0$. Hint: use fact (iii).

THEOREM 4. $E(\hat{\sigma}^2|X) = \sigma^2$.

Proof. We claim that

(13) $$e = (I - H)\epsilon.$$

Indeed, by facts (i) and (vi) about the hat matrix,

$$e = (I - H)Y = (I - H)(X\beta + \epsilon) = (I - H)\epsilon.$$

We write $\tilde{H}$ for $I_{n \times n} - H$, and claim that

(14) $$\|e\|^2 = \epsilon' \tilde{H} \epsilon.$$

Indeed, $\tilde{H}$ is symmetric and idempotent—facts (ii) and (iii) about the hat matrix—so $\|e\|^2 = e'e = \epsilon' \tilde{H}^2 \epsilon = \epsilon' \tilde{H} \epsilon$, proving (14). Check that

$$E(\epsilon' \tilde{H} \epsilon | X) = E\left(\sum_{i=1}^{n} \sum_{j=1}^{n} \epsilon_i \tilde{H}_{ij} \epsilon_j \Big| X \right) = \sum_{i=1}^{n} \sum_{j=1}^{n} E(\epsilon_i \tilde{H}_{ij} \epsilon_j | X).$$

The next step is to simplify the double sum on the right. The matrix $\tilde{H}$ is fixed: we've conditioned on X. Conditioning on X doesn't change the distribution of ϵ, because $\epsilon \perp\!\!\!\perp X$. If $i \neq j$, then $E(\epsilon_i \tilde{H}_{ij} \epsilon_j | X) = 0$ because ϵ_i and ϵ_j are independent with $E(\epsilon_i) = 0$. On the other hand, $E(\epsilon_i \tilde{H}_{ii} \epsilon_i | X) = \sigma^2 \tilde{H}_{ii}$. Thus,

$$E(\epsilon' \tilde{H} \epsilon | X) = \sigma^2 \sum_{i=1}^{n} \tilde{H}_{ii} = \sigma^2 \text{trace}(\tilde{H}).$$

By (14),
$$E(\|e\|^2 | X) = \sigma^2 \text{trace}(\tilde{H}).$$

Now we have to work out the trace. Remember, $H = X(X'X)^{-1}X'$. By exercise 3B11,

$$\text{trace}(H) = \text{trace}\left[(X'X)^{-1} X'X \right] = \text{trace}(I_{p \times p}) = p.$$

Then $\text{trace}(\tilde{H}) = \text{trace}(I_{n \times n} - H) = \text{trace}(I_{n \times n}) - \text{trace}(H) = n - p$. So

$$E(\|e\|^2 | X) = \sigma^2 (n - p).$$

To wrap things up,

$$E(\hat{\sigma}^2 | X) = \frac{1}{n-p} E(\|e\|^2 | X) = \frac{1}{n-p} \sigma^2 (n - p) = \sigma^2,$$

completing the proof of theorem 4.

Things we don't need

Theorems 1–4 show that under certain conditions, OLS is a good way to estimate a model; also see theorem 5 below. There are a lot of assumptions we *don't* need to make. For instance—

- The columns of X don't have to be orthogonal to each other.
- The random errors don't have to be normally distributed.

Exercise set B

The first five exercises concern the regression model (1)–(5), and X_i denotes the ith row of the design matrix X.

1. True or false: $E(Y_i|X) = X_i\beta$.

2. True or false: the sample mean of the Y_i's is $\overline{Y} = n^{-1}\sum_{i=1}^{n} Y_i$. Is $\overline{Y}$ a random variable?

3. True or false: $\text{var}(Y_i|X) = \sigma^2$.

4. True or false: the sample variance of the Y_i's is $n^{-1}\sum_{i=1}^{n}(Y_i - \overline{Y})^2$. Is this a random variable?

5. Conditionally on X, show that the joint distribution of the random vectors $(\hat{\beta} - \beta, e)$ is the same for all values of β. Hint: express $(\hat{\beta} - \beta, e)$ in terms of X and ϵ.

6. Can you put standard errors on the estimated coefficients in Yule's equation (1.2)? Explain briefly. Hint: see exercise A7.

7. In section 2.3, we estimated the intercept and slope for Hooke's law. Can you put standard errors on these estimates? Explain briefly.

8. Here are two equations:
$$\text{(i) } Y = X\beta + \epsilon \qquad \text{(ii) } Y = X\hat{\beta} + e$$
   Which is the regression model? Which equation has the parameters and which has the estimates? Which equation has the random errors? Which has the residuals?

9. We use the OLS estimator $\hat{\beta}$ in the usual regression model, and the unbiased estimator of variance $\hat{\sigma}^2$. Which of the following statements are true, and why?
   (i) $\text{cov}(\beta) = \sigma^2(X'X)^{-1}$.
   (ii) $\text{cov}(\hat{\beta}) = \sigma^2(X'X)^{-1}$.
   (iii) $\text{cov}(\hat{\beta}|X) = \sigma^2(X'X)^{-1}$.
   (iv) $\text{cov}(\hat{\beta}|X) = \hat{\sigma}^2(X'X)^{-1}$.
   (v) $\widehat{\text{cov}}(\hat{\beta}|X) = \hat{\sigma}^2(X'X)^{-1}$.

10. True or false, and explain.
    (a) If you fit a regression equation to data, the sum of the residuals is 0.
    (b) If the equation has an intercept, the sum of the residuals is 0.

11. True or false, and explain.
    (a) In the regression model, $E(\hat{Y}|X) = X\hat{\beta}$.
    (b) In the regression model, $E(\hat{Y}|X) = X\beta$.

12. If X is $n \times n$ with rank n, show that $X(X'X)^{-1}X' = I_{n \times n}$, so $\hat{Y} = Y$.
    Hint: is X invertible?

13. Suppose there is an intercept in the regression model (1), so the first
    column of X is all 1's. Let $\overline{Y}$ be the mean of Y. Let $\overline{X}$ be the mean of
    X, column by column. Show that $\overline{Y} = \overline{X}\hat{\beta}$.

14. (Hard.) Suppose $Y_i = a + bX_i + \epsilon_i$ for $i = 1, \ldots, n$, the ϵ_i being IID
    with mean 0 and variance σ^2, independent of the X_i. (Reminder: IID
    stands for "independent and identically distributed.") Equation (2.5)
    expressed $\hat{a}, \hat{b}$ in terms of five summary statistics: two means, two SDs,
    and r. Derive the formulas for $\hat{a}, \hat{b}$ from equation (6) in this chapter.
    Show also that, conditionally on X,

$$\text{SE}\,\hat{a} = \frac{\sigma}{\sqrt{n}}\sqrt{1 + \frac{\overline{X}^2}{\text{var}(X)}}, \qquad \text{SE}\,\hat{b} = \frac{\sigma}{s_X\sqrt{n}},$$

where $\overline{X} = \frac{1}{n}\sum_{i=1}^n X_i$, $\text{var}(X) = \frac{1}{n}\sum_{i=1}^n (X_i - \overline{X})^2$, and $s_X^2 = \text{var}(X)$.
Hints. The design matrix M will be $n \times 2$. What is the first column? the
second? Find $M'M$. Show that $\det(M'M) = n^2\text{var}(X)$. Find $(M'M)^{-1}$
and $M'Y$.

4.3 Explained variance in multiple regression

After fitting the regression model, we have the equation $Y = X\hat{\beta} + e$.
All the quantities are observable. Suppose the equation has an intercept, so
there is a column of 1's in X. We will show in a bit that

(15) $\text{var}(Y) = \text{var}(X\hat{\beta}) + \text{var}(e)$.

Here, we think of Y as a data variable, and

$$\text{var}(Y) = \frac{1}{n}\sum_{i=1}^n (Y_i - \overline{Y})^2.$$

Variances on the right hand side of (15) are defined in a similar way: $\text{var}(X\hat{\beta})$
is called "explained variance," and $\text{var}(e)$ is "unexplained" or "residual" variance. The fraction of variance "explained" by the regression is

(16) $R^2 = \text{var}(X\hat{\beta})/\text{var}(Y)$.

The proof of (15) takes a bit of algebra. Let u be an $n \times 1$ column of 1's, corresponding to the intercept in the regression equation. Recall that $Y = X\hat{\beta} + e$. As always, $e \perp X$, so $\bar{e} = 0$. Now

$$Y - \bar{Y}u = X\hat{\beta} - \bar{Y}u + e,$$

$$\|Y - \bar{Y}u\|^2 = \|X\hat{\beta} - \bar{Y}u\|^2 + \|e\|^2,$$

$$\bar{Y} = \overline{X\hat{\beta}} = \bar{X}\hat{\beta},$$

$$\|Y - \bar{Y}u\|^2 = n\operatorname{var}(Y), \quad \|X\hat{\beta} - \bar{Y}u\|^2 = n\operatorname{var}(X\hat{\beta}), \quad \|e\|^2 = n\operatorname{var}(e),$$

(17) $$n\operatorname{var}(Y) = n\operatorname{var}(X\hat{\beta}) + n\operatorname{var}(e).$$

Division by n gives equation (15).

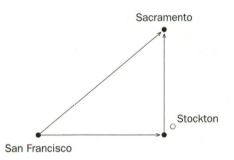

The math is fine, but the concept is a little peculiar. (Many people talk about explained variance, perhaps without sufficient consideration.) First, as a descriptive statistic, variance is the wrong size and has the wrong units. Second, well, let's take an example. Sacramento is about 78 miles from San Francisco, as the crow flies. Or, the crow could fly 60 miles East and 50 miles North, passing near Stockton at the turn. If we take the 60 and 50 as exact, Pythagoras tells us that the squared hypotenuse in the triangle is

$$60^2 + 50^2 = 3600 + 2500 = 6100 \text{ miles}^2.$$

With "explained" as in "explained variance," the geography lesson can be cruelly summarized. The area—squared distance—between San Francisco and Sacramento is 6100 miles^2, of which 3600 is explained by East.

The analogy is exact. Projecting onto East stands for (i) projecting Y and X orthogonally to the vector u that is all 1's, and then (ii) projecting the remainder of Y onto what is left of the column space of X. The hypotenuse of the triangle is $Y - \bar{Y}u$, with squared length $\|Y - \bar{Y}u\|^2 = n\operatorname{var}(Y)$. The horizontal edge is $X\hat{\beta} - \bar{Y}u$, with $\|X\hat{\beta} - \bar{Y}u\|^2 = n\operatorname{var}(X\hat{\beta})$. The vertical edge is e, and $\|e\|^2 = n\,\mathrm{var}(e)$. The theory of explained variance—equations (15)-(16)-(17)—boils down to Pythagoras' theorem on the crow's triangular flight. Explaining the area between San Francisco and Sacramento by East is zany, and explained variance may not be much better.

Although "explained variance" is peculiar terminology, R^2 is a useful descriptive statistic. High R^2 indicates a good fit between the data and the equation: the residuals are small relative to the SD of Y. Conversely, low R^2 indicates a poor fit. In fields like political science and sociology, $R^2 < 1/10$ is commonplace. This may indicate large random effects, difficulties in measurement, and so forth. Or, there may be many important factors omitted from the equation, which might raise questions about confounding.

Association or causation?

R^2 measures goodness of fit, not the validity of any underlying causal model. For example, over the period 1950–1999, the correlation between the purchasing power of the United States dollar each year and the death rate from lung cancer in that year is -0.95. So $R^2 = (-0.95)^2 = 0.9$, which is a lot bigger than what you find in run-of-the-mill regression studies of causation. If you run a regression of lung cancer death rates on the purchasing power of the dollar, the data will follow the line very closely.

Inflation, however, neither causes nor prevents lung cancer. The purchasing power of the dollar was going steadily downhill from 1950 to 1999. Death rates from lung cancer were generally going up (with a peak in 1990). These facts create a high R^2. Death rates from lung cancer were going up because of increases in smoking during the first half of the century. And the value of the dollar was shrinking because, well, let's not go there.

4.4 Generalized least squares

The OLS regression model says that $Y = X\beta + \epsilon$, where Y is an $n \times 1$ vector of observable random variables, X is an $n \times p$ matrix of observable random variables with rank $p < n$, and ϵ is an $n \times 1$ vector of unobservable random variables, IID with mean 0 and variance σ^2, independent of X. In this section, we're going to drop the independence assumptions about ϵ, and make a weaker—less restrictive—set of assumptions:

$$(18) \qquad\qquad E(\epsilon|X) = 0_{n\times 1}, \quad \mathrm{cov}(\epsilon|X) = \sigma^2 I_{n\times n}.$$

Theorems 1–4 continue to hold. The weaker assumptions will be more convenient for comparing GLS (Generalized Least Squares) to OLS. We begin with a theorem about OLS in the case where X is not random. In that case, condition (18) is simpler: $E(\epsilon) = 0_{n \times 1}$, and $\text{cov}(\epsilon) = \sigma^2 I_{n \times n}$.

THEOREM 5. GAUSS-MARKOV. Assume (1) and (18). Suppose X is fixed (i.e., not random). The OLS estimator is BLUE.

The acronym BLUE stands for Best Linear Unbiased Estimator, i.e., the one with the smallest variance. Let $\gamma = c'\beta$, where c is $p \times 1$: the parameter γ is a linear combination of the components of β. Examples would include β_1, or $\beta_2 - \beta_3$. The OLS estimator for γ is $\hat{\gamma} = c'\hat{\beta} = c'(X'X)^{-1}X'Y$. This is unbiased by (18), and $\text{var}(\hat{\gamma}) = \sigma^2 c'(X'X)^{-1}c$. Cf. exercise C2(a) below. Let $\tilde{\gamma}$ be another linear unbiased estimator for γ. Then $\text{var}(\tilde{\gamma}) \geq \text{var}(\hat{\gamma})$, and $\text{var}(\tilde{\gamma}) = \text{var}(\hat{\gamma})$ entails $\tilde{\gamma} = \hat{\gamma}$. That is what the theorem says.

Proof. A detailed proof is beyond our scope, but here is a sketch. Recall that X is fixed. Since $\tilde{\gamma}$ is by assumption a linear function of Y, there is an $n \times 1$ vector d with $\tilde{\gamma} = d'Y = d'X\beta + d'\epsilon$. Then $E(\tilde{\gamma}) = d'X\beta$ by (18). Since $\tilde{\gamma}$ is unbiased, $d'X\beta = c'\beta$ for all β. Therefore,

$$d'X = c'.$$

Let $q = d - X(X'X)^{-1}c$, an $n \times 1$ vector. So

$$q' = d' - c'(X'X)^{-1}X'.$$

(Motivation: $q'Y = \tilde{\gamma} - \hat{\gamma}$.) Multiply on the right by X:

$$q'X = d'X - c'(X'X)^{-1}X'X$$
$$= d'X - c' = 0_{1 \times p}$$

because $d'X = c'$. Thus

$$\text{var}(\tilde{\gamma}) = \text{var}(d'\epsilon)$$
$$= \sigma^2 d'd$$
$$= \sigma^2 [q' + c'(X'X)^{-1}X'][q + X(X'X)^{-1}c]$$
$$= \sigma^2 [q'q + c'(X'X)^{-1}c]$$
$$= \sigma^2 q'q + \text{var}(\hat{\gamma}).$$

The cross-product terms dropped out: $q'X(X'X)^{-1}c = c'(X'X)^{-1}X'q = 0$ because $q'X = 0_{1 \times p}$. Finally, $q'q = \sum_i q_i^2 \geq 0$. The inequality is strict unless $q = 0_{n \times 1}$, i.e., $\tilde{\gamma} = \hat{\gamma}$. This completes the proof.

We now keep the equation $Y = X\beta + \epsilon$, but change assumption (18) to

(19) $E(\epsilon|X) = 0_{n \times 1}, \quad \text{cov}(\epsilon|X) = G,$

where G is a positive definite $n \times n$ matrix: this is the *GLS regression model*. We can still define the OLS estimator $\hat{\beta}_{\text{OLS}}$ by (6). And $\hat{\beta}_{\text{OLS}}$ is still unbiased given X, by (9). However, the formula for $\mathrm{cov}(\hat{\beta}_{\mathrm{OLS}}|X)$ in theorem 3 no longer holds. Instead (exercise C2 below),

(20) $$\mathrm{cov}(\hat{\beta}_{\mathrm{OLS}}|X) = (X'X)^{-1}X'GX(X'X)^{-1}.$$

Moreover, $\hat{\beta}_{\mathrm{OLS}}$ is no longer BLUE. Some people regard this as a fatal flaw.

The fix—if you know G —is to transform equation (1). You multiply on the left by $G^{-1/2}$, getting

(21) $$(G^{-1/2}Y) = (G^{-1/2}X)\beta + (G^{-1/2}\epsilon).$$

(Why does $G^{-1/2}$ make sense? See exercise 3D7.) The transformed model has $G^{-1/2}Y$ as the response vector, $G^{-1/2}X$ as the design matrix, and $G^{-1/2}\epsilon$ as the vector of disturbances. The parameter vector is still β. Condition (18) holds (with $\sigma^2 = 1$) for the transformed model, by exercises 3C3 and 3C4. That was the whole point of the transformation—and the reason for introducing (18).

The *GLS estimator* for β is obtained by applying OLS to (21):

$$\hat{\beta}_{\mathrm{GLS}} = \left[(G^{-1/2}X)'(G^{-1/2}X)\right]^{-1}(G^{-1/2}X)'G^{-1/2}Y.$$

Since $(AB)' = B'A'$ and $G^{-1/2}G^{-1/2} = G^{-1}$,

(22) $$\hat{\beta}_{\mathrm{GLS}} = (X'G^{-1}X)^{-1}X'G^{-1}Y.$$

Exercise 1 in the next section shows that $X'G^{-1}X$ on the right hand side of (22) is invertible; X is $n \times p$, so X' is $p \times n$, while G and G^{-1} are $n \times n$. Thus, $X'G^{-1}X$ is $p \times p$: and $\hat{\beta}_{\mathrm{GLS}}$ is $p \times 1$, as it should be. By theorem 2,

(23) the GLS estimator is conditionally unbiased given X.

By theorem 3 and the tiniest bit of matrix algebra,

(24) $$\mathrm{cov}(\hat{\beta}_{\mathrm{GLS}}|X) = (X'G^{-1}X)^{-1}.$$

There is no σ^2 in the formula: σ^2 is built into G on the right hand side of (24). In the case of fixed X, the GLS estimator is BLUE by theorem 5.

In applications, G is usually unknown, and has to be estimated from the data. Constraints have to be imposed on G. Without constraints, there are too many covariances to estimate and not enough data. The estimate $\hat{G}$ is substituted for G in (22), giving the *feasible GLS* or *Aitken* estimator $\hat{\beta}_{\mathrm{FGLS}}$:

(25a) $$\hat{\beta}_{\mathrm{FGLS}} = (X'\hat{G}^{-1}X)^{-1}X'\hat{G}^{-1}Y.$$

Covariances would be estimated by plugging in $\hat{G}$ for G in (24):

(25b) $$\widehat{\text{cov}}(\hat{\beta}_{\text{FGLS}}|X) = (X'\hat{G}^{-1}X)^{-1}.$$

Sometimes the "plug-in" covariance estimator $\widehat{\text{cov}}$ is a good approximation. But sometimes it isn't—if there are a lot of covariances to estimate and not enough data to do it well (chapter 7). Moreover, feasible GLS is usually nonlinear. Therefore, $\hat{\beta}_{\text{FGLS}}$ is usually biased, at least by a little. Remember,

$$\hat{\beta}_{\text{FGLS}} \neq \hat{\beta}_{\text{GLS}}.$$

4.5 Examples on GLS

We are in the GLS model (1), with assumption (19) on the errors.

Example 2. Suppose Γ is a known positive definite $n \times n$ matrix, and $G = \lambda\Gamma$ where λ is an unknown parameter. We multiply (1) on the left by $\Gamma^{-1/2}$ and apply OLS to the transformed model:

$$(\Gamma^{-1/2}Y) = (\Gamma^{-1/2}X)\beta + (\Gamma^{-1/2}\epsilon).$$

Assumption (18) holds for this transformed model, with $\sigma^2 = \lambda$ to be estimated as part of the OLS fit. This is "weighted" least squares. Given X, the GLS estimator is linear and unbiased—because Γ is known and fixed. OLS is a special case, with $\Gamma = I_{n \times n}$.

Example 3. Suppose n is even, K is a positive definite 2×2 matrix, and

$$G = \begin{pmatrix} K & 0_{2\times2} & \cdots & 0_{2\times2} \\ 0_{2\times2} & K & \cdots & 0_{2\times2} \\ \vdots & \vdots & \ddots & \vdots \\ 0_{2\times2} & 0_{2\times2} & \cdots & K \end{pmatrix}.$$

The $n \times n$ matrix G has K repeated along the main diagonal. Here, K is unknown, to be estimated from the data. Chapter 7 has a case study with this sort of matrix.

Make a first pass at the data, estimating β by OLS. This gives $\hat{\beta}^{(0)}$, with residual vector $e = Y - X\hat{\beta}^{(0)}$. Estimate K using mean products of residuals:

$$\hat{K}_{11} = \frac{2}{n}\sum_{j=1}^{n/2} e_{2j-1}^2, \quad \hat{K}_{22} = \frac{2}{n}\sum_{j=1}^{n/2} e_{2j}^2, \quad \hat{K}_{12} = \hat{K}_{21} = \frac{2}{n}\sum_{j=1}^{n/2} e_{2j-1}e_{2j}.$$

Plug $\hat{K}$ into the formula for G, and then $\hat{G}$ into (22) to get $\hat{\beta}^{(1)}$, which is a feasible GLS estimator called *one-step* GLS. (Division by $n - 2$ is also fine.)

The estimation procedure can be repeated iteratively: get residuals off $\hat{\beta}^{(1)}$, use them to re-estimate K, use the new $\hat{K}$ to get a new $\hat{G}$. Now do feasible GLS again. Voilà: $\hat{\beta}^{(2)}$ is the *two-step* GLS estimator. People usually keep going, until the estimator settles down. This sort of procedure is called "iteratively reweighted" least squares.

Terminology. Consider the model (1), assuming only that $E(\epsilon|X) = 0_{n \times 1}$. Suppose too that the Y_i are uncorrelated given X, i.e., $\text{cov}(\epsilon|X)$ is a diagonal matrix. In this setup, *homoscedasticity* means that $\text{var}(Y_i|X)$ is the same for all i, so that assumption (18) holds—although σ^2 may depend on the design matrix. *Heteroscedasticity* means that $\text{var}(Y_i|X)$ isn't the same for all i, so that assumption (18) fails. Then people fall back on (19) and GLS.

Exercise set C

1. If the $n \times p$ matrix X has rank $p < n$ and G is $n \times n$ positive definite, show that $X'G^{-1}X$ is $p \times p$ positive definite, hence invertible. Hint: see exercise 3D7.

2. Let $\hat{\beta}$ be the OLS estimator in (1), where the design matrix X has full rank $p < n$.

    (a) Suppose c is $p \times 1$. Assume (18). Show that $E(c'\hat{\beta}|X) = c'\beta$ and $\text{var}(c'\hat{\beta}|X) = \sigma^2 c'(X'X)^{-1}c$. Find $\text{var}(\hat{\beta}_1 - \hat{\beta}_2|X)$, where $\hat{\beta}_i$ is the ith component of $\hat{\beta}$.

    (b) Assume (19), i.e., the GLS model. Show that $E(Y|X) = X\beta$ and $\text{cov}(Y|X) = G$. Verify that $E(\hat{\beta}|X) = \beta$ and $\text{cov}(\hat{\beta}|X) = (X'X)^{-1}X'GX(X'X)^{-1}$.

    Hint: look at the proofs of theorems 2 and 3. Bigger hint: look at equations (8) and (9).

3. Suppose U_i are IID for $i = 1, \ldots, m$ with mean α and variance σ^2. Suppose V_i are IID for $i = 1, \ldots, n$ with mean α and variance τ^2. The mean is the same, but variance and sample size are different. Suppose the U's and V's are independent. How would you estimate α if σ^2 and τ^2 are known? if σ^2 and τ^2 are unknown? Hint: get this into the GLS framework by defining $\epsilon_j = U_j - \alpha$ for $j = 1, \ldots, m$, and $\epsilon_j = V_{j-m} - \alpha$ for $j = m + 1, \ldots, m + n$.

4. Suppose $Y = X\beta + \epsilon$. The design matrix X is $n \times p$ with rank $p < n$, and $\epsilon \perp\!\!\!\perp X$. The ϵ_i are independent with $E(\epsilon_i) = 0$. However, $\text{var}(\epsilon_i) = \lambda c_i$. The c_i are known positive constants.

(a) If λ is known, show that the GLS estimator for β is the γ that minimizes

$$\sum_i (Y_i - X_i\gamma)^2/\text{var}\,(Y_i|X).$$

(b) If λ is unknown, show that the GLS estimator for β is the γ that minimizes

$$\sum_i (Y_i - X_i\gamma)^2/c_i.$$

5. (Hard.) There are three observations on a variable Y for each individual $i = 1, 2, \ldots, 800$. There is an explanatory variable Z, which is scalar. Maria thinks that each subject i has a "fixed effect" a_i and there is a parameter b common to all 800 subjects. Her model can be stated this way:

$$Y_{ij} = a_i + Z_{ij}b + \epsilon_{ij} \quad \text{for } i = 1, 2, \ldots, 800 \text{ and } j = 1, 2, 3.$$

She is willing to assume that the ϵ_{ij} are independent with mean 0. She also believes that the ϵ's are independent of the Z's and $\text{var}(\epsilon_{ij})$ is the same for $j = 1, 2, 3$. But she is afraid that $\text{var}(\epsilon_{ij}) = \sigma_i^2$ depends on the subject i. Can you get this into the GLS framework? What would you use for the response vector Y in (1)? The design matrix? (This will get ugly.) With her model, what can you say about G in (19)? How would you estimate her model?

4.6 What happens to OLS if the assumptions break down?

If $E(\epsilon|X) \neq 0_{n\times 1}$, the bias in the OLS estimator is $(X'X)^{-1}X'E(\epsilon|X)$, by equation (9). If $E(\epsilon|X) = 0_{n\times 1}$ but $\text{cov}(\epsilon|X) \neq \sigma^2 I_{n\times n}$, OLS will be unbiased but theorem 3 breaks down. Then $\hat{\sigma}^2(X'X)^{-1}$ may be a misleading estimator of $\text{cov}(\hat{\beta}|X)$. See (20).

If the assumptions behind OLS are wrong, the estimator can be severely biased. Even if the estimator is unbiased, standard errors computed from the data can be way off. Significance levels, which are based on the SEs, would not be trustworthy (section 7 below).

4.7 Normal theory

In this section and the next, we review the conventional theory of the OLS model (1), which conditions on X—an $n\times p$ matrix of full rank $p < n$—and restricts the ϵ_i to be independent $N(0, \sigma^2)$. The principal results are the t-test and the F-test. As usual, $e = Y - X\hat{\beta}$ is the vector of residuals. Fix

$k = 1, \ldots, p$. Write β_k for the kth component of the vector β. To test the null hypothesis that $\beta_k = 0$ against the alternative $\beta_k \neq 0$, we use the t-statistic. Here,

$$t = \hat{\beta}_k / \widehat{SE},$$

with $\widehat{SE}$ equal to $\hat{\sigma}$ times the square root of the kkth element of $(X'X)^{-1}$. We reject the null hypothesis when $|t|$ is large, e.g., $|t| > 2$. For testing at a fixed level, the critical value depends (to some extent) on $n - p$. When $n - p$ is large, people refer to the t-test as the "z-test:" under the null, t is close to $N(0, 1)$. If the terminology is unfamiliar, see the definitions below.

THEOREM 6. With independent $N(0, \sigma^2)$ errors, the OLS estimator $\hat{\beta}$ has a normal distribution with mean β and covariance matrix $\sigma^2 (X'X)^{-1}$. Moreover, $e \perp\!\!\!\perp \hat{\beta}$ and $\|e\|^2 \sim \sigma^2 \chi_d^2$ with $d = n - p$.

COROLLARY. Under the null hypothesis, t is distributed as $U/\sqrt{V/d}$, where $U \perp\!\!\!\perp V$, $U \sim N(0, 1)$, $V \sim \chi_d^2$, and $d = n - p$. In other words, if the null hypothesis is right, the t-statistic follows Student's t-distribution, with $n - p$ degrees of freedom.

Definitions. $U \sim N(0, 1)$, for instance, means that the random variable U is normally distributed with mean 0 and variance 1. Likewise, $U \sim W$ means that U and W have the same distribution. Suppose $U_1, U_2, \ldots$ are IID $N(0, 1)$. We write χ_d^2 for a variable distributed as $\sum_{i=1}^{d} U_i^2$, and say that χ_d^2 has the *chi-squared distribution* with d degrees of freedom. Furthermore,

$$U_{d+1} / \sqrt{d^{-1} \sum_{i=1}^{d} U_i^2}$$

has *Student's t-distribution* with d degrees of freedom.

Sketch of proof. In the leading special case, X vanishes except along the main diagonal of the top p rows, where $X_{ii} = 1$. For instance, if $n = 5$ and $p = 2$,

$$X = \begin{pmatrix} 1 & 0 \\ 0 & 1 \\ 0 & 0 \\ 0 & 0 \\ 0 & 0 \end{pmatrix}.$$

The theorem and corollary are pretty obvious in the special case, because $Y_i = \beta_i + \epsilon_i$ for $i \leq p$ and $Y_i = \epsilon_i$ for $i > p$. Furthermore, $\hat{\beta}$ consists of the first p elements of Y, $\mathrm{cov}(\hat{\beta}) = \sigma^2 I_{p \times p} = \sigma^2 (X'X)^{-1}$, and e consists of p zeros stacked on top of the last $n - p$ elements of Y.

The general case is beyond our scope, but here is a sketch of the argument. There is a $p \times p$ upper triangular matrix M such that the columns of XM are orthonormal. To construct M, regress column j on the previous $j - 1$ columns (the "Gram-Schmidt process"). The residual vector from this regression is the part of column j orthogonal to the previous columns. Since X has rank p, column 1 cannot vanish; nor can column j be a linear combination of columns $1, \ldots, j - 1$. The orthogonal pieces can therefore be normalized to have length 1. A bit of matrix algebra shows this set of orthonormal vectors can be written as XM, where $M_{ii} \neq 0$ for all i and $M_{ij} = 0$ for all $i > j$. In particular, M is invertible.

Let S be the special $n \times p$ matrix discussed above, with the $p \times p$ identity matrix in the top p rows and 0's in the bottom $n - p$ rows. There is an $n \times n$ orthogonal matrix R with $RXM = S$. To get R, take the $p \times n$ matrix $(XM)'$, whose rows are orthonormal. Add $n - p$ rows: keep it all orthonormal....

Consider the transformed regression model $(RY) = (RXM)\gamma + \delta$, where $\gamma = M^{-1}\beta$ and $\delta = R\epsilon$. The δ_i are IID $N(0, \sigma^2)$: see exercise 3E2. Let $\hat{\gamma}$ be the OLS estimates from the transformed model, and let $f = RY - (RXM)\hat{\gamma}$ be the residuals. The special case of the theorem applies to the transformed model.

You can check that $\hat{\beta} = M\hat{\gamma}$. So $\hat{\beta}$ is multivariate normal, as required (section 3.5). The covariance matrix of $\hat{\beta}$ can be obtained from theorem 3. But here is a direct argument: $\mathrm{cov}(\hat{\beta}) = M\mathrm{cov}(\hat{\gamma})M' = \sigma^2 MM'$. We claim that $MM' = (X'X)^{-1}$. Indeed, $RXM = S$, so $XM = R'S$. Then $M'X'XM = S'RR'S = S'S = I_{p \times p}$. Multiply on the left by M'^{-1} and on the right by M^{-1} to see that $X'X = M'^{-1}M^{-1} = (MM')^{-1}$. Invert!

For the residuals, $e = R^{-1}f$, where f was the residual vector from the transformed model. But $R^{-1} = R'$ is orthogonal, so $\|e\|^2 = \|f\|^2 \sim \sigma^2 \chi^2_{n-p}$: cf. exercise 3D3. Independence is the last issue. In our leading special case, $f \perp\!\!\!\perp \hat{\gamma}$. Thus, $R^{-1}f \perp\!\!\!\perp M\hat{\gamma}$, i.e., $e \perp\!\!\!\perp \hat{\beta}$, completing a sketch proof of theorem 6.

Suppose we drop the normality assumption, requiring only that the ϵ_i are independent and identically distributed with mean 0 and finite variance σ^2. If n is a lot larger than p, and the design matrix is not too weird, then $\hat{\beta}$ will be close to normal—thanks to the central limit theorem. Furthermore, $\|e\|^2/(n - p) \doteq \sigma^2$. The *observed significance level*—aka *P-value*—of the two-sided t-test will be essentially the area under the normal curve beyond $\pm \hat{\beta}_k / \widehat{SE}$. Without the normality assumption, however, little can be said about the asymptotic distribution of $\|e\|^2$.

Statistical significance

If $P < 10\%$, then $\hat{\beta}_k$ is statistically significant at the 10% level, or *barely significant*. If $P < 5\%$, then $\hat{\beta}_k$ is statistically significant at the 5% level, or *statistically significant*. If $P < 1\%$, then $\hat{\beta}_k$ is statistically significant at the 1% level, or *highly significant*. When $n - p$ is large, the respective cutoffs for a two-sided t-test are 1.64, 1.96, and 2.58: see page 282 below. If $\hat{\beta}_j$ and $\hat{\beta}_k$ are both statistically significant, the corresponding explanatory variables are said to have *independent effects* on Y: this has nothing to do with statistical independence.

Statistical significance is little more than technical jargon. Over the years, however, the jargon has acquired enormous—and richly undeserved—emotional power. For additional discussion, see Freedman-Pisani-Purves (1998, chapter 29).

Exercise set D

1. We have an OLS model with $p = 1$; and X is a column of 1's. Find $\hat{\beta}$ and $\hat{\sigma}^2$ in terms of Y and n. If the errors are IID $N(0, \sigma^2)$, find the distribution of $\hat{\beta} - \beta$, $\hat{\sigma}^2$, and $\sqrt{n}(\hat{\beta} - \beta)/\hat{\sigma}$. Hint: see exercise 3B16.

2. Lei is a PhD student in sociology. She has a regression equation $Y_i = a + bX_i + Z_i\gamma + \epsilon_i$. Here, X_i is a scalar, while Z_i is a 1×5 vector of control variables, and γ is a 5×1 vector of parameters. Her theory is that $b \neq 0$. She is willing to assume that the ϵ_i are IID $N(0, \sigma^2)$, independent of X and Z. Fitting the equation to data for $i = 1, \dots, 57$ by OLS, she gets $\hat{b} = 3.79$ with $\hat{SE} = 1.88$. True or false and explain—

    (a) For testing the null hypothesis that $b = 0$, $t \doteq 2.02$. (Reminder: the dotted equals sign means "about equal.")

    (b) $\hat{b}$ is statistically significant.

    (c) $\hat{b}$ is highly significant.

    (d) The probability that $b \neq 0$ is about 95%.

    (e) The probability that $b = 0$ is about 5%.

    (f) If the model is right and $b = 0$, there is about a 5% chance of getting $|\hat{b}/\widehat{SE}| > 2$.

    (g) If the model is right and $b = 0$, there is about a 95% chance of getting $|\hat{b}/\widehat{SE}| < 2$.

    (h) Lei can be about 95% confident that $b \neq 0$.

    (i) The test shows the model is right.

    (j) The test assumes the model is right.

(k) If the model is right, the test gives some evidence that $b \neq 0$.

3. A philosopher of science writes,

> "Suppose we toss a fair coin 10,000 times, the first 5000 tosses being done under a red light, and the last 5000 under a green light. The color of the light does not affect the coin. However, we would expect the statistical null hypothesis—that exactly as many heads will be thrown under the red light as the green light—would very likely not be true. There will nearly always be random fluctuations that make the statistical null hypothesis false."

Has the null hypothesis been set up correctly? Explain briefly.

4. An archeologist fits a regression model, rejecting the null hypothesis that $\beta_2 = 0$, with $P < 0.005$. True or false and explain:

    (a) β_2 must be large.

    (b) $\hat{\beta}_2$ must be large.

4.8 The F-test

We are in the OLS model (1). The design matrix X has full rank $p < n$. The ϵ_i are independent $N(0, \sigma^2)$ with $\epsilon \perp\!\!\!\perp X$. We condition on X. Suppose $p_0 \geq 1$ and $p_0 \leq p$. We are going to test the null hypothesis that the last p_0 of the β_i's are 0: that is, $\beta_i = 0$ for $i = p - p_0 + 1, \ldots, p$. The alternative hypothesis is $\beta_i \neq 0$ for at least one $i = p - p_0 + 1, \ldots, p$. The usual test statistic is called F, in honor of Sir R. A. Fisher. To define F, we need to fit a big model and a small model.

   (A) First, we fit the full model. Let $\hat{\beta}$ be the OLS estimate, and e the residual vector.

   (B) Next, we fit the smaller model that satisfies the null hypothesis: $\beta_i = 0$ for all $i = p - p_0 + 1, \ldots, p$. Let $\hat{\beta}^{(s)}$ be the OLS estimate for the smaller model.

In effect, the smaller model just drops the last p_0 columns of X; then $\hat{\beta}^{(s)}$ is a $(p - p_0) \times 1$ vector. Or, think of $\hat{\beta}^{(s)}$ as $p \times 1$, the last p_0 entries being 0. The test statistic is

$$F = \frac{\left(\|X\hat{\beta}\|^2 - \|X\hat{\beta}^{(s)}\|^2 \right)/p_0}{\|e\|^2/(n - p)}.$$

Example 4. We have a regression model

$$Y_i = a + bu_i + cv_i + dw_i + fz_i + \epsilon_i \text{ for } i = 1, \ldots, 72.$$

(The coefficients skip from d to f because e is used for the residual vector in the big model.) The u, v, w, z are just data, and the design matrix has full rank. The ϵ_i are IID $N(0, \sigma^2)$. There are 72 data points and β has 5 components:

$$\beta = \begin{pmatrix} a \\ b \\ c \\ d \\ f \end{pmatrix}.$$

So $n = 72$ and $p = 5$. We want to test the null hypothesis that $d = f = 0$. So $p_0 = 2$ and $p - p_0 = 3$. The null hypothesis leaves the first 3 parameters alone but constrains the last 2 to be 0. The small model would just drop w and z from the equation, leaving $Y_i = a + bu_i + cv_i + \epsilon_i$ for $i = 1, \ldots, 72$.

To say this another way, the design matrix for the big model has 5 columns. The first column is all 1's, for the intercept. There are columns for u, v, w, z. The design matrix for the small model only has 3 columns. The first column is all 1's. Then there are columns for u and v. The small model throws away the columns for w and z. That is because the null hypothesis says $d = f = 0$. The null hypothesis does not allow the columns for w and z to come into the equation. To compute $X\hat{\beta}^{(s)}$, use the smaller design matrix; or, if you prefer, use the original design matrix and pad out $\hat{\beta}^{(s)}$ with two 0's.

THEOREM 7. With independent $N(0, \sigma^2)$ errors, under the null hypothesis,

$$\|X\hat{\beta}\|^2 - \|X\hat{\beta}^{(s)}\|^2 \sim U, \quad \|e\|^2 \sim V, \quad F \sim \frac{U/p_0}{V/(n-p)},$$

where $U \perp\!\!\!\perp V$, $U \sim \sigma^2 \chi^2_{p_0}$, and $V \sim \sigma^2 \chi^2_{n-p}$.

A reminder on the notation: p_0 is the number of parameters that are constrained to 0, while $\hat{\beta}^{(s)}$ estimates the other coefficients. The distribution of F under the null hypothesis is *Fisher's F-distribution*, with p_0 degrees of freedom in the numerator and $n - p$ in the denominator. The σ^2 cancels out. We reject when F is large, e.g., $F > 4$. For testing at a fixed level, the critical value depends on the degrees of freedom in numerator and denominator. See page 282 on finding critical values.

The theorem can be proved like theorem 6; details are beyond our scope. Intuitively, if the null hypothesis is right, numerator and denominator are both estimating σ^2, so F should be around 1. If p_0 and p are fixed while n gets large, and the design matrix behaves itself, the normality assumption is not too important. If p_0, p, and $n - p$ are similar in size, normality may be an issue. A careful (graduate-level) treatment of the t- and F-tests and related theory will be found in Lehmann (1991ab). Also see the comments after lab 7 at the back of the book.

"The" F-test in applied work

In journal articles, a typical regression equation will have an intercept and several explanatory variables. The regression output will usually include an F-test, with $p - 1$ degrees of freedom in the numerator and $n - p$ in the denominator. The null hypothesis will not be stated. The missing null hypothesis is that all the coefficients vanish, except for the intercept.

If F is significant, that is often thought to validate the model. Mistake. The F-test takes the model as given. Significance only means this: *if* the model is right *and* the coefficients are 0, it was very unlikely to get such a big F-statistic. Logically, there are three possibilities on the table. (i) An unlikely event occurred. (ii) Or the model is right and some of the coefficients differ from 0. (iii) Or the model is wrong. So?

Exercise set E

1. Suppose $U_i = \alpha + \delta_i$ for $i = 1, \ldots, n$. The δ_i are independent $N(0, \sigma^2)$. The parameters α and σ^2 are unknown. How would you test the null hypothesis that $\alpha = 0$ against the alternative that $\alpha \neq 0$?

2. Suppose U_i are independent $N(\alpha, \sigma^2)$ for $i = 1, \ldots, n$. The parameters α and σ^2 are unknown. How would you test the null hypothesis that $\alpha = 0$ against the alternative that $\alpha \neq 0$?

3. In exercise 1, what happens if the δ_i are IID with mean 0, but are not normally distributed? if n is small? large?

4. In Yule's model (1.1), how would you test the null hypothesis $c = d = 0$ against the alternative $c \neq 0$ or $d \neq 0$? Be explicit. You can use the metropolitan unions, 1871–81, for an example. What assumptions would be needed on the errors in the equation? (See lab 8 at the back of the book.)

5. There is another way to define the numerator of the F-statistic. Let $e^{(s)}$ be the vector of residuals from the small model. Show that
$$\|X\hat{\beta}\|^2 - \|X\hat{\beta}^{(s)}\|^2 = \|e^{(s)}\|^2 - \|e\|^2.$$
Hint: what is $\|X\hat{\beta}^{(s)}\|^2 + \|e^{(s)}\|^2$?

6. (Hard.) George uses OLS to fit a regression equation with an intercept, and computes R^2. Georgia wants to test the null hypothesis that all the

coefficients are 0, except for the intercept. Can she compute F from R^2, n, and p? If so, what is the formula? If not, why not?

7. (Hard.) For a regression equation with an intercept, show that R^2 is the square of the correlation between $\hat{Y}$ and Y.

4.9 Data snooping

The point of testing is to help distinguish between real effects and chance variation. People sometimes jump to the conclusion that a result which is statistically significant cannot be explained as chance variation. However, even if the null hypothesis is right, there is a 5% chance of getting a "statistically significant" result, and there is 1% chance to get a "highly significant" result. An investigator who makes 100 tests can expect to get five results that are "statistically significant" and one that is "highly significant," even if the null hypothesis is right in every case.

Investigators often decide which hypotheses to test only after they've seen the data. Statisticians call this *data snooping*. To avoid being fooled by statistical artifacts, it would help to know how many tests were run before "statistically significant" differences turned up. Such information is seldom reported.

Replicating studies would be even more useful, so the statistical analysis could be repeated on an independent batch of data. This is commonplace in the physical and health sciences, rare in the social sciences. An easier option is *cross validation*: you put half the data in cold storage, and look at it only after deciding which models to fit. This isn't as good as real replication but it's much better than nothing. Cross validation is standard in some fields, not in others.

Investigators often screen out insignificant variables and refit the equations before publishing their models. What does this data snooping do to P-values?

Example 5. Suppose Y consists of 100 independent random variables, each being $N(0, 1)$. This is pure noise. The design matrix X is 100×50. All the variables are independent $N(0, 1)$. More noise. We regress Y on X. There won't be much to report, although we can expect an R^2 of around $50/100 = 0.5$. (This follows from theorem 7, with $n = 100$ and $p_0 = p = 50$, so $\hat{\beta}^{(s)} = 0_{50 \times 1}$.)

But now, suppose we test each of the 50 coefficients at the 10% level, and keep only the "significant" variables. There will be about $50 \times 0.1 = 5$ keepers. If we just run the regression on the keepers, quietly discarding the other variables, we are likely to get a decent R^2—by social-science standards—and dazzling t-statistics. One simulation, for example, gave 5 keeper columns out of 50 starters in X. In the regression of Y on the keepers, the R^2 was 0.2, and the t-statistics were -1.037, 3.637, 3.668, -3.383, -2.536.

This is just one simulation. Maybe the data set was exceptional? Try it yourself. There is one gotcha. The expected number of keepers is 5, but the SD is over 3, so there is a lot of variability. With more keepers, the R^2 is likely to be better; with fewer keepers, R^2 is worse. There is a small chance of having no keepers at all—in which case, try again. . . .

R^2 *without an intercept.* If there is no intercept in a regression equation, R^2 is defined as

(26) $$\|\hat{Y}\|^2/\|Y\|^2.$$

Exercise set F

1. The number of keeper columns isn't binomial. Why not?

2. In a regression equation without an intercept, show that $1 - R^2 = \|e\|^2/\|Y\|^2$, where $e = Y - \hat{Y}$ is the vector of residuals.

4.10 Discussion questions

Some of these questions cover material from previous chapters.

1. Suppose X_i are independent normal random variables with variance 1, for $i = 1, 2, 3$. The means are $\alpha + \beta$, $\alpha + 2\beta$, and $2\alpha + \beta$, respectively. How would you estimate the parameters α, β?

2. In the OLS regression model—

    (a) Is it the residuals that are independent from one subject to another, or the random errors?

    (b) Is it the residuals that are independent of the explanatory variables, or the random errors?

    (c) Is it the vector of residuals that is orthogonal to the column space of the design matrix, or the vector of random errors?

    Explain briefly.

3. In the OLS regression model, do the residuals always have mean 0? Discuss briefly.

4. True or false, and explain. If the disturbance terms in a regression equation are correlated with each other across subjects, then—

    (a) the OLS estimates are likely to be biased;

    (b) the estimated standard errors are likely to be biased.

5. An OLS regression model is defined by equation (2), with assumptions (4) and (5) on the ϵ's. Are the Y_i independent? identically distributed? Discuss briefly.

6. You are using OLS to fit a regression equation. True or false, and explain:

    (a) If you exclude a variable from the equation, but the excluded variable is orthogonal to the other variables in the equation, you won't bias the estimated coefficients of the remaining variables.

    (b) If you exclude a variable from the equation, and the excluded variable isn't orthogonal to the other variables, your estimates are going to be biased.

    (c) If you put an extra variable into the equation—as long as that extra variable is independent of the error term—you won't bias the estimated coefficients.

    (d) If you put an extra variable into the equation, you are likely to bias the estimated coefficients—if that extra variable is dependent on the error term.

7. True or false, and explain: as long as the design matrix has full rank, the computer can find the OLS estimator $\hat{\beta}$. If so, what are the assumptions good for? Discuss briefly.

8. Does R^2 measure the degree to which a regression equation fits the data? Or does it measure the validity of the model? Discuss briefly.

9. The F-test, like the t-test, assumes something in order to demonstrate something. What needs to be assumed, and what can be demonstrated? To what extent can the model itself be tested using F? Discuss briefly.

10. Suppose $Y_i = au_i + bv_i + \epsilon_i$ for $i = 1, \ldots, 100$. The ϵ_i are independent $N(0, 1)$. The u's and v's are fixed not random; these two data variables have mean 0 and variance 1: the correlation between them is r. If $r = \pm 1$, show that the design matrix has rank 1. Otherwise, let $\hat{a}, \hat{b}$ be the OLS estimator. Find the variance of $\hat{a}$; of $\hat{b}$; of $\hat{a} - \hat{b}$. What happens if $r = 0.99$? What are the implications of collinearity for applied work? For instance, what sort of inferences about a and b are made easier or harder by collinearity? Comments. *Collinearity* sometimes means $r = \pm 1$; more often, it means $r \doteq \pm 1$. A synonym is *multicollinearity*. The case $r = \pm 1$ is better called *exact collinearity*. Also see lab 9 at the back of the book.

11. True or false, and explain:
    (a) Collinearity leads to bias in the OLS estimates.
    (b) Collinearity leads to bias in the estimated standard errors for the OLS estimates.
    (c) Collinearity leads to big standard errors for some estimates.

12. Suppose $Y = X\beta + \epsilon$ where
    (i) X is $n \times p$ of rank p, and
    (ii) $E(\epsilon|X) = \gamma$, a non-random $n \times 1$ vector, and
    (iii) $\text{cov}(\epsilon|X) = G$, a non-random positive definite $n \times n$ matrix.
    Let $\hat{\beta} = (X'X)^{-1}X'Y$. True or false and explain:
    (a) $E(\hat{\beta}|X) = \beta$.
    (b) $\text{cov}(\hat{\beta}|X) = \sigma^2(X'X)^{-1}$.
    In (a), the exceptional case $\gamma \perp X$ should be discussed separately.

13. (This continues question 12.) Suppose $p > 1$, the first column of X is all 1's, and $\gamma_1 = \cdots = \gamma_n$.
    (a) Is $\hat{\beta}_1$ biased or unbiased given X?
    (b) What about $\hat{\beta}_2$?

14. Suppose $Y = X\beta + \epsilon$ where
    (i) X is fixed not random, $n \times p$ of rank p, and
    (ii) the ϵ_i are IID with mean 0 and variance σ^2, but
    (iii) the ϵ_i need not be normal.
    Let $\hat{\beta} = (X'X)^{-1}X'Y$. True or false and explain:
    (a) $E(\hat{\beta}) = \beta$.
    (b) $\text{cov}(\hat{\beta}) = \sigma^2(X'X)^{-1}$.
    (c) If $n = 100$ and $p = 6$, it is probably OK to use the t-test.
    (d) If $n = 100$ and $p = 96$, it is probably OK to use the t-test.

15. Suppose (X_i, W_i, δ_i) are IID as triplets across subjects $i = 1, \ldots, n$, where n is large; $E(X_i) = E(W_i) = E(\delta_i) = 0$, and δ_i is independent of (X_i, W_i). Happily, X_i and W_i have positive variance; they are not perfectly correlated. The response variable Y_i is in truth this:

$$Y_i = aX_i + bW_i + \delta_i.$$

We can recover a and b, up to random error, by running a regression of Y_i on X_i and W_i; no intercept is needed. (Why? And what's wrong with a perfect correlation?) However, Tom elects to run a regression of Y_i on X_i, omitting W_i. He will use the coefficient of X_i to estimate a.
    (a) What happens to Tom if X_i and W_i are independent?

(b) What happens to Tom if X_i and W_i are dependent?

Hint: see exercise 3B15.

16. Suppose (X_i, W_i, δ_i) are IID as triplets across subjects $i = 1, \ldots, n$, where n is large; $E(X_i) = E(W_i) = E(\delta_i) = 0$, and δ_i is independent of X_i. As it happens, $E(X_i^2) = E(\delta_i^2) = 1$, and $E(W_i^2) > 0$. The response variable Y_i is in truth this:

$$Y_i = aX_i + \delta_i .$$

We can recover a, up to random error, by running a regression of Y_i on X_i; no intercept is needed. (Why?)

Dick elects to run a regression of Y_i on X_i and W_i, again without an intercept. Dick will use the coefficient of X_i in his regression to estimate a. If W_i is independent of δ_i, Dick still gets a, up to random error. This is so even if W_i is correlated with X_i, as long as the correlation isn't perfect. (Why? And what's wrong with a perfect correlation?)

Suppose, however, that $W_i = cX_i + d\delta_i$, so W_i is correlated with δ_i. Then Dick has a problem. To see the problem more clearly, assume that n is large. Let $Q = (X \; W)$ be the design matrix, i.e., the first column is the X_i and the second column is the W_i. Show that

$$Q'Q/n \doteq \begin{pmatrix} E(X_i^2) & E(X_i W_i) \\ E(X_i W_i) & E(W_i^2) \end{pmatrix}, \quad Q'Y/n \doteq \begin{pmatrix} E(X_i Y_i) \\ E(W_i Y_i) \end{pmatrix}.$$

(a) Suppose $a = 1$, $c = 1$, $d = 1$. What will Dick estimate for the coefficient of X_i in his regression?

(b) Suppose $a = 1$, $c = 1$, $d = -1$. What will Dick estimate for the coefficient of X_i in his regression?

(c) A textbook on regression advises that, when in doubt, put more explanatory variables into the equation, rather than fewer. What do you think?

17. Suppose that $X_1, X_2, \ldots, X_n, \delta_1, \delta_2, \ldots, \delta_n$ are independent $N(0, 1)$ variables, and $Y_i = X_i^2 - 1 + \delta_i$. However, Harry regresses Y_i on X_i. What will he conclude about the relationship between X_i and Y_i?

18. You are thinking about a regression model $Y = X\beta + \epsilon$, with the usual assumptions. A friend suggests adding a column Z to the design matrix. If you do it, the bigger design matrix still has full rank. What are the arguments for putting Z into the equation? Against putting it in?

19. Suppose U and $V_1, \ldots, V_n$ are IID $N(0, 1)$ variables; μ is a real number. Let $X_i = \mu + U + V_i$. Let $\overline{X} = n^{-1} \sum_{i=1}^{n} X_i$ and $s^2 = (n-1)^{-1} \sum_{i=1}^{n} (X_i - \overline{X})^2$.

   (a) What is the distribution of X_i?
   (b) Do the X_i have a common distribution?
   (c) Are the X_i independent?
   (d) What is the distribution of $\overline{X}$? of s^2?

20. Suppose X_i are $N(\mu, \sigma^2)$ for $i = 1, \ldots, n$, where n is large. We use $\overline{X}$ to estimate μ. True or false and explain:

   (a) If the X_i are independent, then $\overline{X}$ will be around μ, being off by something like $s/\sqrt{n}$; the chance that $|\overline{X} - \mu| < s/\sqrt{n}$ is about 68%.

   (b) Even if the X_i are dependent, $\overline{X}$ will be around μ, being off by something like $s/\sqrt{n}$; the chance that $|\overline{X} - \mu| < s/\sqrt{n}$ is about 68%.

   What are the implications for applied work? For instance, how would dependence affect your ability to make statistical inferences about μ? (Notation: $\overline{X}$ and s^2 are defined in question 19.)

21. Suppose X_i has mean μ and variance σ^2 for $i = 1, \ldots, n$, where n is large. These random variables have a common distribution, which is not normal. We use $\overline{X}$ to estimate μ. True or false and explain:

   (a) If the X_i are IID, then $\overline{X}$ will be around μ, being off by something like $s/\sqrt{n}$; the chance that $|\overline{X} - \mu| < s/\sqrt{n}$ is about 68%.

   (b) Even if the X_i are dependent, $\overline{X}$ will be around μ, being off by something like $s/\sqrt{n}$; the chance that $|\overline{X} - \mu| < s/\sqrt{n}$ is about 68%.

   What are the implications for applied work? (Notation: $\overline{X}$ and s^2 are defined in question 19.)

22. There is a population consisting of N subjects, with data variables x and y. A simple regression equation can in principle be fitted by OLS to the population data: $y_i = a + bx_i + u_i$, where $\sum_{i=1}^{N} u_i = \sum_{i=1}^{N} x_i u_i = 0$. Although Julia does not have access to data on the full population, she can take a sample of size $n < N$, at random with replacement: n is moderately large, but small relative to N. She will estimate the parameters a and b by running a regression of y_i on x_i for i in the sample. She will have an intercept in the equation.

   (a) Are the OLS estimates biased or unbiased? Why?

(b) Should she believe the standard errors printed out by the computer? Discuss briefly.

23. Over the period 1950–99, the correlation between the size of the population in the United States and the death rate from lung cancer was 0.92. Does population density cause lung cancer? Discuss briefly.

24. (Hard.) Suppose $X_1, \ldots, X_n$ are dependent random variables. They have a common mean, $E(X_i) = \alpha$. They have a common variance, $\mathrm{var}(X_i) = \sigma^2$. Let r_{ij} be the correlation between X_i and X_j for $i \neq j$. Let

$$r = \frac{1}{n(n-1)} \sum_{1 \leq i \neq j \leq n} r_{ij}$$

be the average correlation. Let $S_n = X_1 + \cdots + X_n$.

(a) Show that $\mathrm{var}(S_n) = n\sigma^2 + n(n-1)\sigma^2 r$.

(b) Show that $\mathrm{var}\left(\dfrac{S_n}{n}\right) = \dfrac{1}{n}\sigma^2 + \dfrac{n-1}{n}\sigma^2 r$.

Hint for (a):

$$\left[\sum_{i=1}^{n} (X_i - \alpha)\right]^2 = \sum_{i=1}^{n} (X_i - \alpha)^2 + \sum_{1 \leq i \neq j \leq n} (X_i - \alpha)(X_j - \alpha).$$

Notes. (i) There are $n(n-1)$ pairs of indices (i, j) with $1 \leq i \neq j \leq n$. (ii) If $n = 100$ and $r = 0.05$, say, $\mathrm{var}(S_n/n)$ will be a lot bigger than σ^2/n. Small correlations are hard to spot, so casual assumptions about independence can be quite misleading.

25. A random sample of size 25 is taken from a population with mean μ. The sample mean is 105.8 and the sample variance is 110. The computer makes a t-test of the null hypothesis that $\mu = 100$. It doesn't reject the null. Comment briefly.

26. Discussing an application like example 3 in section 5, a social scientist says "one-step GLS is very problematic because it simply downweights observations that do not fit the OLS model."

(a) Does one-step GLS downweight observations that do not fit the OLS model?

(b) Would this be a bug or a feature?

Hint: look at exercises C3–4.

27. Let Δ stand for the percentage difference from 1871 to 1881 and let i range over the 32 metropolitan unions. Yule's model (section 1.4) explains ΔPaup_i in terms of ΔOut_i, ΔOld_i, and ΔPop_i.

   (a) Is option (i) below the OLS regression model, or the fitted equation? What about (ii)?

   (b) In (i), is b a parameter or an estimate? What about 0.755 in (ii)?

   (c) In (i), is ϵ_i an observable residual or an unobservable error term? What about e_i in (ii)?

   (i) $\Delta\text{Paup}_i = a + b \times \Delta\text{Out}_i + c \times \Delta\text{Old}_i + d \times \Delta\text{Pop}_i + \epsilon_i$, the ϵ_i's being IID with mean 0 and variance σ^2, independent of the explanatory variables.

   (ii) $\Delta\text{Paup}_i = 13.19 + 0.755\Delta\text{Out}_i - 0.022\Delta\text{Old}_i - 0.322\Delta\text{Pop}_i + e_i$, the e_i's having mean 0 with e orthogonal to the explanatory variables.

28. A box has N numbered tickets; N is known; the mean μ of the numbers in the box is an unknown parameter; the variance σ^2 of the numbers in the box is another unknown parameter. We draw n tickets at random with replacement: X_1 is the first draw, X_2 is the second draw, ..., X_n is the nth draw. Fill in the blanks, using the options below: _____ is an unbiased estimator for _____. Options:

   (i) n (ii) σ^2 (iii) $E(X_1)$

   (iv) $\dfrac{X_1 + X_2 + \cdots + X_n}{n}$

   (v) None of the above

29. (This continues question 28.) Let

$$\overline{X} = \frac{X_1 + X_2 + \cdots + X_n}{n}.$$

   True or false:

   (a) The X_i are IID.

   (b) $E(X_i) = \mu$ for all i.

   (c) $E(X_i) = \overline{X}$ for all i.

   (d) $\text{var}(X_i) = \sigma^2$ for all i.

   (e) $\dfrac{(X_1 - \overline{X})^2 + (X_2 - \overline{X})^2 + \cdots + (X_n - \overline{X})^2}{n} = \sigma^2$.

   (f) $\dfrac{(X_1 - \overline{X})^2 + (X_2 - \overline{X})^2 + \cdots + (X_n - \overline{X})^2}{n} \doteq \sigma^2$ if n is large.

30. Labrie et al (2004) report on a randomized controlled experiment to
     see whether routine screening for prostate cancer reduces the death rate
     from that disease. The experimental subjects consisted of 46,486 men
     age 45–80 who were registered to vote in Quebec City. The investigators
     randomly selected 2/3 of the subjects, inviting them to annual screening.
     The other 1/3 of the subjects were used as controls. Among the 7,348
     men who accepted the invitation to screening, 10 deaths from prostate
     cancer were observed during the first 11 years following randomization.
     Among the 14,231 unscreened controls, 74 deaths from prostate cancer
     were observed during the same time period. The ratio of death rates
     from prostate cancer is therefore

$$\frac{10/7{,}348}{74/14{,}231} = 0.26,$$

     i.e., screening cuts the death rate by 74%. Is this analysis convincing?
     Answer yes or no, and explain briefly.

4.11 End notes for chapter 4

Conditional vs unconditional expectations. The OLS estimate involves
an inverse, $(X'X)^{-1}$. If everything is integrable, then OLS is uncondition-
ally unbiased. Integrability apart, conditionally unbiased is the stronger and
more useful property. In many conventional models, $X'X$ is relatively constant when n is large. Then there is little difference between conditional and
unconditional inference.

Consistency and asymptotic normality. Consider the OLS estimator $\hat{\beta}$
in the usual model (1)–(5). One set of regularity conditions that guarantees
consistency and asymptotic normality of $\hat{\beta}$ is the following: p is fixed, n is
large, the elements of X are uniformly $o(\sqrt{n})$, and $X'X = nV + o(n)$ with V a
positive definite $p \times p$ matrix. Furthermore, under this set of conditions, the F-
statistic is asymptotically $\chi^2_{p_0}/p_0$ when the null hypothesis holds (section 8).
For additional discussion, see

http://www.stat.berkeley.edu/users/census/Ftest.pdf

White's correction for heteroscedasticity. Also called the "Huber-White
correction." It may seem natural to estimate the covariance of $\hat{\beta}_{OLS}$ given
X as $(X'X)^{-1}X'\hat{G}X(X'X)^{-1}$ where $\hat{G}_{ij} = e_i e_j$ and $e = Y - X\hat{\beta}_{OLS}$ is the
vector of residuals: see (20). This $\hat{G}$ has rank 1. Worse, $e \perp X$: so $X'e = e'X = 0$ and the estimator is identically 0. If the ϵ_i are assumed independent,
the off-diagonal elements of $\hat{G}$ would be set to 0. That often works. With
dependence, smoothing can be tried. A key reference is White (1980).

Explained variance. One point was elided in section 3. If Q projects orthogonally to the constant vectors, we must show that the projection of QY on QX is $X\hat{\beta} - \overline{Y}$. To begin with, $QY = Y - \overline{Y}$ and $QX = X - \overline{X}$. Now $Y - \overline{Y} = X\hat{\beta} - \overline{Y} + e = (X - \overline{X})\hat{\beta} + e = (QX)\hat{\beta} + e$ because $\overline{Y} = \overline{X}\hat{\beta}$. Plainly, $e \perp QX$. QED

BLUEness. If X is random, the OLS estimator is linear in Y but not X. Furthermore, the set of unbiased estimators is much larger than the set of conditionally unbiased estimators. Restricting to fixed X makes life easier. There is a more elegant (although perhaps more opaque) matrix form of the theorem; see, e.g.,

http://www.stat.berkeley.edu/users/census/GaussMar.pdf

Data snooping. The simulation discussed in section 9 was run another 1000 times. There were 19 runs with no keepers. Otherwise, the simulations gave a total of 5213 t-statistics whose distribution is shown in the histogram. A little bit of data-snooping goes a long way: t-statistics with $|t| > 2$ are the rule not the exception—in regressions on the keeper columns. If we add an intercept to the model, "the" F-test will give off-scale P-values.

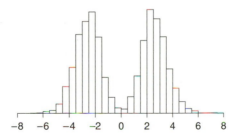

Replication is the best antidote (Ehrenberg and Bound 1993), but replication is unusual (Dewald et al 1986, Hubbard et al 1998). Many texts actually recommend data snooping. See, e.g., Hosmer and Lemeshow (2000, pp. 95ff); they suggest a preliminary screen at the 25% level, which will inflate R^2 and F even beyond our example.

Fixed-effects models. These are now widely used, as are "random-effects models" (where subjects are viewed as a random sample from some super-population). One example of a fixed-effects model, which illustrates the strengths and weaknesses of the technique, is Grogger (1995).

The discussion questions. Questions 8–9 are about the interpretation of R^2 and F. Questions 10–11 are about collinearity: the general point is that some linear combinations of the β's will be easy to estimate, and some— the $c'\beta$ with $Xc \doteq 0$—will be very hard. (Collinearity can also make results more sensitive to omitted variables and to data entry errors.) Questions 12–14 look at assumptions in the regression model. Question 15 gives an example of *omitted-variables bias* when W is correlated with X. In question 16, if W is correlated with δ, then including W creates "endogeneity bias" (also called "simultaneity bias"). Question 22 is a nice test case: do the regression assumptions hold in a sampling model? Question 26 is based on Beck (2001, pp. 276–77).

Questions 4, 19–21, and 24 show in various ways that independence is the key to estimating precision of estimates from internal evidence; homoscedasticity is perhaps of lesser importance. Of course, if the mode of dependence is known, adjustments can be made. Generally, such things are hard to know; on the other hand, assumptions are easy to make.

In question 17, the true regression is nonlinear: $E(Y_i|X_i) = X_i^2 - 1$. Linear approximation is awful. On the other hand, if $Y_i = X_i^3$, linear approximation is pretty good, on average. (If you want local behavior, say at 0, linear approximation is a bad idea; it is also bad for large x; nor should you trust the usual formulas for the SE.) We need the moments of X_i to work on this; see below. The regression of X_i^3 on X_i is just $3X_i$. The correlation between X_i and X_i^3 is $3/\sqrt{15} = 0.77$, which isn't bad at all. With a little more work, the moments can be used to get explicit formulas for asymptotic bias and variance; the latter will differ from the nominal variance.

Normal moments. Let Z be $N(0, 1)$. The odd moments of Z vanish. The even moments can be computed recursively. Integration by parts shows that $E(Z^{2n+2}) = (2n + 1)E(Z^{2n})$. So

$$E(X^2) = 1, \quad E(Z^4) = 3, \quad E(Z^6) = 5 \times 3 = 15, \quad E(Z^8) = 7 \times 15 = 105 \ldots .$$

Data sources. In section 3 and discussion question 23, lung cancer death rates are for males, age standardized to the United States population in 1970, from the American Cancer Society. Purchasing power of the dollar is based on the Consumer Price Index: *Statistical Abstract of the United States*, 2000, table 767. Total population is from *Statistical Abstract of the United States*, 1994, 2000, table 2; the earlier edition was used for the period 1950–59.

Spurious correlations. Hendry (1980, figure 8) reports an R^2 of 0.998 for predicting inflation by cumulative rainfall over the period 1964–75: both variables were increasing steadily. (The equation is quadratic, with an adjustment for autocorrelation.) Yule (1926) reports an R^2 of 0.9 between English mortality rates and the percentage of marriages performed in the Church of England over the period 1886–1911: both variables were declining. Hans Melberg provided the citations.

5

Path Models

5.1 Stratification

A *path model* is a graphical way to represent a regression equation or several linked regression equations. These models, developed by the geneticist Sewell Wright, are often used to make causal inferences. We will look at a couple of examples and then explain the logic, which involves *response schedules* and the idea of *stability under interventions*.

Blau and Duncan (1967) are thinking about the stratification process in the United States. According to Marxist scholars of the time, the US is a highly stratified society. Status is determined by family background, and transmitted through the school system. Blau and Duncan have data in their chapter 2, showing that family background variables do influence status— but the system is far from deterministic. The US has a permeable social structure, with many opportunities to succeed or fail. Blau and Duncan go on to develop the path model shown in figure 1 on the next page, in order to answer questions like these:

> "how and to what degree do the circumstances of birth condition subsequent status? and, how does status attained (whether by ascription or achievement) at one stage of the life cycle affect the prospects for a subsequent stage?" [p. 164]

Figure 1. Path model. Stratification, US, 1962.

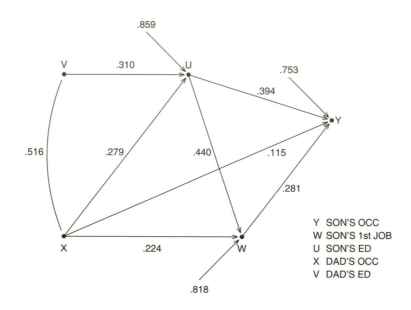

The five variables in the diagram are son's occupation, son's first job, son's education, father's occupation, and father's education. Data come from a special supplement to the March 1962 Current Population Survey. The respondents are the sons (age 20–64), who answer questions about current job, first job, and parents. There are 20,000 respondents. Education is measured on a scale from 0 to 8, where 0 means no schooling, 1 means 1–4 years of schooling, ..., 8 means some post-graduate education. Occupation is measured on Duncan's prestige scale from 0 to 96. The scale takes into account income, education, and raters' opinions of job prestige. Hucksters and peddlers are near the bottom of the pyramid, with clergy in the middle and judges at the top.

The path diagram uses standardized variables. Before running regressions, you subtract the mean from each data variable, and divide by the standard deviation. After standardization, means are 0 and variances are 1; furthermore, variables pretty much fall in the range from −3 to 3. Table 1 shows the correlation matrix for the data.

How is figure 1 to be read? The diagram unpacks to three regression equations:

$$(1) \qquad\qquad U = aV + bX + \delta,$$

Table 1. Correlation matrix for variables in Blau and Duncan's path model.

		Y Son's occ	W Son's 1st job	U Son's ed	X Dad's occ	V Dad's ed
Y	Son's occ	1.000	.541	.596	.405	.322
W	Son's 1st job	.541	1.000	.538	.417	.332
U	Son's ed	.596	.538	1.000	.438	.453
X	Dad's occ	.405	.417	.438	1.000	.516
V	Dad's ed	.322	.332	.453	.516	1.000

$$(2) \qquad W = cU + dX + \epsilon,$$

$$(3) \qquad Y = eU + fX + gW + \eta.$$

Equations are estimated by least squares. No intercepts are needed because the variables are standardized. (See exercise C6 for the reasoning on the intercepts; statistical assumptions will be discussed in section 5 below.)

In figure 1, the arrow from V to U indicates a causal link, and V is entered by Blau and Duncan on the right hand side of the regression equation (1) that explains U. The path coefficient 0.310 next to the arrow is the estimated coefficient $\hat{a}$ of V. The number 0.859 on the "free arrow" (that points into U from outside the diagram) is the estimated standard deviation of the error term δ in (1). The free arrow itself represents δ.

The other arrows in figure 1 are interpreted in a similar way. There are three equations because three variables in the diagram (U, W, Y) have arrows pointing into them. The curved line joining V and X is meant to indicate association rather than causation: V and X influence each other, or are influenced by some common causes not represented in the diagram. The number on the curved line is just the correlation between V and X (table 1).

The Census Bureau (which conducts the Current Population Survey used by Blau and Duncan) would not release raw data, due to confidentiality concerns. The Bureau did provide the correlation matrix in table 1. As it turns out, the correlations are all that is needed to fit the standardized equations. We illustrate the process on equation (1), which can be rewritten in matrix form as

$$(4) \qquad U = M \begin{pmatrix} a \\ b \end{pmatrix} + \delta,$$

where U and δ are $n \times 1$ vectors, while M is the $n \times 2$ "partitioned matrix"

$$M = \begin{pmatrix} V & X \end{pmatrix}.$$

In other words, the design matrix has one row for each subject, one column for the variable V, and a second column for X. Initially, father's education is in the range from 0 to 8. After it is standardized to have mean 0 and variance 1 across respondents, V winds up (with rare exceptions) in the range from -3 to 3. Similarly, father's occupation starts in the range from 0 to 96, but X winds up between -3 and 3. Algebraically, the standardization implies

$$(5) \qquad \frac{1}{n}\sum_{i=1}^{n} V_i = 0, \qquad \frac{1}{n}\sum_{i=1}^{n} V_i^2 = 1.$$

Similarly for X and U. In particular,

$$(6) \qquad r_{VX} = \frac{1}{n}\sum_{i=1}^{n} V_i X_i$$

is the data-level correlation between V and X, computed across respondents $i = 1, \ldots, n$. See equation (2.4).

   To summarize the notation, the sample size n is about 20,000. Next, V_i is the education of the ith respondent's father, standardized. And X_i is the father's occupation, scored on Duncan's prestige scale from 0 to 96, then standardized. So,

$$M'M = \begin{pmatrix} \sum_{i=1}^{n} V_i^2 & \sum_{i=1}^{n} V_i X_i \\ \sum_{i=1}^{n} V_i X_i & \sum_{i=1}^{n} X_i^2 \end{pmatrix} = n\begin{pmatrix} 1 & r_{VX} \\ r_{VX} & 1 \end{pmatrix} = n\begin{pmatrix} 1.000 & 0.516 \\ 0.516 & 1.000 \end{pmatrix}.$$

(You can find the 0.516 in table 1.) Similarly,

$$M'U = \begin{pmatrix} \sum_{i=1}^{n} V_i U_i \\ \sum_{i=1}^{n} X_i U_i \end{pmatrix} = n\begin{pmatrix} r_{VU} \\ r_{XU} \end{pmatrix} = n\begin{pmatrix} 0.453 \\ 0.438 \end{pmatrix}.$$

Now we can use equation (4.6) to get the OLS estimates:

$$\begin{pmatrix} \hat{a} \\ \hat{b} \end{pmatrix} = (M'M)^{-1}M'U = \begin{pmatrix} 0.309 \\ 0.278 \end{pmatrix}.$$

These differ in the 3rd decimal place from path coefficients in figure 1, probably due to rounding.

   What about the numbers on the free arrows? The *residual variance* in a regression equation—the mean square of the residuals—is used to estimate the variance of the disturbance term. Let $\hat{\sigma}^2$ be the residual variance in (1). We're going to derive an equation that can be solved for $\hat{\sigma}^2$. As a first step, let $\hat{\delta}$ be the residuals after fitting (1) by OLS. Then

$$(7) \quad 1 = \frac{1}{n} \sum_{i=1}^{n} U_i^2 \quad \text{because } U \text{ is standardized}$$

$$= \frac{1}{n} \sum_{i=1}^{n} \left(\hat{a} V_i + \hat{b} X_i + \hat{\delta}\right)^2$$

$$= \hat{a}^2 \frac{1}{n} \sum_{i=1}^{n} V_i^2 + \hat{b}^2 \frac{1}{n} \sum_{i=1}^{n} X_i^2 + 2\hat{a}\hat{b} \frac{1}{n} \sum_{i=1}^{n} V_i X_i + \frac{1}{n} \sum_{i=1}^{n} \hat{\delta}_i^2.$$

Two cross-product terms were dropped in (7). This is legitimate because the residuals are orthogonal to the design matrix, so

$$2\hat{a} \frac{1}{n} \sum_{i=1}^{n} V_i \hat{\delta}_i = 2\hat{b} \frac{1}{n} \sum_{i=1}^{n} X_i \hat{\delta}_i = 0.$$

Because V and X were standardized,

$$\frac{1}{n} \sum_{i=1}^{n} V_i^2 = 1, \quad \frac{1}{n} \sum_{i=1}^{n} X_i^2 = 1, \quad \frac{1}{n} \sum_{i=1}^{n} V_i X_i = r_{VX}.$$

Substitute back into (7). Since $\hat{\sigma}^2$ is the mean square of the residuals $\hat{\delta}$,

$$(8) \qquad\qquad 1 = \hat{a}^2 + \hat{b}^2 + 2\hat{a}\hat{b} r_{VX} + \hat{\sigma}^2.$$

Equation (8) can be solved for $\hat{\sigma}^2$. Take the square root to get the SD. The SDs are shown on the free arrows in figure 1. With a small sample, this isn't such a good way to estimate σ^2, because it doesn't take degrees of freedom into account. The fix would be to multiply $\hat{\sigma}^2$ by $n/(n-p)$. When $n = 20{,}000$ and $p = 3$ or 4, this is not an issue. If n were a lot smaller, in standardized equations like (1) and (2) with two variables, the best choice for p is 3. Behind the scenes, there is an intercept being estimated. That is the third parameter. In an equation like (3), with three variables, take $p = 4$. The sample size n cancels when computing the path coefficients, but is needed for standard errors.

The large SDs in figure 1 show the permeability of the social structure. Since variables are standardized, the SDs cannot exceed 1. (See exercise 4 below.) So 0.753 is a big number. Even if we know your family background and your education and your first job, the variation in the social status of your current job is 75% of the variation in the full sample. (Variation is measured by SD not variance: variance is on the wrong scale.)

The big SDs are a good answer to the Marxist argument, and so is the data analysis in Blau and Duncan (1967, chapter 2). As social physics, however, figure 1 leaves something to be desired. Why linearity? Why the same coefficients for everybody? What about variables like intelligence or motivation? And where are the mothers??

Now let's return to standardization. Standardizing might be sensible if (i) units are meaningful only in comparative terms (e.g., prestige points), or (ii) the meaning of units changes over time (e.g., years of education) while correlations are stable.

For descriptive statistics, with only one data set at issue, standardizing is really a matter of taste: do you like pounds, kilograms, or standard units? All variables should be similar in scale after standardization, which may make it easier to compare regression coefficients. That could be why social scientists like to standardize.

If the object is to find laws of nature that are stable under intervention, standardizing may be a bad idea, because estimated parameters would depend on irrelevant details of the study design (section 2 below). Generally, the intervention idea gets muddier with standardization. It will be difficult to hold the standard deviations constant when individual values are manipulated. If the SDs change too, what is supposed to be invariant and why? (*Manipulation* means an intervention, as in an experiment, to set a variable at a certain value: there is no connotation of unfairness.)

The terminology is peculiar. "Standardized regression coefficients" are just coefficients that come from fitting the equation to standardized variables. Similarly, "unstandardized regression coefficients" come from fitting the equation to the "raw"—unstandardized—variables. It is not coefficients that get standardized, but variables.

Exercise set A

1. Fit the equations in figure 1; find the SDs. (Cf. lab 4, back of book.)

2. Is *a* in equation (1) a parameter or an estimate? 0.322 in table 1? 0.310 in figure 1? How is 0.753 in figure 1 related to equation (3)?

3. True or false, and explain: after fitting equation (1), the mean square of the residuals equals their variance.

4. Prove that the SDs in a path diagram cannot exceed 1, if variables are standardized.

5. When we consider what figure 1 says about permeability of the social system, should we measure variation by the SD, or variance? Why?

6. In figure 1, why is there no arrow from V to W or V to Y? In principle, could there be an arrow from Y to U?

7. What are some important variables omitted from equation (3)?

8. The education variable in figure 1 takes values $0, 1, \ldots, 8$. Does that have any implications for linearity in (1)? What if the education variable only took values $0, 1, 2, 3, 4$? If the education variable only took values 0 and 1?

5.2 Hooke's law revisited

According to Hooke's law (section 2.3), if weight x is hung on a spring, and x is not too large, the length of the spring is $a + bx + \epsilon$. (Near the elastic limit of the spring, the physics will be more complicated.) In this equation, a and b are physical constants that depend on the spring not the weights. The parameter a is the length of the spring with no load. The parameter b is the length added to the spring by each additional unit of weight. The ϵ is random measurement error, with the usual assumptions.

If we were to standardize, the crucial slope parameter would depend on the weights, and on the accuracy of the device used to measure the length of the spring. To see this, let $v > 0$ be the variance of the weights used in the experiment. Let σ^2 be the variance of ϵ. Let s^2 be the mean square of the residuals. The standardized regression coefficient is

$$(9) \qquad \hat{b}\sqrt{\frac{v}{\hat{b}^2 v + s^2}} \doteq b\sqrt{\frac{v}{b^2 v + \sigma^2}} \,.$$

(See exercise 2 below.) The dotted equals sign means "approximately equal"; s^2 is the sum of the squared residuals, divided by n.

The standardized regression coefficient tells us about a parameter—the right hand side of (9)—that depends on v and σ^2. But v and σ^2 are features of the measurement procedure not the spring. The parameter we want to estimate is b, which tells us how the spring responds when the load is manipulated (by Hooke's law). The unstandardized $\hat{b}$ works like a charm.

If a regression coefficient is stable under interventions, standardizing is not a good idea—stability will get lost in the shuffle. That is what (9) shows.

Standardize coefficients only if there is a good reason to do so.

Exercise set B

1. Is v in equation (9) the variance of a data variable, or a random variable? What about σ^2?

2. Check that the left hand side of (9) is the standardized slope. Hint: work out the correlation coefficient between the weights and the lengths.

3. What happens to (9) if $\sigma^2 \doteq 0$? What would that tell us about springs and weights?

5.3 Political repression during the McCarthy era

Gibson (1988), reprinted at the back of the book, is about the causes of McCarthyism in the United States—the great witch-hunt for Reds in public life, particularly in Hollywood and the State Department. (With the opening of Soviet archives, it became pretty clear there had been many agents of influence in the US, but McCarthy probably did more harm than all of them put together.) Was repression due to the masses or the elites? Gibson argues that elite intolerance is the root cause. His chief piece of empirical evidence is the path diagram in figure 2, which is redrawn from the paper.

Figure 2. Path model. The causes of McCarthyism. The free arrow pointing into repression is not shown.

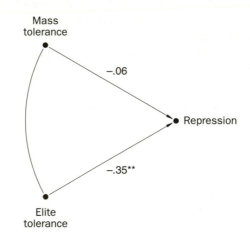

The unit of analysis is the state. The dependent variable is a measure of repressive legislation in each state (table 1 in the paper, and note 4). The independent variables are mean tolerance scores for each state, derived from the "Stouffer survey of masses and elites" (table A1 in the paper, and note 8).

The "masses" are just ordinary people who turn up in a sample of the population. "Elites" include school board presidents, commanders of the American Legion, bar association presidents, labor union leaders. Data on masses were available for 36 states; on elites, for 26 states. Gibson computes correlations from the available data, then estimates a standardized regression equation. He says,

> "Generally, it seems that elites, not masses, were responsible for the repression of the era.... The beta for mass opinion is $-.06$; for elite opinion, it is $-.35$ (significant beyond .01)."

His equation for legislative scores is

(10)   Repression $= \beta_1$ Mass tolerance $+ \beta_2$ Elite tolerance $+ \delta$.

Variables are standardized. The two straight arrows in figure 2 represent causal links: mass and elite tolerance affect repression. The estimated coefficients are $\hat{\beta}_1 = -0.06$ and $\hat{\beta}_2 = -0.35$. The curved line in figure 2 represents an association between mass and elite tolerance scores. Each one can influence the other, or both can have some common cause. The association is not analyzed in the diagram.

Gibson is looking at an interesting qualitative question: was it the masses or the elites who were responsible for McCarthyism? To address this issue by regression, he has to quantify everything—tolerance, repression, the causal effects, and statistical significance. The quantification is problematic.

As social physics, the path model is weak. Too many crucial issues are left dangling. What intervention is contemplated? Are there other variables in the system? Why are relationships linear? Signs apart, for example, why does a unit increase in tolerance have the same effect on repression as a unit decrease? Why are coefficients the same for all states? Why are states statistically independent? Such questions are not addressed in the paper. (The paper is not unique in this respect.)

McCarthy became a force in national politics with a speech attacking the State Department in 1950. The turning point came in 1954, with public humiliation in the Army-McCarthy hearings. Censure by the Senate followed in 1957. Gibson scores repressive legislation over the period 1945–65, long before McCarthy mattered, and long after (note 4 in the paper). The Stouffer survey was done in 1954, when the McCarthy era was ending. The timetable does not hang together.

Even if all such issues are set aside, and we allow Gibson the statistical assumptions, there is a big problem. Gibson finds that $\hat{\beta}_2$ is significant and $\hat{\beta}_1$ is insignificant. But this does not impose much of a constraint on the difference $\hat{\beta}_2 - \hat{\beta}_1$. The standard error for the difference can be computed from data in the paper (exercise 4 below). The difference is not significant. Since $\beta_2 = \beta_1$ is a viable null hypothesis, the data are not strong enough to distinguish elites from masses.

Note 9 in the paper is also worth some attention. Gibson used generalized least squares because he "could not assume that the variances of the observations were equal"; instead, he "weighted the observations by the square root of the numbers of respondents within the state." This confuses the variance of Y with the variance of X. When observations are independent, but $\mathrm{var}(Y_i|X)$ differs from one i to another, $\hat{\beta}$ should be chosen (exercise 4C4) to minimize

$$\sum_i (Y_i - X_i\hat{\beta})^2/\mathrm{var}(Y_i|X).$$

Weighting by sample size makes little sense. It is the variance of the mean response to the Stouffer survey that depends on the number of respondents. Gibson's Y_i is the repression score. The variance of Y_i has nothing to do with the Stouffer survey.

Exercise set C

1. Is the -0.35 in figure 2 a parameter or an estimate? How is it related to equation (10)?

2. The correlation between mass and elite tolerance scores is 0.52; between mass tolerance scores and repression scores, -0.26; between elite tolerance scores and repression scores, -0.42. Compute the path coefficients in figure 2.

   Note. Apparently, Gibson used weighted correlations, but this makes almost no difference to the estimates. Exercises 2–4 can be done on a pocket calculator, but it's easier with a computer: see lab 5 at the back of the book, and exercise 4C2(a).

3. Estimate the SD of δ in equation (10). You may assume the correlations are based on 36 states but you need to decide if p is 2 or 3. (See text for Gibson's sample sizes.)

4. Find the SEs for the path coefficients and their difference.

5. The repression scale is lumpy: scores go from 0 to 3.5 in steps of 0.5 (table 1 in the paper). Does this make the linearity assumption more plausible, or less plausible?

6. Suppose we run a regression of Y on U and V, getting

$$Y = \hat{a} + \hat{b}U + \hat{c}V + e,$$

where e is the vector of residuals. Express the standardized coefficients in terms of the unstandardized coefficients and the sample variances of U, V, Y.

5.4 Inferring causation by regression

The key to making causal inferences by regression is a *response schedule*. This is a new idea, and a complicated one. We'll start with a mathematical example to illustrate the idea of a "place holder." Logarithms can be defined by the equation

$$(11) \qquad \log x = \int_1^x \frac{1}{z} \, dz \text{ for } 0 < x < \infty.$$

The symbol ∞ stands for "infinity." But what does the x stand for? Not much. It's a place holder. You could change the x's in (11) to u's without changing the content, namely, the equality between the two sides of the equation. Similarly, z is a place holder inside the integral. You could change both z's to v's without changing the value of the integral. (Mathematicians refer to place holders as "dummy variables," but statisticians use the language differently: section 6 below.)

Now let's take an example that's closer to regression—Hooke's law (section 2). Suppose we're going to hang some weights on a spring. We do this on n occasions, indexed by $i = 1, \ldots, n$. Fix an i. If we put weight x on the spring on occasion i, our physics consultant assures us that the length of the spring will be

$$(12) \qquad Y_{i,x} = 439 + 0.05x + \epsilon_i .$$

If we put a 5-unit weight on the spring, the length will be $439 + 0.05 \times 5 + \epsilon_i = 439.25 + \epsilon_i$. If instead we put a 6-unit weight on the spring, the length will be $439.30 + \epsilon_i$. A 1-unit increase in x makes the spring longer, by 0.05 units—causation has come into the picture. The random disturbance term ϵ_i represents measurement error. These random errors are IID for $i = 1, \ldots, n$, with mean 0 and known variance σ^2. The units for x are kilograms; the units for length are centimeters, so ϵ_i and σ must be in centimeters too. (Reminder: IID is shorthand for independent and identically distributed.)

Equation (12) looks like a regression equation, but it isn't. It is a response schedule that describes a theoretical relationship between weight and length. Conceptually, x is a weight that you could hang on the spring. If you did, equation (12) tells you what the spring would do. This is all in the subjunctive.

Formally, x is a place holder. The equation gives length $Y_{i,x}$ as a function of weight x, with a bit of random error. For any particular i, we can choose *one* x, electing to observe $Y_{i,x}$ for that x and that x only. The rest of the response schedule—the other potential outcomes—would be lost to history.

Let's make the example a notch closer to social science. We might not know (12), but only

$$(13) \qquad\qquad Y_{i,x} = a + bx + \epsilon_i \,,$$

where the ϵ_i are IID with mean 0 and variance σ^2. This time, a, b, and σ^2 are unknown. These parameters have to be estimated. More troublesome: we can't do an experiment. However, observational data are available. On occasion i, weight X_i is found on the spring, we just don't quite know how it got there. The length of the spring is measured as Y_i. We're still in business, *if*

(i) Y_i was determined from the response schedule (13), so $Y_i = Y_{i,X_i} = a + bX_i + \epsilon_i$, and

(ii) the X_i's were chosen at random by Nature, independent of the ϵ_i's.

Condition (i) ties the observational data to the response schedule (13), and gives us most of the statistical conditions we need on the random errors: these errors are IID with mean 0 and variance σ^2. Condition (ii) is *exogeneity*. Exogeneity—$X \perp\!\!\!\perp \epsilon$—is the rest of what we need. With these assumptions, OLS gives unbiased estimates for a and b. Example 4.1 explains how to set up the design matrix. Conditions (4.1–5) are all satisfied.

The response schedule tells us that the parameter b we're estimating has a causal interpretation: if we intervene and change x to x', then y is expected to change by $b(x' - x)$. The response schedule tells us that the relation is linear rather than quadratic or cubic or . . . . It tells us that interventions won't affect a or b. It tells us the errors are IID. It tells us there is no confounding: X causes Y without any help from any other variable. The exogeneity condition says that Nature ran the observational study just the way we would run an experiment. We don't have to randomize. Nature did it for us. Nice.

What would happen without exogeneity? Suppose Nature puts a big weight X_i on the spring whenever ϵ_i is large and positive. Nasty. Now OLS over-estimates b. In this hypothetical, the spring doesn't stretch as much as you might think: measurement error gets mixed up with stretch. (This is "selection bias" or "endogeneity bias," to be discussed in chapters 6 and 8.) The response schedule is a powerful assumption, and so is exogeneity. For Hooke's law, the response schedule and exogeneity are reasonably convincing. With typical social science applications, there might be some harder questions to answer.

The discussion so far is about a one-dimensional x, but the generalization is easy. The response schedule would be

$$(14) \qquad\qquad Y_{i,x} = x\beta + \epsilon_i,$$

where x is $1 \times p$ vector of treatments and β is a $p \times 1$ parameter vector. Again, the errors ϵ_i are IID with mean 0 and variance σ^2. In the next section, we'll see that path models put together several response schedules like (14).

> A response schedule says how one variable would respond, if you intervened and manipulated other variables. Together with the exogeneity assumption, the response schedule is a theory of how the data were generated. If the theory is right, causal effects can be estimated from observational data by regression. If the theory is wrong, regression coefficients measure association not causation, and causal inferences can be quite misleading.

Exercise set D

1. (This is a hypothetical; SAT stands for Scholastic Achievement Test, widely used for college admissions in the US.) Dr. Sally Smith is doing a study on coaching for the Math SAT. She assumes the response schedule $Y_{i,x} = 450 + 3x + \delta_i$. In this equation, $Y_{i,x}$ is the score that subject i would get on the Math SAT with x hours of coaching. The error term δ_i is normal, with mean 0 and standard deviation 100.

   (a) If subject #77 gets 10 hours of coaching, what does Dr. Smith expect for this subject's Math SAT score?

   (b) If subject #77 gets 20 hours of coaching, what does Dr. Smith expect for this subject's Math SAT score?

   (c) If subject #99 gets 10 hours of coaching, what does Dr. Smith expect for this subject's Math SAT score?

   (d) If subject #99 gets 20 hours of coaching, what does Dr. Smith expect for this subject's Math SAT score?

2. (This continues exercise 1; it is still a hypothetical.) After thinking things over, Dr. Smith still believes that the response schedule is linear: $Y_{i,x} = a + bx + \delta_i$, the δ_i being IID $N(0, \sigma^2)$. But she decides that her values for a, b, and σ^2 are unrealistic. (They probably are.) She wants to estimate these parameters from data.

   (a) Does she need to do an experiment, or can she get by with an observational study? (The latter would be much easier to do.)

(b) If she can use observational data, what else would she have to assume, beyond the response schedule?

(c) And, how would she estimate the parameters from the observational data?

5.5 Response schedules for path diagrams

Path models are often held out as rigorous statistical engines for inferring causation from association. Statistical techniques can indeed be rigorous— given their assumptions. But the assumptions are usually imposed on the data by the analyst, and this is not a rigorous process. The assumptions behind the models are of two kinds: (i) causal and (ii) statistical. This section will lay out the assumptions in more detail. A relatively simple path model is shown in figure 3, where a hypothesized causal relationship between Y and Z is confounded by X.

Figure 3. Path model. The relationship between Y and Z is confounded by X. Free arrows leading into Y and Z are not shown.

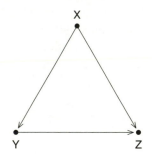

This sort of diagram is used to draw causal conclusions from observational data. The diagram is therefore more complicated than it looks: causation is a complicated business. Let's assume that Dr. Alastair Arbuthnot has collected data on X, Y, and Z in an observational study. He draws the diagram shown in figure 3, and fits the two regression equations suggested by the figure:

$$Y = \hat{a} + \hat{b}X + \text{error}, \quad Z = \hat{c} + \hat{d}X + \hat{e}Y + \text{error}$$

Estimated coefficients are positive and significant. He is now trying to explain the findings to his colleague, Dr. Beverly Braithwaite.

Dr. A   So you see, Dr. Braithwaite, if X goes up by one unit, then Y goes up by $\hat{b}$ units.

Dr. B   Quite.

Dr. A   Furthermore, if X goes up by one unit with Y held fixed, then Z goes up by $\hat{d}$ units. This is the direct effect of X on Z. ["Held fixed" means, kept the same.]

Dr. B   But Dr. Arbuthnot, you just told me that if X goes up by one unit, then Y will go up by $\hat{b}$ units.

Dr. A   Moreover, if Y goes up by one unit with X held fixed, the change in Y makes Z go up by $\hat{e}$ units. The effect of Y on Z is $\hat{e}$.

Dr. B   Dr. Arbuthnot, hello, why would Y go up unless X goes up? "Effects"? "Makes"? How did you get into causation?? And what about my first point?!?

Dr. Arbuthnot's explanation is not unusual. But Dr. Braithwaite has some good questions. Our objective in this section is to answer her, by developing a logically coherent set of assumptions which—if true—would justify Dr. Arbuthnot's data analysis and his interpretations. On the other hand, as we will see, Dr. Braithwaite has good reason for her skepticism.

At the back of his mind, Dr. Arbuthnot has two response schedules describing hypothetical experiments. In principle, these two experiments are unrelated to one another. But, to model the observational study, the experiments have to be linked in a complicated way. We will describe the two experiments first, and then explain how they are put together to model Dr. Arbuthnot's data.

(i) *First hypothetical experiment.* Treatment at level x is applied to a subject. A response Y is observed, corresponding to the level of treatment. There are two parameters, a and b, that describe the response. With no treatment ($x = 0$), the response level for each subject will be a, up to random error. All subjects are assumed to have the same value for a. Each additional unit of treatment adds b to the response. Again, b is the same for all subjects at all levels of x, by assumption. Thus, when treatment is applied at level x, the response Y is assumed to be

(15) $a + bx + \text{random error}.$

For example, colleges send students with weak backgrounds to summer boot-camp with mathematics drill. In an evaluation study of such a program, x might be hours spent in math drill, and Y might be test scores.

(ii) *Second hypothetical experiment.* In the second experiment, there are two treatments and a response variable Z. There are two treatments because there are two arrows leading into Z. The treatments are labeled X and Y in figure 3. Both treatments may be applied to a subject. In Experiment #1, Y was the response variable. But in Experiment #2, Y is one of the treatment variables: the response variable is Z.

There are three parameters, c, d, and e. With no treatment at all ($x = y = 0$), the response level for each subject will be c, up to random error. Each additional unit of treatment X adds d to the response. Likewise, each additional unit of treatment Y adds e to the response. (Here, e is a parameter not a residual vector.) The constancy of parameters across subjects and levels of treatment is an assumption. Thus, when the treatments are applied at levels x and y, the response Z is assumed to be

(16) $c + dx + ey +$ random error.

Three parameters are needed because it takes three parameters to specify the linear relationship (16), an intercept and two slopes.

Random errors in (15) and (16) are assumed to be independent from subject to subject, with a distribution that is constant across subjects: the expectation is zero and the variance is finite. The errors in (16) are assumed to be independent of the errors in (15). Equations (15) and (16) are *response schedules*: they summarize Dr. Arbuthnot's ideas about what would happen *if* he could do the experiments.

Putting the experiments together. Dr. Arbuthnot collected the data on X, Y, Z in an observational study. He wants to use the observational data to figure out what would have happened *if* he could have intervened and manipulated the variables. There is a price to be paid.

To begin with, he has to assume the response schedules (15) and (16). He also has to assume that the X's are independent of the random errors in the two hypothetical experiments—"exogeneity." Thus, Dr. Arbuthnot is pretending that Nature randomized subjects to levels of X. If so, there is no need for experimental manipulation on his part, which is convenient. The exogeneity of X has a graphical representation: arrows come out of X in figure 3, but no arrows lead into X.

Dr. Arbuthnot also has to assume that Nature generates Y from X as if by substituting X into (15). Then Nature generates Z as if by substituting X and Y—the very same X that was the input to (15), and the Y that was the output from (15)—into (16). Using the output from (15) as an input to (16) links the two equations together.

Let's take another look at this linkage. In principle, the experiments described by the two response schedules are separable from one another. There is no a priori connection between the value of x in (15) and the value of x in (16). There is no a priori connection between outputs from (15) and inputs to (16). However, to model his observational study, Dr. Arbuthnot links the equations "recursively." He assumes that one value of X is chosen and used as an input for both equations; that the Y generated from (15) is used as an input to (16); and there is no feedback from (16) to (15).

Given all these assumptions, the parameters a, b can be estimated by regression of Y on X. Likewise, c, d, e can be estimated by regression of Z on X and Y. Moreover, the regression estimates have legitimate causal interpretations. This is because causation is built into the assumptions, via the response schedules (15) and (16). If causation were not assumed, causation would not be demonstrated by running the regressions.

One point of Dr. Arbuthnot's regressions is to estimate the direct effect of X on Z. The direct effect is d in (16). If X is increased by one unit with Y held fixed—i.e., kept at its old value—then Z is expected to go up by d units. This is shorthand for the mechanism in the second experiment. The response schedule (16) says what happens to Z when x and y are manipulated. In particular, y can be held at an old value while x is made to increase.

Dr. Arbuthnot imagines that he can keep the Y generated by Nature, while replacing X by $X + 1$. He just substitutes his values ($X + 1$ and Y) into the response schedule (16), getting

$$c + d(X + 1) + eY + \text{error} = (c + dX + eY + \text{error}) + d.$$

This is what Z would have been, if X had been increased by 1 unit with Y held fixed: Z would have been d units bigger.

Dr. Arbuthnot also wants to estimate the effect e of Y on Z. If Y is increased by one unit with X held fixed, then Z is expected to go up by e units. Dr. Arbuthnot thinks he can keep Nature's value for X, while replacing Y by $Y + 1$. He just substitutes X and $Y + 1$ into the response schedule (16), getting

$$c + dX + e(Y + 1) + \text{error} = (c + dX + eY + \text{error}) + e.$$

This is what Z would have been, if Y had been increased by 1 unit with X kept unchanged: Z would have been e units bigger. Of course, even Dr. Arbuthnot has to replace parameters by estimates. If $e = 0$—or could be 0 because $\hat{e}$ is statistically insignificant—then manipulating Y should not affect Z, and Y would not be a cause of Z after all. This is a qualitative inference. Again, the inference depends on the response schedule (16).

In short, Dr. Arbuthnot uses the observational data to estimate parameters. But when he interprets the results—for instance, when he talks about the "effects" of X and Y on Z—he's thinking about the hypothetical experiments described by the response schedules (15)-(16), not about the observational data themselves. His causal interpretations depend on a rather subtle model: the same response schedules, with the same parameter values, apply to his observational data and to his hypothetical experiments.

To state the model more formally, we would index the subjects by a subscript i in the range from 1 to n, where n is the number of subjects. In this notation, X_i is the value of X for subject i. The level of treatment #1 is denoted by x, and $Y_{i,x}$ is the response for variable Y when treatment at level x is applied to subject i, as in (15). Similarly, $Z_{i,x,y}$ is the response for variable Z when treatment #1 at level x and treatment #2 at level y are applied to subject i, as in (16). The response schedules are interpreted causally.

- $Y_{i,x}$ is what Y_i would be if X_i were set to x by intervention.

- $Z_{i,x,y}$ is what Z_i would be if X_i were set to x and Y_i were set to y by intervention.

Figure 3 unpacks into two equations, which are more precise versions of (15) and (16), with subscripts for the subjects. Greek letters are used for the random error terms:

(17) $$Y_{i,x} = a + bx + \delta_i,$$
(18) $$Z_{i,x,y} = c + dx + ey + \epsilon_i.$$

The parameters a, b, c, d, e and the error terms δ_i, ϵ_i are not observed. The parameters are assumed to be the same for all subjects. There are assumptions about the error terms—the statistical component of the assumptions behind the path diagram:

(i) δ_i and ϵ_i are independent of each other within each subject i.
(ii) These error terms are independent across subjects i.
(iii) The distribution of δ_i is constant across subjects i; so is the distribution of ϵ_i. (However, δ_i and ϵ_i need not have the same distribution.)
(iv) δ_i and ϵ_i have expectation zero and finite variance.
(v) The X_i's are independent of the δ_i's and ϵ_i's, where X_i is the value of X for subject i in the observational study.

Assumption (v) says that Nature chooses X_i for us as if by randomization. In other words, the X_i's are "exogenous." By further assumption, Nature determines the response Y_i for subject i as if by substituting X_i into (17):

$$Y_i = Y_{i,X_i} = a + bX_i + \delta_i.$$

The rest of the response schedule— $Y_{i,x}$ for $x \neq X_i$—is not observed. After all, even in an experiment, subject i would be assigned to one level of treatment. The response at other levels would not be observed.

Similarly, we observe $Z_{i,x,y}$ only for $x = X_i$ and $y = Y_i$. The response for subject i is determined by Nature, as if by substituting X_i and Y_i into (18):

$$Z_i = Z_{i,X_i,Y_i} = c + dX_i + eY_i + \epsilon_i.$$

The rest of the response schedule remains unobserved, namely, the responses $Z_{i,x,y}$ for all the other possible values of x and y.

Economists call the unobserved $Y_{i,x}$ and $Z_{i,x,y}$ *potential outcomes*. The model specifies unobservable response schedules, not just regression equations. The model has another feature worth noticing: each subject's responses are determined by the levels of treatment for that subject only. Treatments applied to subject j do not affect the responses of subject i. For treating infectious diseases, this is not such a good model. (If one subject sneezes, another will catch the flu: stop the first sneeze, prevent the second flu.) There may be similar problems with social experiments, when subjects interact with each other.

Figure 4. The path diagram as a box model.

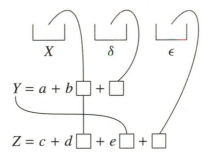

The box model in figure 4 illustrates the statistical assumptions. Independent random errors with constant distributions are represented as draws made at random with replacement from a box of potential errors (Freedman-Pisani-Purves 1998). Since the box remains the same from one draw to another, the probability distribution of one draw is the same as the distribution of any other. The distribution is constant. Furthermore, the outcome of one draw cannot affect the distribution of another. That is independence.

Figure 4 also shows how the two hypothetical causal mechanisms—response schedules (17) and (18)—are linked together to model the observational data. Let's take this apart and put it back together. We can think about each response schedule as a little machine, which accepts inputs and makes output. There are two of these little machines at work.

- *First causal mechanism.* You feed an x—any x that you like—into machine #1. The output from the machine is $Y = a + bx$, plus a random draw from the δ-box.

- *Second causal mechanism.* You feed x and y—any x and y that you like—into machine #2. The output from the machine is $Z = c + dx + ey$, plus a random draw from the ϵ-box.

- *Linkage.* You don't feed anything into anything. Nature chooses X at random from an X-box, independent of the δ's and ϵ's. She puts X into machine #1, to generate a Y. She puts the same X—and the Y she just generated—into machine #2, to generate Z. You get to see (X, Y, Z) for each subject. This is Dr. Arbuthnot's model for his observational data.

- *Estimation.* You estimate a, b, c, d, e by OLS, from the observational data, namely, triples of observed values on (X, Y, Z) for many subjects.

- *Causal inference.* You can say what would happen if you got your hands on the machines and put an x into machine #1. You can also say what would happen if you put x and y into machine #2.

You never do touch the machines. (After all, these are purely theoretical entities.) Still, you seem to be free to use your own x's and y's, rather than the ones generated by Nature, as inputs. You can say what the machines will do in response to your inputs. That is causal inference from observational data. Causal inference is legitimate because—by assumption—you know the social physics: response schedules (17) and (18).

What about the assumptions? Checking (17) and (18), which involve potential outcomes, is going to be hard work. Checking the statistical assumptions will not be much easier. The usual point of running regressions is to make causal inferences without doing real experiments. On the other hand, without the real experiments, the assumptions behind the models are going to be iffy. Inferences get made by ignoring the iffiness of the assumptions. That is the paradox of causal inference by regression, and a good reason for Dr. Braithwaite's skepticism.

Path models do not infer causation from association. Instead, path models *assume* causation through response schedules, and—using additional statistical assumptions—estimate causal effects from observational data. The statistical assumptions (independence, expectation zero, constant variance) justify estimating coefficients by ordinary least squares. With large samples, standard errors, confidence intervals, and significance tests would follow. With small samples, the errors would have to follow a normal distribution in order to justify t-tests.

Evaluating the statistical models in chapters 1–5. Earlier in the book, we discussed several examples of causal inference by regression—Yule on poverty, Blau and Duncan on stratification, Gibson on McCarthyism. We found serious problems. These studies are among the strongest in the social sciences, in terms of clarity, interest, and data analysis. (Gibson, for example, won a prize for best paper of the year—and is still viewed as a landmark study in political behavior.) The problems are built into the assumptions behind the statistical models.

> Typically, a regression model assumes causation and uses the data to estimate the size of a causal effect. If the estimate isn't statistically significant, lack of causation is inferred. Estimation and significance testing require statistical assumptions. Therefore, you need to think about the assumptions—both causal and statistical— behind the models. If the assumptions don't hold, the conclusions don't follow from the statistics.

Selection vs intervention

The conditional expectation of Y given $X = x$ is the average of Y for subjects with $X = x$. (This ignores sampling error: chapter 4.) The response-schedule formalism connects two very different ideas of conditional expectation: (i) selecting the subjects with $X = x$, versus (ii) intervening to set $X = x$. The first is something you can actually do with observational data. The second would require manipulation. Response schedules crystallize the assumptions you need to get from selection to intervention. (*Intervention* means interrupting the natural flow of events in order to manipulate a variable—as in an experiment; the contrast is with passive observation.)

> Selection is one thing, intervention is another.

Structural equations and stable parameters

In econometrics, "structural" equations describe causal relationships. Response schedules give a clearer meaning to this idea, and to the idea of "stability under intervention." The parameters in a path diagram, for instance, are defined through response schedules like (17) and (18), separately from the data. These parameters are constant across subjects and levels of treatment, and stay the same whether you intervene or just observe the natural course of events—by assumption, of course. Response schedules bundle up these assumptions for us, along with similar assumptions on the error distributions. (Stability is also called "constancy" or "invariance" under intervention.)

> Regression equations are structural, with parameters that are stable under intervention, when the equations derive from response schedules.

Ambiguity in notation

Look back at figure 3. In the observational study, there is an X_i for each subject i. In some contexts, X just means the X_i for a generic subject. In other contexts, X is the vector whose ith component is X_i. Often, X is the design matrix. This sort of ambiguity is commonplace. You have to pay attention to context, and figure out what is meant each time.

Exercise set E

1. In the path diagram below, free arrows are omitted. How many free arrows should there be, where do they go, and what do they mean? What does the curved line mean? The diagram represents some regression equations. What are the equations? the parameters? State the assumptions that would be needed to estimate the parameters by OLS. What data would you need? What additional assumptions would be needed to make causal inferences? Give an example of a qualitative causal inference that could be made from one of the equations. Give an example of a quantitative causal inference.

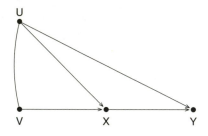

2. With the assumptions of this section, show that a regression of Y_i on X_i gives unbiased estimates, conditionally on the X_i's, of a and b in (17). Show also that a regression of Z_i on X_i and Y_i gives unbiased estimates,

conditionally on the X_i's and Y_i's, of c, d, and e in (18). Hints. What are the design matrices in the two regressions? Can you verify assumptions (4.2)–(4.5)? [Cross-references: (4.2) is equation (2) in chapter 4.]

3. Suppose you are only interested in the effects of X and Y on Z; you are not interested in the effect of X on Y. You are willing to assume the response schedule (18), with IID errors ϵ_i, independent of the X_i's and Y_i's. How would you estimate c, d, e? Do the estimates have a causal interpretation? Why?

4. True or false, and explain.
   (a) In figure 1, father's education has a direct influence on son's occupation.
   (b) In figure 1, father's education has an indirect influence on son's occupation through son's education.
   (c) In exercise 1, U has a direct influence on Y.
   (d) In exercise 1, V has a direct influence on Y.

5. Suppose Dr. Arbuthnot's models are correct; and in his data, $X_{77} = 12$, $Y_{77} = 2$, $Z_{77} = 29$.
   (a) How much bigger would Y_{77} have been, if Dr. Arbuthnot had intervened, setting X_{77} to 13?
   (b) How much bigger would Z_{77} have been, if Dr. Arbuthnot had intervened, setting X_{77} to 13 and Y_{77} to 5?

6. An investigator writes, "Statistical tests are a powerful tool for deciding whether effects are large." Do you agree or disagree? Discuss briefly.

5.6 Dummy variables

A "dummy variable" takes the value 0 or 1. Dummy variables are used to represent the effects of qualitative factors in a regression equation. Sometimes, dummies are even used to represent quantitative factors, in order to weaken linearity assumptions. (Dummy variables are also called "indicator" variables or "binary" variables; programmers call them "flags.")

Example. A company is accused of discriminating against female employees in determining salaries. The company counters that male employees have more job experience, which explains the salary differential. To explore that idea, a statistician might fit the equation

$$Y = a + b\,\text{MAN} + c\,\text{EXPERIENCE} + \text{error}.$$

Here, MAN is a dummy variable, taking the value 1 for men and 0 for women. EXPERIENCE would be years of job experience. A significant positive value for b would be taken as evidence of discrimination.

Objections could be raised to the analysis. For instance, why does EX-PERIENCE have a linear effect? To meet that objection, some analysts would put in a quadratic term:

$$Y = a + b\,\text{MAN} + c\,\text{EXPERIENCE} + d\,\text{EXPERIENCE}^2 + \text{error}.$$

Others would break up EXPERIENCE into categories, e.g.,

category 1   under 5 years
category 2   5–10 years (inclusive)
category 3   over 10 years

Then dummies for the first two categories could go into the equation:

$$Y_i = a + b\,\text{MAN} + c_1\,\text{CAT}_1 + c_2\,\text{CAT}_2 + \text{error}.$$

For example, CAT_1 is 1 for all employees who have less than 5 years of experience, and 0 for the others. Don't put in all three dummies: if you do, the design matrix won't have full rank.

The coefficients are a little tricky to interpret. You have to look for the missing category, because effects are measured relative to the missing category. For MAN, it's easy. The baseline is women. The equation says that men earn b more than women, other things equal (experience). For CAT_1, it's less obvious. The baseline is the third category, over 10 years of experience. The equation says that employees in category 1 earn c_1 more than employees in category 3. Furthermore, employees in category 2 earn c_2 more than employees in category 3.

We expect c_1 and c_2 to be negative, because long-term employees get higher salaries. Similarly, we expect $c_1 < c_2$. Other things are held equal in these comparisons, namely, gender. (Saying that Harriet earns $-\$5,000$ more than Harry is a little perverse; ordinarily, we would talk about earning $\$5,000$ less: but this is statistics.)

Of course, the argument would continue. Why these categories? What about other variables? If people compete with each other for promotion, how can error terms be independent? And so forth. The point here was just to introduce the idea of dummy variables.

Types of variables

A *qualitative* or *categorical* variable is not numerical. Examples include gender and marital status, values for the latter being never-married, married, widowed, divorced, separated. By contrast, a *quantitative* variable takes numerical values. If the possible values are few and relatively widely separated, the variable is *discrete*; otherwise, *continuous*. These are useful distinctions, but the boundaries are a little blurry. A dummy variable, for instance, can be seen as converting a categorical variable with two values into a numerical variable taking the values 0 and 1. The discrimination example broke EXPE-RIENCE down into three categories, and two dummies were needed in the regression equation.

5.7 Discussion questions

Some of these questions cover material from previous chapters.

1. A regression of wife's educational level (years of schooling) on husband's educational level gives the equation

$$\text{WifesEdLevel} = 5.60 + 0.57 \times \text{HusbandsEdLevel} + \text{residual}.$$

   (Data are from the Current Population Survey in 2001.) If Mr. Wang's company sends him back to school for a year to catch up on the latest developments in his field, do you expect Mrs. Wang's educational level to go up by 0.57 years? If not, what does the 0.57 mean?

2. In equation (10), δ is a random error; there is a δ for each state. Gibson finds that $\hat{\beta}_1$ is statistically insignificant, while $\hat{\beta}_2$ is highly significant (two-tailed). Suppose that Gibson computed his P-values from the standard normal curve; the area under the curve between -2.6 and $+2.6$ is 0.99. True or false and explain—

   (a) The absolute value of $\hat{\beta}_2$ is more than 2.6 times its standard error.

   (b) The statistical model assumes that the random errors are independent across states.

   (c) However, the estimated standard errors are computed from the data.

   (d) The computation in (c) can be done whether or not the random errors are independent across states: the computation uses the tolerance scores and repression scores, but does not use the random errors themselves.

   (e) Therefore, Gibson's significance tests are fine, even if the random errors are dependent across states.

3. Timberlake and Williams (1984) offer a regression model to explain political oppression (PO) in terms of foreign investment (FI), energy development (EN), and civil liberties (CV). High values of PO correspond to authoritarian regimes that exclude most citizens from political participation. High values of CV indicate few civil liberties. Data were

collected for 72 countries. The equation proposed by Timberlake and Williams is

$$PO = a + b\,FI + c\,EN + d\,CV + \text{random error},$$

with the usual assumptions about the random errors. The estimated coefficient $\hat{b}$ of FI is significantly positive, and is interpreted as measuring the effect of foreign investment on political oppression.

(a) There is one random error for each _____ so there are _____ random errors in all. Fill in the blanks.

(b) What are the "usual assumptions" on the random errors?

(c) From the data in the table below, can you estimate the coefficient a in the equation? If so, how? If not, why not? What about b?

(d) How can $\hat{b}$ be positive, given that $r(FI, PO)$ is negative?

(e) From the data in the table, can you tell whether $\hat{b}$ is significantly different from 0? If so, how? If not, why not?

(f) Comment briefly on the statistical logic used by Timberlake and Williams. Do you agree that foreign investment causes political oppression? You might consider the following points. (i) Does CV belong on the right hand side of the equation? (ii) If not, and you drop it, what happens? (iii) What happens if you run a regression of CV on PO, FI, and EN?

The Timberlake and Williams data. 72 countries. Correlation matrix for political oppression (PO), foreign investment (FI), energy development (EN), and civil liberties (CV).

	PO	FI	EN	CV
PO	1.000	−.175	−.480	+.868
FI	−.175	1.000	+.330	−.391
EN	−.480	+.330	1.000	−.430
CV	+.868	−.391	−.430	1.000

Note. Regressions can be done with a pocket calculator, but it's easier with a computer. We're using different notation from the paper.

4. Alba and Logan (1993) develop a regression model to explain residential integration. The equation is $Y_i = X_i\beta + \delta_i$, where i indexes individuals and X_i is a vector of some three dozen dummy variables describing various characteristics of subject i, including—

AGE GROUP   under 5, 5–17, ...
HOUSEHOLD TYPE   married couple, ...
INCOME LEVEL   under $5,000, $5,000–$10,000, ...
EDUCATIONAL LEVEL grammar school, some high school, ...

The parameter vector β is taken as constant across subjects within each of four demographic groups (Asians, Hispanics, non-Hispanic blacks, non-Hispanic whites). The dependent variable Y_i is the percentage of non-Hispanic whites in the town where subject i resides, and is the same for all subjects in that town. Four equations are estimated, one for each of the demographic groups. Estimation is by OLS, with census data on 674 suburban towns in the New York metropolitan area. The R^2's range from 0.04 to 0.29. Some coefficients are statistically significant for certain groups but not others, which is viewed as evidence favoring one theory of residential integration rather than another. Do the OLS assumptions apply? If not, how would this affect statistical significance? Discuss briefly.

5. Rodgers and Maranto (1989) developed a model for

> "the complex causal processes involved. . . . [in] the determinants of publishing success. . . . the good news is that academic psychologists need not attend a prestigious graduate program to become a productive researcher. . . . the bad news is that attending a nonprestigious PhD program remains an impediment to publishing success."

The Rodgers-Maranto model (figure 7 in the paper) is shown in the diagram below.

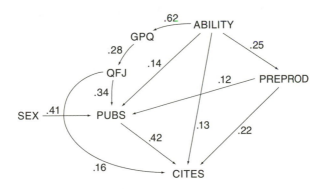

The investigators sent questionnaires to a probability sample of 932 members of the American Psychological Association who were currently working as academic psychologists, and obtained data on 241 men and

244 women. Cases with missing data were deleted, leaving 86 men and 76 women. Variables include—

SEX
: respondent's gender (a dummy variable).

ABILITY
: measures selectivity of respondent's undergraduate institution, respondent's membership in Phi Beta Kappa, etc.

GPQ
: measures the quality of respondent's graduate institution, using national rankings, publication rates of faculty, etc.

QFJ
: measures quality of respondent's first job.

PREPROD
: respondent's quality-weighted number of publications before the PhD. (Mean is 0.8, SD is 1.6.)

PUBS
: number of respondent's publications within 6 years after the PhD. (Mean is 7, SD is 6.)

CITES
: number of times PUBS were cited by others. (Mean is 20, SD is 44.)

Variables were standardized before proceeding with the analysis. Six models were developed but considered inferior to the model shown in the diagram. What does the diagram mean? What are the numbers on the arrows? Where do you see the good news/bad news? Do you believe the news? Discuss briefly.

6. A balance gives quite precise measurements for the difference between weights that are nearly equal. A, B, C, D each weigh about 1 kilogram. The weight of A is known exactly: it is 53 μg above a kilogram, where a μg is a millionth of a gram. (A kilogram is 1000 grams.) The weights of B, C, D are determined through a "weighing design" that involves 6 comparisons shown in the table below.

Comparison	Difference in μg
A and B vs C and D	+42
A and C vs B and D	−12
A and D vs B and C	+10
B and C vs A and D	−65
B and D vs A and C	−17
C and D vs A and B	+11

According to the first line in the table, for instance, A and B are put on the left hand pan of the balance; C and D on the right hand pan. The difference in weights (left hand pan minus right hand pan) is 42 μg.

(a) Are these data consistent or inconsistent?

    (b) What might account for the inconsistencies?

    (c) How would you estimate the weights of B, C, and D?

    (d) Can you put standard errors on the estimates?

    (e) What assumptions are you making?

Explain your answers.

7. (Hard.) There is a population of N subjects, indexed by $i = 1, \ldots, N$. Associated with subject i there is a number v_i. A sample of size n is chosen at random without replacement.

    (a) Show that the sample average of the v's is an unbiased estimate of the population average. (There are hints below.)

    (b) If the sample v's are denoted $V_1, V_2, \ldots, V_n$, show that the probability distribution of $V_2, V_1, \ldots, V_n$ is the same as the probability distribution of $V_1, V_2, \ldots, V_n$. In fact, the probability distribution of any permutation of the V's is the same as any other. In short, the sample is *exchangeable*.

Hints. If you're starting from scratch, it might be easier to do part (b) first. For (b), a permutation π of $\{1, \ldots, N\}$ is a 1–1 mapping of this set onto itself. There are $N!$ permutations. You can choose a sample of size n by choosing π at random, and taking the subjects with index numbers $\pi(1), \ldots, \pi(n)$ as the sample.

8. There is a population of N subjects, indexed by $i = 1, \ldots, N$. A treatment x can be applied at level 0, 10, or 50. Each subject will be assigned to treatment at one of these levels. Subject i has a response $y_{i,x}$ if assigned to treatment at level x. For instance, with a drug to reduce cholesterol levels, x would be the dose and y the cholesterol level at the end of the experiment. Note: $y_{i,x}$ is fixed, not random.

Each subject i has a $1 \times p$ vector of personal characteristics w_i, unaffected by assignment. In the cholesterol experiment, these characteristics might include weight and cholesterol level just before the experiment starts. If you assign subject i to treatment at level 10, say, you observe $y_{i,10}$ but not $y_{i,0}$ or $y_{i,50}$. You can always observe w_i. Population parameters of interest are

$$\alpha_0 = \frac{1}{N} \sum_{i=1}^{N} y_{i,0}, \quad \alpha_{10} = \frac{1}{N} \sum_{i=1}^{N} y_{i,10}, \quad \alpha_{50} = \frac{1}{N} \sum_{i=1}^{N} y_{i,50}.$$

[Question continues on next page.]

The parameter α_0 is the average result we would see if all subjects were put into treatment at level 0. We could measure this directly, by assigning all the subjects to treatment at level 0, but would then lose our chance to learn about the other parameters.

Suppose n_0, n_1, n_2 are positive numbers whose sum is N. In a "randomized controlled experiment," n_0 subjects are chosen at random without replacement and assigned to treatment at level 0. Then n_1 subjects are chosen at random without replacement from the remaining subjects and assigned to treatment at level 10. The last n_2 subjects are assigned to treatment at level 50. From the experimental data—

(a) Can you estimate the three population parameters of interest?

(b) Can you estimate the average response if all the subjects had been assigned to treatment at level 75?

Explain briefly.

9. (This continues question 8.) Let $X_i = x$ if subject i is assigned to treatment at level x. A simple regression model says that given the assignments, the response Y_i of subject i is $\alpha + X_i\beta + \epsilon_i$, where α, β are scalar parameters and the ϵ_i are IID with mean 0 and variance σ^2. Does randomization justify the model? If the model is true, can you estimate the average response if all the subjects had been assigned to treatment at level 75? Explain.

10. (This continues questions 8 and 9.) Let Y_i be the response of subject i. According to a multiple regression model, given the assignments, $Y_i = \alpha + X_i\beta + w_i\gamma + \epsilon_i$, where w_i is a vector of personal characteristics for subject i (question 8); α, β are scalar parameters, γ is a vector of parameters, and the ϵ_i are IID with mean 0 and variance σ^2. Does randomization justify the model? If the model is true, can you estimate the response if a subject with characteristics w_j is assigned to treatment at level 75? Explain.

11. Suppose (X_i, ϵ_i) are IID as pairs for $i = 1, \ldots, n$, with $E(\epsilon_i) = 0$ and $\mathrm{var}(\epsilon_i) = \sigma^2$. Here X_i is a $1 \times p$ random vector and ϵ_i is a random variable (unobservable). Suppose $E(X_i'X_i)$ is $p \times p$ positive definite. Finally, $Y_i = X_i\beta + \epsilon_i$ where β is a $p \times 1$ vector of unknown parameters. Is OLS biased or unbiased? Explain.

12. To demonstrate causation, investigators have used (i) natural experiments, (ii) randomized controlled experiments, and (iii) regression models, among other methods. What are the strengths and weaknesses of

methods (i), (ii), and (iii)? Discuss, preferably giving examples to illustrate your points.

13. True or false, and explain: if the OLS assumptions are wrong, the computer can't fit the model to data.

14. An investigator fits the linear model $Y = X\beta + \epsilon$ to data. The OLS estimate for β is $\hat{\beta}$, and the fitted values are $\hat{Y}$. The investigator writes down the equation $\hat{Y} = X\hat{\beta} + \hat{\epsilon}$. What is $\hat{\epsilon}$?

15. Suppose the X_i are IID $N(0, 1)$. Let $\epsilon_i = 0.025(X_i^4 - 3X_i^2)$ and $Y_i = X_i + \epsilon_i$. An investigator does not know how the data were generated, and runs a regression of Y on X.

   (a) Show that R^2 is about 0.97. (This is hard.)

   (b) Do the OLS assumptions hold?

   (c) Should the investigator trust the usual regression formulas for standard errors?

   Hints. Part (a) can be done by calculus—see the end notes to chapter 4 for the moments of the normal distribution—but it gets a little intricate. A computer simulation may be easier. Assume there is a large sample, e.g., $n = 500$.

16. Assume the response schedule $Y_{i,x} = a + bx + \epsilon_i$. The ϵ_i are IID $N(0, \sigma^2)$. The variables X_i are IID $N(0, \tau^2)$. In fact, the pairs (ϵ_i, X_i) are IID in i, and jointly normal. However, the correlation between (ϵ_i, X_i) is ρ, which may not be 0. The parameters $a, b, \sigma^2, \tau^2, \rho$ are unknown. You observe X_i and $Y_i = Y_{i,X_i}$ for $i = 1, \ldots, 500$.

   (a) If you run a regression of Y_i on X_i, will you get unbiased estimates for a and b?

   (b) Is the relationship between X and Y causal?

   Explain briefly.

17. A statistician fits a regression model ($n = 107$, $p = 6$) and tests whether the coefficient she cares about is 0. Choose one or more of the options below. Explain briefly.

   (i) The null hypothesis says that $\beta_2 = 0$.

   (ii) The null hypothesis says that $\hat{\beta}_2 = 0$.

   (iii) The null hypothesis says that $t = 0$.

   (iv) The alternative hypothesis says that $\beta_2 \neq 0$.

   (v) The alternative hypothesis says that $\hat{\beta}_2 \neq 0$.

(vi) The alternative hypothesis says that $t \neq 0$.

(vii) The alternative hypothesis says that $\hat{\beta}_2$ is statistically significant.

18. Doctors often use body mass index (BMI) to measure obesity. BMI is weight/height^2, where weight is measured in kilograms and height in meters. A BMI of 30 is getting up there. For American women age 18–24, the mean BMI is 24.6 and the variance is 29.4. Although the BMI for a typical woman in this group is something like _____, the BMI of a typical woman will deviate from that central value by something like _____. Fill in the blanks; explain briefly.

19. An epidemiologist says that "randomization does not exclude confounding . . . confounding is very likely if information is collected—as it should be—on a sufficient number of baseline characteristics. . . ." Do you agree or disagree? Discuss briefly.

Notes. "Baseline characteristics" are characteristics of subjects measured at the beginning of the study, i.e., just before randomization. The quote, slightly edited, is from Victora et al (2004).

20. A political scientist is studying a regression model with the usual assumptions, including IID errors. The design matrix X is fixed, with full rank $p = 5$, and $n = 57$. The chief parameter of interest is $\beta_2 - \beta_4$. One possible estimator is $\hat{\beta}_2 - \hat{\beta}_4$, where $\hat{\beta} = (X'X)^{-1}X'Y$. Is there another linear unbiased estimator with smaller variance? Explain briefly.

5.8 End notes for chapter 5

Discussion questions. In question 5, some details of the data analysis are omitted. Question 6 is hypothetical. Two references on weighing designs are Banerjee (1975) and Cameron et al (1977); these are fairly technical. Background for question 19: epidemiologists like to adjust for imbalance in baseline characteristics by statistical modeling, on the theory that they're getting more power—as they would, *if* their models were right. See question 10; also see discussion question 11 in chapter 6.

Measurement error. This is an important topic, not covered in the text. In brief, random error in Y can be incorporated into ϵ, as in the example on Hooke's law. Random error in X usually biases the coefficient estimates. The bias can go either way. For example, random error in a confounder can make an estimated effect too big; random error in measurements of a putative cause can dilute the effect. Biased measurements of X or Y create other problems. There are ways to model the impact of errors, both random and systematic. Such correctives would be useful if the supplementary models were good approximations. Arguments get very complicated very quickly, and benefits remain doubtful (Freedman 1987, 2005). For a broader discussion of measurement issues in the social sciences, see Adcock and Collier (2001).

Dummy variable. The term starts popping up in the statistical literature around 1950: see Oakland (1950) or Klein (1951). The origins are unclear, but the *Oxford English Dictionary* notes related usage in computer science around 1948.

Current Population Survey. This survey is run by the US Bureau of the Census for the Bureau of Labor Statistics, and is the principal source of employment data in the US. There are supplementary questionnaires on other topics of interest, including computer use, demographics, and electoral participation. For information on the design of the survey, see Freedman-Pisani-Purves (1998, chapter 22).

Path diagrams. The choice of variables and arrows in a path diagram is up to the analyst, as are the directions in which the arrows point, although some choices may fit the data less well, and some choices may be illogical. If the graph is "complete"—every pair of nodes joined by an arrow—the direction of the arrows is not constrained by the data (Freedman 1997, pp. 138, 142). Ordering the variables in time may reduce the number of options. There are some algorithms that claim to be able to induce the path diagram from the data, but the track record is not good (Freedman 1997, 2004; Humphreys and Freedman 1996, 1999). Achen (1977) is critical of standardization; also see Blalock (1989).

Response schedules provide a rationale for the usual statistical analysis of path diagrams, and there seems to be no alternative that is much simpler. The statistical assumptions can be weakened a little; see, e.g., (4.18). Figure 4 suggests that the X's are IID. This is the best case for path diagrams, especially when variables are standardized, but all that is needed is exogeneity. Setting up parameters when non-IID data are standardized is a little tricky; see, e.g.,

http://www.stat.berkeley.edu/users/census/standard.pdf

The phrase "response schedule" combines "response surface" from statistics with "supply and demand schedules" from economics (chapter 8). One of the first papers to mention response schedules is Bernheim, Shleifer, and Summers (1985, p. 1051). Some economists have started to write "supply response schedule" and "demand response schedule."

Invariance. The discussion in sections 4–5 assumes that errors are invariant under intervention. It might make more sense to assume that the error distributions are invariant, rather than the errors themselves (Freedman 2004).

Ideas of causation. Embedded in the response-schedule formalism is the conditional distribution of Y, if we were to intervene and set the value of X. This conditional distribution is a counter-factual, at least when the study is observational. The conditional distribution answers the question, what would have happened if we had intervened and set X to x, rather than letting Nature take its course? The idea is best suited to experiments or hypothetical experiments. (The latter are also called "thought experiments" or "gedanken experiments.") The formalism applies less well to non-manipulationist ideas of causation: the moon causes the tides, earthquakes cause property values to go down, time heals all wounds. Time is not manipulable; neither are earthquakes or the moon.

Investigators may hope that regression equations are like laws of motion in classical physics: if position and momentum are given, you can determine the future of the system and discover what would happen with different initial conditions. Some other formalism may be needed to make this non-manipulationist account more precise. Evans (1993) has an interesting survey of causal ideas in epidemiology, with many examples. In the legal context, the survey to read is Hart and Honoré (1985).

Otis Dudley Duncan was one of the great empirical social scientists of the 20th century. Blau and Duncan (1967) were optimistic about the use of statistical models in the social sciences, but Duncan's views darkened after 20 years of experience—

> "Coupled with downright incompetence in statistics, paradoxically, we often find the syndrome that I have come to call *statisticism*: the notion that computing is synonymous with doing research, the naive faith that statistics is a complete or sufficient basis for scientific methodology, the superstition that statistical formulas exist for evaluating such things as the relative merits of different substantive theories or the 'importance' of the causes of a 'dependent variable'; and the delusion that decomposing the covariations of some arbitrary and haphazardly assembled collection of variables can somehow justify not only a 'causal model' but also, praise the mark, a 'measurement model.' There would be no point in deploring such caricatures of the scientific enterprise if there were a clearly identifiable sector of social science research wherein such fallacies were clearly recognized and emphatically out of bounds." (Duncan 1984, p. 226)

6

Maximum Likelihood

6.1 Introduction

Maximum likelihood is a general (and, with large samples, very power-ful) method for estimating parameters in a statistical model. The maximum likelihood estimator is usually called the MLE. Here, we begin with textbook examples like the normal, binomial, and Poisson. Then comes the probit model, with a real application—the effects of Catholic schools (Evans and Schwab 1995, reprinted at the back of the book). This application will show the strengths and weaknesses of the probit model in action.

Example 1. $N(\mu, 1)$ with $-\infty < \mu < \infty$. The density at x is

$$\frac{1}{\sqrt{2\pi}} \exp\left[-\frac{1}{2}(x-\mu)^2\right], \quad \text{where } \exp(x) = e^x.$$

See section 3.5. For n independent $N(\mu, 1)$ variables $X_1, \ldots, X_n$, the density at $x_1, \ldots, x_n$ is

$$\left(\frac{1}{\sqrt{2\pi}}\right)^n \exp\left[-\frac{1}{2}\sum_{i=1}^{n}(x_i-\mu)^2\right].$$

The *likelihood function* is the density evaluated at the data $X_1, \ldots, X_n$, viewed as a function of the parameter μ. The log likelihood function is more useful:

$$L_n(\mu) = -\frac{1}{2} \sum_{i=1}^{n} (X_i - \mu)^2 - n \log\left(\sqrt{2\pi}\right).$$

The notation makes it explicit that $L_n(\mu)$ depends on the sample size n and the parameter μ. There is also dependence on the data, because the likelihood function is evaluated at the X_i: look at the right hand side of the equation.

The MLE is the parameter value $\hat{\mu}$ that maximizes $L_n(\mu)$. To find the MLE, you can start by differentiating $L_n(\mu)$ with respect to μ:

$$L'_n(\mu) = \sum_{i=1}^{n} (X_i - \mu).$$

Set $L'_n(\mu)$ to 0 and solve. The unique μ with $L'_n(\mu) = 0$ is $\hat{\mu} = \overline{X}$, the sample mean. Check that

$$L''_n(\mu) = -n.$$

Thus, $\overline{X}$ is the maximum not the minimum. (Here, L'_n means the derivative not the transpose, and L''_n is the second derivative.)

Example 2. Binomial$(1, p)$ with $0 < p < 1$. Let X_i be independent. Each X_i is 1 with probability p and 0 with remaining probability $1 - p$, so X_i has the Binomial$(1, p)$ distribution. Let $x_i = 0$ or 1. The probability that $X_i = x_i$ for $i = 1, \ldots, n$ is

$$\prod_{i=1}^{n} p^{x_i} (1 - p)^{1 - x_i}.$$

The reasoning: due to independence, the probability is the product of n factors. If $x_i = 1$, the ith factor is $P_p(X_i = 1) = p = p^{x_i}(1 - p)^{1 - x_i}$, because $(1 - p)^0 = 1$. If $x_i = 0$, the factor is $P_p(X_i = 0) = 1 - p = p^{x_i}(1 - p)^{1 - x_i}$, because $p^0 = 1$. (Here, P_p is the probability that governs the X_i's when the parameter is p.) Let $S = X_1 + \cdots + X_n$. Check that

$$L_n(p) = \sum_{i=1}^{n} \left[X_i \log p + (1 - X_i) \log(1 - p) \right]$$
$$= S \log p + (n - S) \log(1 - p).$$

Now

$$L'_n(p) = \frac{S}{p} - \frac{n - S}{1 - p}.$$

and

$$L_n''(p) = -\frac{S}{p^2} - \frac{n - S}{(1 - p)^2}.$$

The MLE is $\hat{p} = S/n$.

If $S = 0$, the likelihood function is maximized at $\hat{p} = 0$. This is an "endpoint maximum." Similarly, if $S = n$, the likelihood function has an endpoint maximum at $p = 1$. In the first case, $L_n' < 0$ on $(0, 1)$. In the second case, $L_n' > 0$ on $(0, 1)$. Either way, the equation $L_n'(p) = 0$ has no solution.

Example 3. Poisson(λ) with $0 < \lambda < \infty$. Let X_i be independent Poisson(λ). If $j = 0, 1, \ldots$ then

$$P_\lambda(X_i = j) = e^{-\lambda}\frac{\lambda^j}{j!}$$

and

$$P_\lambda(X_i = j_i \text{ for } i = 1, \ldots, n) = e^{-n\lambda}\lambda^{j_1 + \cdots + j_n}\prod_{i=1}^{n}\frac{1}{j_i!},$$

where P_λ is the probability distribution that governs the X_i's when the parameter is λ. Let $S = X_1 + \cdots + X_n$. So

$$L_n(\lambda) = -n\lambda + S\log\lambda - \sum_{i=1}^{n}\log(X_i!).$$

Now

$$L_n'(\lambda) = -n + \frac{S}{\lambda}$$

and

$$L_n''(\lambda) = -\frac{S}{\lambda^2}.$$

The MLE is $\hat{\lambda} = S/n$. (This is an endpoint maximum if $S = 0$.)

Example 4. Let X be a positive random variable, with $P_\theta(X > x) = \theta/(\theta + x)$ for $0 < x < \infty$, where the parameter θ is a positive real number. The distribution function of X is $x/(\theta + x)$. The density is $\theta/(\theta + x)^2$. Let $X_1, \ldots, X_n$ be independent, with density $\theta/(\theta + x)^2$. Then

$$L_n(\theta) = n\log\theta - 2\sum_{i=1}^{n}\log(\theta + X_i).$$

Now

$$L'_n(\theta) = \frac{n}{\theta} - 2\sum_{i=1}^{n} \frac{1}{\theta + X_i}$$

and

$$L''_n(\theta) = -\frac{n}{\theta^2} + 2\sum_{i=1}^{n} \frac{1}{(\theta + X_i)^2}.$$

There is no explicit formula for the MLE, but you can find it by numerical methods on the computer. (Computer labs 10–12 at the back of the book will get you started on numerical maximization, or see the end notes for the chapter; a detailed treatment is beyond our scope.) This example is a little artificial; it will be used to illustrate some features of the MLE.

REMARKS. In example 1, the sample mean $\overline{X}$ is $N(\mu, 1/n)$. In example 2, the sum is Binomial(n, p):

$$P_p(S = j) = \binom{n}{j} p^j (1 - p)^{n-j}.$$

In example 3, the sum is Poisson$(n\lambda)$:

$$P_\lambda(S = j) = e^{-n\lambda} \frac{(n\lambda)^j}{j!}.$$

DEFINITION. There is a statistical model parameterized by θ. The *Fisher information* is $I_\theta = -E_\theta[L''_1(\theta)]$, namely, the negative of the expected value of the second derivative of the log likelihood function, for a sample of size 1.

THEOREM 1. Suppose $X_1, \ldots, X_n$ are IID with probability distribution governed by the parameter θ. Let θ_0 be the true value of θ. Under regularity conditions (which are omitted here), the MLE for θ is asymptotically normal. The asymptotic mean of the MLE is θ_0. The asymptotic variance can be computed in three ways:

(i) $I_{\theta_0}^{-1}/n$,

(ii) $I_{\hat\theta}^{-1}/n$,

(iii) $[-L''_n(\hat\theta)]^{-1}$.

If $\hat\theta$ is the MLE and v_n is the asymptotic variance, the theorem says that $(\hat\theta - \theta_0)/\sqrt{v_n}$ is nearly $N(0, 1)$ when the sample size n is large—and we're sampling from θ_0. ("Asymptotic" results are nearly right for large samples.)

The $[-L_n''(\hat{\theta})]$ in (iii) is often called "observed information." With option (iii), the sample size n is built into L_n: there is no division by n.

The MLE can be used in multi-dimensional problems, and theorem 1 generalizes. When the parameter vector θ is p-dimensional, $L'(\theta)$ is a p-vector. The jth component of $L'(\theta)$ is $\partial L/\partial \theta_j$. Furthermore, $L''(\theta)$ is a $p \times p$ matrix. The ijth component of $L''(\theta)$ is

$$\frac{\partial^2 L}{\partial \theta_i \, \partial \theta_j} = \frac{\partial^2 L}{\partial \theta_j \, \partial \theta_i}.$$

We're assuming that L is smooth. Then the matrix L'' is symmetric. We still define $I_\theta = -E_\theta\big[L_1''(\theta)\big]$. This is now a $p \times p$ matrix. The diagonal elements of I_θ^{-1}/n give asymptotic variances for the components of $\hat{\theta}$; the off-diagonal elements, the covariances. Similar comments apply to $-L_n''(\hat{\theta})^{-1}$.

The examples. The normal, binomial, and Poisson are "exponential families" where the theory is especially attractive (although it is beyond our scope). Among other things, the likelihood function generally has a unique maximum. With other kinds of models, there are usually several local maxima and minima.

Caution. Ordinarily, the MLE is biased—although the bias is small with large samples. The asymptotic variance is also an approximation. Moreover, with small samples, the distribution of the MLE is often far from normal.

Software. Some numerical analysis software will maximize L as a function of the parameter, in which case you need to write some code to compute L. Other software looks for the minimum of $-L$. Some routines use L' or L''; others compute derivatives for you. When there are several maxima, the software tries to find a local extremum near a starting point you set. With more dimensions, numerical maximization becomes trickier.

Exercise set A

1. In example 1, the log likelihood function is a sum—as it is in examples 2, 3, and 4. Is this a coincidence? If not, what is the principle?

2. (a) Suppose $X_1, X_2, \ldots, X_n$ are IID $N(\mu, 1)$. Find the mean and variance of the MLE for $\hat{\mu}$. Find the distribution of the MLE and compare to the theorem. Show that $-L_n''(\hat{\mu})/n \to I_\mu$. Comment: for the normal, the asymptotics are awfully good.

   (b) If U is $N(0, 1)$, show that U is *symmetric*: namely, $P(U < y) = P(-U < y)$. Hints. (i) $P(-U < y) = P(U > -y)$, and (ii) $\exp(-x^2/2)$ is a symmetric function of x.

3. Repeat 2(a) for the binomial in example 2. Is the MLE normally distributed? Or is it only approximately normal?

4. Repeat 2(a) for the Poisson in example 3. Is the MLE normally distributed? Or is it only approximately normal?

5. Find the density of $\theta U/(1-U)$, where U is uniform on $[0,1]$ and $\theta > 0$.

6. Suppose the $X_i > 0$ are independent, and their common density is $\theta/(\theta + x)^2$ for $i = 1, \ldots, n$, as in example 4. Show that $\theta L'_n(\theta) = -n + 2\sum_{i=1}^{n} X_i/(\theta + X_i)$. Deduce that $\theta \to \theta L'_n(\theta)$ decreases from n to $-n$ as θ increases from 0 to ∞. Conclude that L_n has a unique maximum. (Reminder: L'_n means the derivative not the transpose.)

7. What is the median of X in example 4?

8. Show that the Fisher information in example 4 is $1/(3\theta^2)$.

9. Suppose X_i are independent for $i = 1, \ldots, n$, with a common Poisson distribution. Suppose $E(X_i) = \lambda > 0$, but the parameter of interest is $\theta = \lambda^2$. Find the MLE for θ. Is the MLE biased or unbiased?

10. As in exercise 9, but the parameter of interest is $\theta = \sqrt{\lambda}$. Find the MLE for θ. Is the MLE biased or unbiased?

11. Let β be a positive real number, which is unknown. Suppose X_i are independent Poisson random variables, with $E(X_i) = \beta i$ for $i = 1, 2, \ldots, 20$. How would you estimate β?

12. Suppose X, Y, Z are independent normal random variables, each having variance 1. The means are $\alpha + \beta$, $\alpha + 2\beta$, $2\alpha + \beta$, respectively: α, β are parameters to be estimated. Show that the MLE and the OLS estimator are the same. Note: this won't usually be true—the result depends on the normality assumption.

13. Let θ be a positive real number, which is unknown. Suppose the X_i are independent for $i = 1, \ldots, n$, with a common distribution P_θ that depends on θ: $P_\theta\{X_i = j\} = c(\theta)(\theta + j)^{-1}(\theta + j + 1)^{-1}$ for $j = 0, 1, 2, \ldots$. What is $c(\theta)$? How would you estimate θ? Hints on finding $c(\theta)$. What is $\sum_{j=0}^{\infty} (a_j - a_{j+1})$? What is $(\theta + j)^{-1} - (\theta + j + 1)^{-1}$?

14. Suppose X_i are independent for $i = 1, \ldots, n$, with common density $\frac{1}{2}\exp(-|x - \theta|)$, where θ is a parameter, x is real, and n is odd. Show that the MLE for θ is the sample median. Hint: see exercise 2B18.

6.2 Probit models

The probit model explains a 0–1 response variable Y_i for subject i in terms of a row vector of covariates X_i. Let X be the matrix whose ith row is X_i. Each row in X represents the covariates for one subject, and each column represents one covariate. Given X, the responses Y_i are assumed to be independent random variables, taking values 0 or 1, with

$$P(Y_i = 1|X) = \Phi(X_i\beta).$$

Here, Φ is the standard normal distribution function and β is a parameter vector. Any distribution function could be used: Φ is what makes it a probit model rather than a logit model or an xxxit model.

Let's try some examples. About half of Americans age 25+ read a book last year. Strange but true. Probabilities vary with education, income, and gender, among other things. In a (hypothetical) study on this issue, subjects are indexed by $i = 1, \ldots, n$. The response variable Y_i is defined as 1 if subject i read a book last year, else $Y_i = 0$. The vector of explanatory variables for subject i is $X_i = [1, ED_i, INC_i, MAN_i]$:

ED_i is years of schooling completed by subject i.

INC_i is the annual income of subject i, in US dollars.

MAN_i is 1 if subject i is a man, else 0. (This is a dummy variable: section 5.6.)

The parameter vector β is 4×1. Given the covariate matrix X, the Y_i's are assumed to be independent with $P(Y_i = 1) = \Phi(X_i\beta)$, where Φ is the standard normal distribution function.

Example 5. Suppose we know that $\beta_1 = -0.35$, $\beta_2 = 0.02$, $\beta_3 = 1/100,000$, and $\beta_4 = -0.1$. A man has 12 years of education and makes $40,000 a year. His $X_i\beta$ is

$$-0.35 + 12 \times 0.02 + 40,000 \times \frac{1}{100,000} - 0.1$$
$$= -0.35 + 0.24 + 0.4 - 0.1 = 0.19.$$

The probability he read a book last year is $\Phi(0.19) = 0.58$.

A similarly situated woman has $X_i\beta = -0.35 + 0.24 + 0.4 = 0.29$. The probability she read a book last year is $\Phi(0.29) = 0.61$, a bit higher than the 0.58 for her male counterpart in example 5. The point of the dummy variable is to add β_4 to $X_i\beta$ for male subjects but not females. Here, β_4 is negative. (Adding a negative number is what most people would call subtraction.)

Estimation. We turn to the case where β is unknown, to be estimated from the data by maximum likelihood. The probit model makes the independence assumption, so the likelihood function is a product with a factor for each subject. Let's compute this factor for two subjects.

Example 6. Subject is male, with 18 years of education and a salary of $60,000. Not a reader, he watches TV or goes to the opera for relaxation. His factor in the likelihood function is

$$1 - \Phi\big(\beta_1 + 18\beta_2 + 60{,}000\beta_3 + \beta_4\big).$$

It's $1 - \Phi$ because he doesn't read. There's $+\beta_4$ in the equation, because it's him not her. TV and opera are irrelevant.

Example 7. Subject is female, with 16 years of education and a salary of $45,000. She reads books, has red hair, and loves scuba diving. Her factor in the likelihood function is

$$\Phi\big(\beta_1 + 16\beta_2 + 45{,}000\beta_3\big).$$

It's Φ because she reads books. There is no β_4 in the equation: her dummy variable is 0. Hair color and underwater activities are irrelevant.

Since the likelihood is a product—we've conditioned on X—the log likelihood is a sum, with a term for each subject:

$$L_n(\beta) = \sum_{i=1}^{n} \Big(Y_i \log\big[P(Y_i = 1|X_i)\big] + (1 - Y_i)\log\big[1 - P(Y_i = 1|X_i)\big]\Big)$$

$$= \sum_{i=1}^{n} \Big(Y_i \log\big[\Phi(X_i\beta)\big] + (1 - Y_i)\log\big[1 - \Phi(X_i\beta)\big]\Big).$$

The readers contribute terms with $\log \Phi$: the $\log[1 - \Phi]$ drops out, because $Y_i = 1$ if subject i is a reader. It's the reverse for non-readers: $Y_i = 0$, so $\log \Phi$ drops out and $\log[1 - \Phi]$ stays in. If in doubt, review the binomial example in section 1.

Given X, the Y_i are independent. They are not identically distributed: $P(Y_i = 1|X) = \Phi(X_i\beta)$ differs from one i to another. Theorem 1 can be extended to cover this case, although options (i) and (ii) for asymptotic variance get a little more complicated. For instance, (i) becomes $\{-E_{\theta_0}[L_n''(\theta_0)]\}^{-1}$.

We estimate β by maximizing $L_n(\beta)$. In the reading model, this would be impossible by calculus, so it's done on the computer. The asymptotic covariance matrix is $[-L_n''(\hat{\beta})]^{-1}$: observed information is used because it isn't feasible to compute the Fisher information matrix analytically. To get standard errors, take square roots of the diagonal elements.

Why not regression?

You probably don't want to tell the world that $Y = X\beta + \epsilon$. The reason: $X_i\beta$ is going to produce numbers other than 0 or 1, and $X_i\beta + \epsilon_i$ is even worse. The next option might be $P(Y_i = 1|X) = X_i\beta$, the Y_i being assumed conditionally independent across subjects. That's a "linear probability model." Chapter 8 has an example, with additional complications.

Given data from a linear probability model, you can estimate β by feasible GLS. However, there are likely to be some subjects with $X_i\hat{\beta} > 1$, and other subjects with $X_i\hat{\beta} < 0$. A probability of 1.5 is a jolt; so is -0.3. The probit model respects the constraint that probabilities are between 0 and 1.

Regression isn't useless in the probit context. To maximize the likelihood function on the computer, it helps to have a reasonable starting point. Regress Y on X, and start the search from there.

The latent-variable formulation

The probit model is one analog of regression for binary response variables; the logit model, discussed below, is another. So far, there is no error term in the picture. However, the model can be set up with something like an error term. To see how, let's go back to the probit model for reading books.

Subject i has a *latent* (hidden) variable U_i. These are IID $N(0, 1)$ across subjects, independent of the covariates. (Reminder: IID = independent and identically distributed.) Subject i reads books if $X_i\beta + U_i > 0$. However, if $X_i\beta + U_i < 0$, then subject i is not a reader. We don't have to worry that $X_i\beta + U_i = 0$: this is an event with probability 0.

Given the covariate matrix X, the probability that subject i reads books is

$$P(X_i\beta + U_i > 0) = P(U_i > -X_i\beta) = P(-U_i < X_i\beta).$$

Because U_i is symmetric (exercise A2),

$$P(-U_i < X_i\beta) = P(U_i < X_i\beta) = \Phi(X_i\beta).$$

So $P(X_i\beta + U_i > 0) = \Phi(X_i\beta)$. The new formulation with latent variables gives the right probabilities.

The probit model now has something like an error term, namely, the latent variable. But there is an important difference between latent variables and error terms. You can't estimate latent variables. At most, the data tell you $X_i\beta$ and the sign of $X_i\beta + U_i$. That is not enough to determine U_i. By contrast, error terms in a regression model can be approximated by residuals.

The latent-variable formulation does make the assumptions clearer. The probit model requires the U_i's to be independent of the X_i's, and IID across subjects. The U_i's need to be normal. The response for subject i depends only on that subject's covariates. (Look at the formulas!)

The hard questions about probit models are usually ducked. Is IID realistic for reading books? Not if there's word-of-mouth: "Hey, you have to read this book, it's great." Why are the β's the same for everybody? e.g., for men and women? Why is the effect of income the same for all educational levels? What about other variables?

> If the assumptions in the model break down, the MLE will be biased—even with large samples. The bias may be severe. Also, estimated standard errors will not be reliable.

Exercise set B

1. Let Z be $N(0, 1)$ with density function ϕ and distribution function Φ (section 3.5). True or false, and explain:

   (a) The slope of Φ at the real number x is $\phi(x)$.
   (b) The area to the left of x under ϕ is $\Phi(x)$.
   (c) $P(Z = x) = \phi(x)$.
   (d) $P(Z < x) = \Phi(x)$.
   (e) $P(Z \le x) = \Phi(x)$.
   (f) $P(x < Z < x + h) \doteq \phi(x)h$ if h is small and positive.

2. In brief, the probit model for reading says that subject i read a book last year if $X_i \beta + U_i > 0$.

   (a) What are X_i and β?

   (b) The U_i is a _____ variable. Options (more than one may be right):

   > data   random   latent   dummy   observable

   (c) What are the assumptions on U_i?

   (d) The log likelihood function is a _____, with one _____ for each _____. Fill in the blanks using the options below, and explain briefly.

   > sum   product   quotient   matrix   term
   > subject   factor   entry   book   variable

3. As in example 5, suppose we know $\beta_1 = -0.35$, $\beta_2 = 0.02$, $\beta_3 = 1/100{,}000$, $\beta_4 = -0.1$. George has 12 years of education and makes \$40,000 a year. His brother Harry also has 12 years of education but makes \$50,000 a year. True or false, and explain: according to the

model, the probability that Harry read a book last year is 0.1 more than George's probability. If false, compute the difference in probabilities.

Identification vs estimation

Two very technical ideas are coming up: *identifiability* and *estimability*. Take identifiability first. Suppose P_θ is the probability distribution that governs X. The distribution depends on the parameter θ. Think of X as observable, so P_θ is something we can determine. The function $f(\theta)$ is identifiable if $f(\theta_1) \neq f(\theta_2)$ implies $P_{\theta_1} \neq P_{\theta_2}$ for every pair (θ_1, θ_2) of parameter values. In other words, $f(\theta)$ is identifiable if changing $f(\theta)$ changes the distribution of an observable random variable.

Now for the second idea: the function $f(\theta)$ is estimable if there is a function g with $E_\theta[g(X)] = f(\theta)$ for all values of θ, where E_θ stands for expected value computed from P_θ. This is a cold mathematical definition: $f(\theta)$ is estimable if there is an unbiased estimator for it. Nearly unbiased won't do, and variance doesn't matter.

PROPOSITION 1. If $f(\theta)$ is estimable, then $f(\theta)$ is identifiable.

Proof. If $f(\theta)$ is estimable, there is a function g with $E_\theta[g(X)] = f(\theta)$ for all θ. If $f(\theta_1) \neq f(\theta_2)$, then $E_{\theta_1}[g(X)] \neq E_{\theta_2}[g(X)]$. So $P_{\theta_1} \neq P_{\theta_2}$: i.e., θ_1 and θ_2 generate different distributions for X.

The converse to proposition 1 is false. A parameter—or a function of a parameter—can be identifiable without being estimable. That is what the next example shows.

Example 8. Suppose $0 < p < 1$ is a parameter; X is a random variable with $P_p(X = 1) = p$ and $P_p(X = 0) = 1 - p$. Then $\sqrt{p}$ is identifiable but not estimable. To prove identifiability, $\sqrt{p_1} \neq \sqrt{p_2}$ implies $p_1 \neq p_2$, and then $P_{p_1}(X = 1) \neq P_{p_2}(X = 1)$. What about estimating $\sqrt{p}$, for instance, by $g(X)$—where g is some suitable function? Well, $E_p[g(X)] = (1 - p)g(0) + pg(1)$. This is a linear function of p. But $\sqrt{p}$ isn't linear. So $\sqrt{p}$ isn't estimable: there is no g with $E_p[g(X)] = \sqrt{p}$ for all p. In short, $\sqrt{p}$ is identifiable but not estimable, as advertised.

For the binomial, the parameter is one-dimensional. However, the definitions apply also to multi-dimensional parameters. Identifiability is an important concept, but it may seem a little mysterious. Let's say it differently.

Something is identifiable if you can get it from the joint distribution of observable random variables.

Example 9. There are three parameters, a, b, and σ^2. Suppose $Y_i = a + bx_i + \delta_i$ for $i = 1, 2, \ldots, 100$. The x_i are fixed and known; in fact, all the x_i happen to be 2. The unobservable δ_i are IID $N(0, \sigma^2)$. Is a identifiable? estimable? How about b? $a + 2b$? σ^2? To begin with, the Y_i are IID $N(a + 2b, \sigma^2)$. The sample mean of the Y_i's estimates $a + 2b$. Thus, $a + 2b$ is estimable and identifiable. The sample variance of the Y_i's estimates σ^2—if you divide by 99 rather than 100. Thus, σ^2 is estimable and identifiable.

However, a and b are not separately identifiable. For instance, if $a = 0$ and $b = 1$, the Y_i would be IID $N(2, \sigma^2)$. If $a = 1$ and $b = 0.5$, the Y_i would be IID $N(2, \sigma^2)$. If $a = \sqrt{17}$ and $b = (2 - \sqrt{17})/2$, the Y_i would be IID $N(2, \sigma^2)$. Infinitely many combinations of a and b generate exactly the same joint distribution for the Y_i. That is why information about the Y_i can't help you break $a + 2b$ apart, into a and b. If you want to identify a and b separately, you need some variation in the x_i.

Example 10. Suppose U and V are independent random variables: U is $N(a, 1)$ and V is $N(b, 1)$, where a and b are parameters. Although the sum $U + V$ is observable, U and V themselves are not observable. Is $a + b$ identifiable? How about a? b? To begin with, $E(U + V) = a + b$. So $a + b$ is estimable, hence, identifiable. On the other hand, if we increase a by some amount and decrease b by the same amount, $a + b$ is unchanged. The distribution of $U + V$ is also unchanged. Hence, a and b themselves are not identifiable.

What if the U_i are $N(\mu, \sigma^2)$?

Let's go back to the probit model for reading books, and try $N(\mu, 1)$ latent variables. Then β_1—the intercept—is mixed up with μ. You can identify $\beta_1 + \mu$, but can't get the pieces β_1, μ. What about $N(0, \sigma^2)$ for the latents? Without some constraint, parameters are not identifiable. For instance, the combination $\sigma = 1$ and $\beta = \gamma$ produces the same probability distribution for the Y_i given the X_i as $\sigma = 2$ and $\beta = 2\gamma$. Setting $\sigma = 1$ makes the other parameters identifiable. But this is an assumption, and there would be trouble if the distribution of the latent variables changed from one subject to another.

Exercise set C

1. If X is $N(\mu, \sigma^2)$, show that μ is estimable and σ^2 is identifiable.

2. Suppose X_1, X_2, and X_3 are independent normal random variables. Each has variance 1. The means are α, $\alpha + 9\beta$, and $\alpha + 99\beta$, respectively. Are α and β identifiable? estimable?

3. Suppose $Y = X\beta + \epsilon$, where X is a fixed $n \times p$ matrix, β is a $p \times 1$ parameter vector, the ϵ_i are IID with mean 0 and variance σ^2. Is β identifiable if the rank of X is p? if the rank of X is $p - 1$?

4. Suppose $Y = X\beta + \epsilon$, where X is a fixed $n \times p$ matrix of rank p, and β is a $p \times 1$ parameter vector. The ϵ_i are independent with common variance σ^2 and $E(\epsilon_i) = \mu_i$, where μ is an $n \times 1$ parameter vector. Is β identifiable?

5. Suppose X_1 and X_2 are IID, with $P_p(X_1 = 1) = p$ and $P_p(X_1 = 0) = 1 - p$; the parameter p is between 0 and 1. Is p^3 identifiable? estimable?

6. Suppose U and V are independent; U is $N(0, \sigma^2)$ and V is $N(0, \tau^2)$, where σ^2 and τ^2 are parameters. However, U and V are not observable. Only $U + V$ is observable. Is $\sigma^2 + \tau^2$ identifiable? How about σ^2? τ^2?

7. If X is distributed like the absolute value of an $N(\mu, 1)$ variable, show that:
   (a) $|\mu|$ is identifiable. Hint: what is $E(X^2)$?
   (b) μ itself is not identifiable. Hint: μ and $-\mu$ lead to the same distribution for X.

8. For incredibly many bonus points: suppose X is $N(\mu, \sigma^2)$. Is $|\mu|$ estimable? What about σ^2? Comments. We only have one observation X, not many observations. A rigorous solution to this exercise might involve the dominated convergence theorem, or the uniqueness theorem for Laplace transforms.

6.3 Logit models

Logits are often used instead of probits. The specification is the same, except that the logistic distribution function Λ is used instead of the normal Φ:

$$\Lambda(x) = \frac{e^x}{1 + e^x} \quad \text{for} \ -\infty < x < \infty.$$

The *odds ratio* is $p/(1 - p)$. People write *logit* for the log odds ratio:

$$\text{logit } p = \log \frac{p}{1 - p} \quad \text{for} \ 0 < p < 1.$$

The logit model says that the response variables Y_i are independent given the covariates X, and $P(Y_i = 1|X) = \Lambda(X_i\beta)$, that is,

$$\text{logit } P(Y_i = 1|X) = X_i\beta.$$

(See exercise 6 below.) From the latent-variables perspective,

$$Y_i = 1 \text{ if } X_i\beta + U_i > 0, \quad \text{but } Y_i = 0 \text{ if } X_i\beta + U_i < 0.$$

The latent variables U_i are independent of the covariate matrix X, and the U_i are IID, but now the common distribution function of the U_i is Λ. The logit model uses Λ where the probit uses Φ. That's the difference. "Logistic regression" is a synonym for logit models.

Exercise set D

1. Suppose the random variable X has a continuous, strictly increasing distribution function F. Show that $F(X)$ is uniform on $[0,1]$. Hints. Show that F has a continuous, strictly increasing inverse F^{-1}. So $F(X) < y$ if and only if $X < F^{-1}(y)$.

2. Conversely, if U is uniform on $[0,1]$, show that $F^{-1}(U)$ has distribution function F. (This idea is often used to simulate IID picks from F.)

On the logit model

3. Check that the logistic distribution function Λ is monotone increasing. Hint: if $1 - \Lambda$ is decreasing, you're there.

4. Check that $\Lambda(-\infty) = 0$ and $\Lambda(\infty) = 1$.

5. Check that the logistic distribution is symmetric, i.e., $1 - \Lambda(x) = \Lambda(-x)$. Appearances can be deceiving....

6. (a) If $P(Y_i = 1|X) = \Lambda(X_i\beta)$, show that logit $P(Y_i = 1|X) = X_i\beta$.
   (b) If logit $P(Y_i = 1|X) = X_i\beta$, show that $P(Y_i = 1|X) = \Lambda(X_i\beta)$.

7. What is the distribution of $\log U - \log(1 - U)$, where U is uniform on $[0, 1]$? Hints. Show that $\log u - \log(1 - u)$ is a strictly increasing function of u. Then compute the chance that $\log U - \log(1 - U) > x$.

8. For $\theta > 0$, suppose X has the density $\theta/(\theta+x)^2$ on the positive half-line $(0, \infty)$. Show that $\log(X/\theta)$ has the logistic distribution.

9. Show that $\varphi(x) = -\log(1 + e^x)$ is strictly concave on $(-\infty, \infty)$. Hint: check that $\varphi''(x) = -e^x/(1 + e^x)^2 < 0$.

10. Suppose that, conditional on the covariates X, the Y's are independent 0–1 variables, with logit $P(Y_i = 1|X) = X_i\beta$, i.e., the logit model holds. Show that the log likelihood function can be written as

$$L_n(\beta) = -\left(\sum_{i=1}^{n} \log\left[1 + \exp(X_i\beta)\right]\right) + \left(\sum_{i=1}^{n} X_i Y_i\right)\beta.$$

11. (This continues exercises 9 and 10: hard.) Show that $L_n(\beta)$ is a concave function of β, and strictly concave if X has full rank. Hints. Let the parameter vector β be $p \times 1$. Let c be a $p \times 1$ vector with $\|c\| > 0$. You need to show $c'L_n''(\beta)c \leq 0$, with strict inequaltiy if X has full rank. Let X_i be the ith row of X, a $1 \times p$ vector. Confirm that $L_n''(\beta) = \sum_i X_i'X_i\varphi''(X_i\beta)$, where φ was defined in exercise 9. Check that $c'X_i'X_ic \geq 0$ and $\varphi''(X_i\beta) \leq m < 0$ for all $i = 1, \ldots, n$, where m is a real number that depends on β.

On the probit model

12. Let Φ be the standard normal distribution function (mean 0, variance 1). Let $\phi = \Phi'$ be the density. Show that $\phi'(x) = -x\phi(x)$. If $x > 0$, show that

$$\int_x^\infty z\phi(z)\,dz = \phi(x) \quad \text{and} \quad 1 - \Phi(x) < \int_x^\infty \frac{z}{x}\phi(z)\,dz.$$

Conclude that $1 - \Phi(x) < \phi(x)/x$ for $x > 0$. If $x < 0$, show that $\Phi(x) < \phi(x)/|x|$. Show that $\log \Phi$ and $\log(1 - \Phi)$ are strictly concave, because their second derivatives are strictly negative. Hint: do the cases $x > 0$ and $x < 0$ separately.

13. (This continues exercise 12: hard.) Show that the log likelihood for the probit model is concave, and strictly concave if X has full rank. Hint: this is like exercise 11.

6.4 The effect of Catholic schools

Catholic schools in the United States seem to be more effective than public schools. Graduation rates are higher and more of the students get into college. But maybe this is because of student characteristics. For instance, richer students might be more likely to go to Catholic schools, and richer kids tend to do better academically. That could explain the apparent effect. Evans and Schwab (1995) use a probit model to adjust for student characteristics like family income. They use a two-equation model to adjust for selection effects based on unmeasured characteristics, like intelligence and motivation. For example, Catholic schools might look better because they screen out less-intelligent, less-motivated students; or, students who are more intelligent and better motivated might self-select into Catholic schools. (The paper, reprinted at the back of the book, rejects these alternative explanations.)

Data are from the "High School and Beyond" survey of high schools. Evans and Schwab look at students who were sophomores in the original 1980

survey and who responded to followup surveys in 1982 and 1984. Students
who dropped out are excluded. So are a further 389 students who attended
private non-Catholic schools, or whose graduation status was unknown. That
leaves 13,294 students in the sample. Table 1 in the paper summarizes the
data: 97% of the students in Catholic schools graduated, compared to 79%
in public schools—an impressive difference.

Table 1 also demonstrates confounding. For instance, 14% of the students in Catholic schools had family incomes above \$38,000, compared to
7% in public schools. (These are 1980s dollars; \$38,000 then was equivalent
to maybe \$80,000 at the beginning of the 21st century.) Generally, however,
confounding is not prominent. Table 2 has further details on outcomes by
school type. The probit results are in table 3. The bottom line: confounding
by measured variables does not seem to explain the different success rates for
Catholic schools and public schools.

To define the model behind table 3, let the response variable Y_i be 1 if
student i graduates, otherwise Y_i is 0. Given the covariates, the model says
that graduation is independent across students. For student i,

$$(1) \qquad\qquad P(Y_i = 1 \,|\, C, X) = \Phi(C_i \alpha + X_i \beta),$$

where $C_i = 1$ if student i attends Catholic school, while $C_i = 0$ if student i
attends public school. Next, X_i is a vector of dummy variables describing personal characteristics of student i—gender, race, ethnicity, family income....
The matrix X has a row for each student and a column for each variable: X_i
is the ith row of X. Similarly, C is the vector whose ith component is C_i.
As usual, Φ is the standard normal distribution function. The parameters α
and β are estimated by maximum likelihood: α is a scalar, and β is a vector.
(We're not using the same notation as the paper.)

For Evans and Schwab, the interesting parameter in (1) is α, which
measures the effect of the Catholic schools relative to the public schools,
all else equal—gender, race, etc. (It is the assumptions behind the model
that do the equalizing; "all else equal" is not a phrase to be treated lightly.)
The Catholic-school effect on graduation is positive and highly significant:
$\hat{\alpha} = 0.777$, with an SE of 0.056, so $t = 0.777/0.056 \doteq 14$. (See table 3;
this is out of sight, but remember, it's a big sample.) The SE comes from the
observed information, $[-L_n''(\hat{\alpha}, \hat{\beta})]^{-1}$.

For each type of characteristic, effects are relative to an omitted category.
(If you put in all the categories all the time, the design matrix will not have
full rank and parameters will not be identifiable.) For example, there is a
dummy variable for attending Catholic schools, but no dummy variable for

public schools. Attending public school is the omitted category. The effect of attending Catholic schools is measured relative to public schools.

Family income is represented in the model, but not as a continuous variable. Instead, there is a set of dummies to describe family income— missing, below $7000, $7000–$12,000, ... (Respondents ticked a box on the questionnaire to indicate a range for family income; some didn't answer the question.) For each student, one and only one of the income dummies kicks in and takes the value 1; the others are all 0. The omitted category in table 3 is $38,000+. You have to look back at table 1 in the paper to spot the omitted category.

A student whose family income is missing has a smaller chance of graduating than a student whose family income is $38,000+, other things being equal. The difference is -0.111 on the probit scale: you see -0.111 in the "probit coefficient" column for the dummy variable "family income missing" (Evans and Schwab, table 3). The negative sign is no surprise. Generally, missing data is a bad symptom. Similarly, a student whose family income is below $7000 has a smaller chance of graduating than a student whose family income is $38,000+, other things being equal. The difference is -0.300 on the probit scale. The other coefficients in table 3 can be interpreted in the same way.

"Marginal effects" are reported in table 3 of the paper. The marginal effect of Catholic schools is obtained by taking the partial derivative of $\Phi(C_i\hat{\alpha} + X_i\hat{\beta})$ with respect to C_i:

$$(2a) \qquad \frac{\partial}{\partial C_i}\Phi(C_i\hat{\alpha} + X_i\hat{\beta}) = \phi(C_i\hat{\alpha} + X_i\hat{\beta})\hat{\alpha},$$

where $\phi = \Phi'$ is the standard normal density. The marginal effect of the jth component of X_i is the partial derivative with respect to X_{ij}:

$$(2b) \qquad \phi(C_i\hat{\alpha} + X_i\hat{\beta})\hat{\beta}_j .$$

But $\phi(C_i\hat{\alpha} + X_i\hat{\beta})$ depends on C_i, X_i. So, which values do we use? See note 10 in the paper. We're talking about a 17-year-old white female, living with both parents, attending a public school. ...

Marginal effects are interpretable if you believe the model, and the variables are continuous. Even if you take the model at face value, however, there is a big problem for categorical variables. Are you making students a little more female? Catholic schools a little more Catholic??

The average treatment effect (at the end of table 3) is

$$(3) \qquad \frac{1}{n}\sum_{i=1}^{n}\left[\Phi(\hat{\alpha} + X_i\hat{\beta}) - \Phi(X_i\hat{\beta})\right].$$

The formula compares students to themselves in two scenarios: (i) attends Catholic school, (ii) attends public school. You take the difference in graduation probabilities for each student. Then you average over the students in the study: students are indexed by $i = 1, \ldots, n$.

For each student, one scenario is factual; the other is counter-factual. After all, the student can't go to both Catholic and public high schools, or at least, not for long. Graduation is observed in the factual scenario only. The calculation does not use observable outcomes. Instead, the calculation uses probabilities computed from the model. This is OK if the model can be trusted. Otherwise, "average treatment effect" is a misleading phrase, and numbers computed from (3) don't mean very much.

More on table 3

A lot of the coefficient estimates make sense. For instance, the probability of a successful outcome goes up with parental education. The probability of success is higher if the family is intact. And so forth. Some of the results are puzzling. Were blacks and Hispanics more likely to graduate in the 1980s, after controlling for the variables in table 3 of the paper? Compare, e.g., Jencks and Phillips (1998). It is also hard to see why there is no income effect on graduation beyond $20,000 a year, although there is an effect on attending college. (The results in table 2 weaken this objection; the problems with income may be in the data.) It is unclear why the test scores discussed in table 2 are excluded from the model. Indeed, many of the variables discussed in Coleman et al (1982) are ignored by Evans and Schwab, for reasons that are not explained.

Coleman et al (1982, pp. 8, 103–15, 171–78) suggest that a substantial part of the difference in outcomes for students at Catholic and public schools is due to differences in the behavior of student peer groups. If so, independence of outcomes is in question. So is the basic causal model, because changing the composition of the student body may well change the effectiveness of the school. Then responses depend on the treatment of groups, not the treatment of individuals: compare section 5.5. Evans and Schwab have a partial response to problems created by omitted variables and peer groups: see table 4 in their paper.

Latent variables

Equation (1) is equivalent to the following. Student i will graduate if

$$(4) \qquad C_i\alpha + X_i\beta + V_i > 0;$$

otherwise, i does not graduate. Remember, $Y_i = 1$ in the first case, and 0 in the second. Often, the recipe gets shortened rather drastically: $Y_i = 1$ if $C_i\alpha + X_i\beta + V_i > 0$, else $Y_i = 0$. Given the C's and X's, the latent (unobservable) variables V_i are assumed to be IID $N(0, 1)$ across subjects. Latent variables are supposed to capture the effects of unmeasured variables like intelligence, aptitude, motivation, parental attitudes. Evans and Schwab derive equation (1) above from (4), but the "net benefit" talk justifying their version of (4) is, well, just talk.

Response schedules

Evans and Schwab treat Catholic school attendance along with gender, race, . . . as manipulable. This makes little sense. It would be better to think of Catholic school attendance as manipulable, the other measured variables being personal characteristics that cannot be manipulated. Apart from the measured covariates X_i, student i has the latent variable V_i introduced above.

The response schedule behind (4) is this. Student i graduates if

$$(5) \qquad\qquad c\alpha + X_i\beta + V_i > 0;$$

otherwise, no graduation. Here, c can be set to 0 (send the kid to public school) or 1 (send to Catholic school). Manipulating c doesn't affect α, β, X_i, V_i— which is quite an assumption.

There are also statistical assumptions:

$$(6) \qquad\qquad V_i \text{ are IID } N(0, 1) \text{ across students } i,$$

$$(7) \qquad\qquad \text{the } V\text{'s are independent of the } C\text{'s and } X\text{'s}.$$

If (7) holds, then Nature is randomizing students to different combinations of C and X, independently of their V's—which is very nice of her.

There is another way to write the response schedule. Given the covariate matrix X, the conditional probability that i graduates is $\Phi(c\alpha + X_i\beta)$. This function of c says what the graduation probability would be if we intervened and set c to 0. The probability would be $\Phi(X_i\beta)$. The function also says what the graduation probability would be if we intervened and set c to 1. The probability would be $\Phi(\alpha + X_i\beta)$. The normal distribution function Φ is relevant because—by assumption—the latent variable V_i is $N(0, 1)$.

The response schedule is theory. Nobody intervened to set c. High School and Beyond was a sample survey, not an experiment. Nature took its course, and the survey recorded what happened. Thus, C_i is the value for c chosen by Nature for student i.

The response schedule may be just theory, but it's important. The theory is what bridges the gap between association and causation. Without (5), it would be hard to draw causal conclusions from observational data. Without (6) and (7), the statistical procedures would be questionable. Parameter estimates and standard errors might be severely biased.

Evans and Schwab are concerned that C may be *endogenous*, that is, related to V. Endogeneity would bias the study. For instance, Catholic schools might look good because they select good students. Evans and Schwab offer a two-equation model—our next topic—to take care of this problem.

The second equation

The two-equation model is shown in figure 1. The first equation—in its response-schedule form—says that student i graduates if

$$(8) \qquad\qquad c\alpha + X_i\beta + V_i > 0;$$

otherwise, no graduation. This is just (5), repeated for convenience.

We could in principle set c to 1, i.e., put the kid in Catholic school. Or, we could set c to 0, i.e., put him in public school. In fact, Nature chooses c. Nature does it as if by using the second equation in the model. That's the novelty.

To state the second equation, let $\text{IsCat}_i = 1$ if student i is Catholic, else $\text{IsCat}_i = 0$. Then student i attends Catholic school ($C_i = 1$) if

$$(9) \qquad\qquad \text{IsCat}_i a + X_i b + U_i > 0;$$

otherwise, public school ($C_i = 0$). Equation (9) is the second equation in the model: a is a new parameter, and b is a new parameter vector.

Nature proceeds as if by generating C_i from (9), and substituting this C_i for c in (8) to decide whether student i graduates. That is what ties the two equations together. The latent variables U_i and V_i in the two equations might be correlated, as indicated by the dashed curve in figure 1. The correlation is yet another new parameter, to be denoted by ρ.

The statistical assumptions in the two-equation model are as follows.

(10) (U_i, V_i) are IID, as pairs, across students i.

(11) (U_i, V_i) are bivariate normal; U_i has mean 0 and variance 1; so does V_i: the correlation between U_i and V_i is ρ.

(12)   The U's and V's are independent of the IsCat's and X's.

Figure 1. The two-equation model.

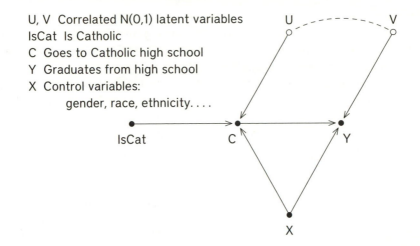

U, V Correlated N(0,1) latent variables
IsCat Is Catholic
C Goes to Catholic high school
Y Graduates from high school
X Control variables:
        gender, race, ethnicity. . . .

Condition (12) makes IsCat and X exogenous (sections 5.4–5). The correlation ρ in (11) is a key parameter. If $\rho = 0$, then C_i is independent of V_i and we don't need the second equation after all. If $\rho \neq 0$, then C_i is dependent on V_i, because V_i is correlated with U_i, and U_i comes into the formula (9) that determines C_i. Then assumption (7) in the single-equation model breaks down. The two-equation model (also called the "bivariate probit") is supposed to take care of the breakdown. That is the whole point of the second equation.

This isn't a simple model, so let's guide Nature through the steps she has to take. (Remember, we don't have access to the parameters α, β, a, b, or ρ—but Nature does.)

1. Choose IsCat_i and X_i.
2. Choose (U_i, V_i) from a bivariate normal distribution, with mean 0, variance 1, and correlation ρ. The (U_i, V_i) are independent of the IsCat's and X's. They are independent across students.
3. Check to see if inequality (9) holds. If so, set C_i to 1 and send student i to Catholic school. Else set C_i to 0 and send i to public school.
4. Set c in (8) to C_i.
5. Check to see if inequality (8) holds. If so, set Y_i to 1 and make student i graduate. Else set Y_i to 0 and prevent i from graduating.
6. Reveal IsCat_i, X_i, C_i, Y_i.
7. Shred U_i and V_i. (Hey, they're latent.)

There has to be at least one exogenous variable that influences C but has no direct influence on Y. That variable is called an "instrument" or "instrumental variable." Here, IsCat is the instrument: IsCat is 1 if the student is Catholic, else 0. IsCat comes into the model (9) for choosing schools, but is excluded, by assumption, from the graduation model (8). Economists call this sort of assumption an "identifying restriction."

In figure 1, there is no arrow from IsCat to Y. That is the graphical tipoff to an identifying restriction. The exogeneity of IsCat and X is another big assumption. In figure 1, there are no arrows or dotted lines connecting IsCat and X to U and V. That is how the graph represents exogeneity. Without the identifying restriction and the exogeneity assumptions, the model breaks down: parameters would not be identifiable. (Verifying this assertion would take some work.)

The two-equation model—equations (8) and (9), with assumptions (10)-(11)-(12) on the latent variables—is estimated by maximum likelihood. Results are shown in line (2), table 6 of the paper. They are similar—at least for school effects—to the single-equation model (table 3). This is because the estimated value for ρ is negligible.

Exogeneity. This term has several different meanings. Here, we use it in a fairly weak sense: exogenous variables are independent of the latent variables. By contrast, *endogenous* variables are dependent on the latent variables. Technically, exogeneity has to be defined relative to a model, which makes the concept even more confusing. For example, take the two-equation model (8)-(9). In this model, C is endogenous, because it is influenced by the latent U. In (4), however, C could be exogenous: if $\rho = 0$, then $C \perp\!\!\!\perp V$. We return to endogeneity in chapter 8.

Mechanics: bivariate probit

In this section, we'll see how to write down the likelihood function for the bivariate probit model. Condition on all the exogenous variables. The likelihood function is a product, with one factor for each student. That comes from the independence assumptions, (10) and (12). Take student i. There are $2 \times 2 = 4$ cases to consider: $C_i = 0$ or 1, and $Y_i = 0$ or 1.

Let's start with $C_i = 1$, $Y_i = 1$. These are facts about student i recorded in the High School and Beyond survey, as are the values for IsCat$_i$ and X_i; what you won't find on the questionnaire is U_i or V_i. We need to compute the chance that $C_i = 1$ and $Y_i = 1$, given all the exogenous variables, including IsCat. According to the model—see (8) and (9)—$C_i = 1$ and $Y_i = 1$ if

$$U_i > -\text{IsCat}_i a - X_i b \text{ and } V_i > -\alpha - X_i \beta.$$

So the chance that $C_i = 1$ and $Y_i = 1$ is

(13) $\qquad P\{U_i > -\text{IsCat}_i a - X_i b \text{ and } V_i > -\alpha - X_i\beta\}.$

The kid contributes the factor (13) to the likelihood. Notice that α appears in (13), because $C_i = 1$.

Let's do one more case: $C_i = 0$ and $Y_i = 1$. The model says that $C_i = 0$ and $Y_i = 1$ if

$$U_i < -\text{IsCat}_i a - X_i b \text{ and } V_i > -X_i\beta.$$

So the chance is

(14) $\qquad P\{U_i < -\text{IsCat}_i a - X_i b \text{ and } V_i > -X_i\beta\}.$

This kid contributes the factor (14) to the likelihood. Notice that α does not appear in (14), because $C_i = 0$. The random element in (14) is (U_i, V_i), while IsCat_i and X_i are treated as data. Ditto for (13).

Now we have to evaluate (13) and (14). Don't be hasty. Multiplying chances in (13), for instance, would not be a good idea—because of the correlation between U_i and V_i:

$$P\{U_i > -\text{IsCat}_i a - X_i b \text{ and } V_i > -\alpha - X_i\beta\} \neq$$
$$P\{U_i > -\text{IsCat}_i a - X_i b\} \bullet P\{V_i > -\alpha - X_i\beta\}.$$

The probabilities can be worked out from the bivariate normal density— assumption (11). The formula will involve ρ, the correlation between U_i and V_i. The bivariate normal density for (U_i, V_i) is

(15) $\qquad \phi(u, v) = \dfrac{1}{2\pi\sqrt{1 - \rho^2}} \exp\left[-\dfrac{u^2 - 2\rho uv + v^2}{2(1 - \rho^2)}\right].$

(This is a special case of the formula in theorem 3.2: the means are 0 and the variances are 1.) So the probability in (13), for example, is

$$\int_{-\alpha - X_i\beta}^{\infty} \int_{-\text{IsCat}_i a - X_i b}^{\infty} \phi(u, v) \, du \, dv.$$

The integral cannot be done in closed form by calculus. Instead, we would have to use numerical methods ("quadrature") on the computer. Numerical methods are beyond our scope, but see the chapter end notes for hints

and references. After working out the likelihood, we would have to maximize it—which means working it out a large number of times. All in all, the
bivariate probit is a big mess to code from scratch. But there is software that
does the whole thing for you, e.g., SHAZAM, Probit.g in GAUSS, Biprob in
STATA, PROC QLIM in SAS, or the VGAM library for R.

Why a model rather than a cross-tab?

There are 2 sexes, 3 racial groups (white, black, other), 2 ethnicities
(Hispanic or not), 8 income categories, 5 educational levels, 5 types of family
structure, 4 age groups, 3 levels of attending religious service: Evans and
Schwab, table 1. The notes to table 3 suggest 3 place types (urban, suburban,
rural) and 4 regions (northeast, midwest, south, west). Altogether that makes

$$2 \times 3 \times 2 \times 8 \times 5 \times 5 \times 4 \times 3 \times 3 \times 4 = 345,600$$

types of students. Each student might or might not attend Catholic school,
which gives another factor of 2. Even with a huge sample, a cross-tab can be
very, very sparse. A probit model enables you to handle a sparse table. This
is good. However, the model assumes—without warrant—that probabilities
are linear and additive (on the probit scale) in the selected variables. Bad.

Let's look more closely at linearity and additivity. The model assumes
that income has the same effect at all levels of education. Effects are the
same for all types of families, wherever they live. And so forth. Especially,
Catholic schools have the same additive effect (on the probit scale) for all
types of students.

Effects are assumed to be constant inside each of the bins that define a
dummy variable. For instance, "some college" is a bin for parent education
(Evans and Schwab, table 3). According to the model, one year of college
for the parents has the same effect on graduation rates as would two years of
college. Similar comments apply to the other bins.

Interactions

To weaken the assumptions of linearity and additivity, people sometimes put *interactions* into the model. Interactions are usually represented
as products. With dummy variables, that's pretty simple. For instance, the
interaction of a dummy variable for male and a dummy for white gives you
a dummy for male whites. A "three-way interaction" between male, white,
and Hispanic gives you a dummy for male white Hispanics. And so forth.

If x, z, and the interaction term xz go into the model as explanatory
variables, and you intervene to change x, you need to think about how the
interaction term will change when x changes. This will depend on the value

of z. The whole point of putting the interaction term into the equation was to get away from linearity and additivity.

If you put in all the interactions, you're back in the cross-tab, and don't have nearly enough data. With finer categories, there could also be a shortage of data. In effect, the model substitutes assumptions (e.g., no interactions) for data. If the assumptions are good, we're making progress. Otherwise, we may only be assuming that progress has been made. Evans and Schwab test their model in several ways, but with 13,000 observations and a few hundred thousand possible interactions, power is limited.

More on the second equation

What the second equation is supposed to do is to take care of a possible correlation between attending Catholic school and the latent variable V in (8). The latent variable represents unmeasured characteristics like intelligence, aptitude, motivation, parental attitudes. Such characteristics are liable to be correlated with some of the covariates, which are then endogenous. Student age is a covariate, and a high school student who is 19+ is probably not the most intelligent and best motivated of people. Student age is likely to be endogenous. So is place of residence, because many parents will decide where to live based on the educational needs of their children. These kinds of endogeneity, which would also bias the MLE, are not addressed in the paper.

There was a substantial non-response rate for the survey: 30% of the sample schools refused to participate in the study. If, e.g., low-achieving Catholic schools are less likely to respond than other schools, the effect of Catholic schools on outcomes will be overstated. If low-achieving public schools are the missing ones, the effect of Catholic schools will be understated.

Within participating schools, about 15% of the students declined to respond in 1980. There were also dropouts—students in the 1980 survey but not the 1982/1984 followup. The dropout rate was in the range 10%–20%. In total, half the data are missing. If participation in the study is endogenous, the MLE is biased. The paper does not address this problem.

The identifying restriction (excluding IsCat from the graduation model) is troublesome. Evans and Schwab use alternative instrumental variables to address some of the modeling issues. In the end, a lot of question marks are left. There seems to be little advance over the analysis in Coleman et al (1982, 1987).

Exercise set E

1. In table 3 of Evans and Schwab, is 0.777 a parameter or an estimate? How is this number related to equation (1)? Is this number on the probability

scale or the probit scale? Repeat for 0.041, in the FEMALE line of the table. (The paper is reprinted at the back of the book.)

2. What does the -0.204 for PARENT SOME COLLEGE in table 3 mean?

3. Here is the two-equation model in brief: student i goes to Catholic school $(C_i = 1)$ if
$$\text{IsCat}_i a + X_i b + U_i > 0,$$
and graduates if
$$C_i \alpha + X_i \beta + V_i > 0.$$

   (a) Which parameter tells you the effect of Catholic schools?
   (b) The U_i and V_i are _____ variables. Options (more than one may be right):

         data   random   latent   dummy   observable

   (c) What are the assumptions on U_i and V_i?

4. In line (2) of table 6 in Evans and Schwab, is 0.859 a parameter or an estimate? How is it related to the equations in exercise 3? What about the -0.053? What does the -0.053 tell you about selection effects in the one-equation model?

5. In the two-equation model, the log likelihood function is a _____, with one _____ for each _____. Fill in the blanks using the options below, and explain briefly.

       sum   product   quotient   matrix   term
                     factor   entry   student   school   variable

6. Student #77 is Presbyterian, went to public school, and graduated. What does this subject contribute to the likelihood function? Write your answer using ϕ in equation (15).

7. Student #4039 is Catholic, went to public school, and failed to graduate. What does this subject contribute to the likelihood function? Write your answer using ϕ in equation (15).

8. Does the correlation between the latent variables in the two equations turn up in your answers to exercises 6 and 7? If so, where?

9. Table 1 in Evans and Schwab shows the total sample as 10,767 in the Catholic schools and 2527 in the public schools. Is this reasonable? Discuss briefly.

10. Table 1 shows that 0.97 of the students at Catholic schools graduated. Underneath the 0.97 is the number 0.17. What is this number, and how is it computed? Comment briefly.

11. For bonus points: suppose the two-equation model is right, and you had a really big sample. Would you get accurate estimates for α? β? the V_i?

6.5 Discussion questions

Some of these questions cover material from previous chapters.

1. Is the MLE biased or unbiased?

2. In the usual probit model, are the response variables independent from one subject to another? Or conditionally independent given the explanatory variables? Do the explanatory variables have to be statistically independent? Do they have to be linearly independent? Explain briefly.

3. Here is the two-equation model of Evans and Schwab, in brief. Student i goes to Catholic school ($C_i = 1$) if

$$\text{IsCat}_i a + X_i b + U_i > 0,$$

otherwise $C_i = 0$. Student i graduates ($Y_i = 1$) if

$$C_i \alpha + X_i \beta + V_i > 0,$$

otherwise $Y_i = 0$. IsCat$_i$ is 1 if i is Catholic, and 0 otherwise; X_i is a vector of dummy variables describing subject i's characteristics, including gender, race, ethnicity, family income, and so forth. Evans and Schwab estimate the parameters by maximum likelihood, finding that $\hat{\alpha}$ is large and highly significant. True or false and explain—

   (a) The statistical model makes a number of assumptions about the latent variables.

   (b) However, the parameter estimates and standard errors are computed from the data.

   (c) The computation in (b) can be done whether or not the assumptions about the latent variables hold true. Indeed, the computation uses IsCat$_i$, X_i, C_i, Y_i for $i = 1, \ldots, n$ and the bivariate normal density but does not use the latent variables themselves.

   (d) Therefore, the statistical calculations in Evans and Schwab are fine, even if the assumptions about the latent variables are not true.

4. To what extent do you agree or disagree with the following statements about the paper by Evans and Schwab?

   (a) The paper demonstrates causation using the data: Catholic schools have an effect on student graduation rates, other things being equal.

(b) The paper assumes causation: Catholic schools have an effect on student graduation rates, other things being equal. The paper assumes a specific functional form to implement the idea of causation and other things being equal—the probit model. The paper uses the data to estimate the size of the Catholic school effect.

(c) The second equation in the model tests for interactions among explanatory variables in the first equation.

(d) The second equation in the model assumes there are no interactions.

(e) The computer derives the bivariate probit model from the data.

(f) The computer is told to assume the bivariate probit model. What the computer derives from the data is estimates for parameters in the model.

5. Suppose high school students work together in small groups to study the material in the courses. Some groups have a strong positive effect, helping the students get on top of the course work. Some groups have a negative effect. And some groups have no effect. Are study groups consistent with the model used by Evans and Schwab? If not, which assumptions are contradicted?

6. Powers and Rock (1999) consider a two-equation model for the effect of coaching on SAT scores:

assignment $X_i = 1$ if $U_i \alpha + \delta_i > 0$, else $X_i = 0$;

response $Y_i = c X_i + V_i \beta + \sigma \epsilon_i$.

Here, $X_i = 1$ if subject i is coached, else $X_i = 0$. The response variable Y_i is subject i's SAT score; U_i and V_i are vectors of personal characteristics for subject i, treated as data. The latent variables (δ_i, ϵ_i) are IID bivariate normal with mean 0, variance 1, and correlation ρ; they are independent of the U's and V's. (In this problem, U and V are observable, δ and ϵ are latent.)

(a) Which parameter measures the effect of coaching? How would you estimate it?

(b) State the assumptions carefully. Do you find them plausible?

(c) Why do Powers and Rock need two equations, and why do they need ρ?

(d) Why can they assume that the disturbance terms have variance 1?

Hint: look at section 6.4.

7. Shaw (1999) uses a regression model to study the effect of TV ads and candidate appearances on votes in the presidential elections of 1988, 1992, and 1996. With three elections and 51 states (DC counts for this purpose), there are 153 data points, i.e., pairs of years and states. Each variable in the model is determined at all 153 points. In a given year and state, the volume *TV* of television ads is measured in 100s of GRPs (gross rating points). Rep.*TV*, for example, is the volume of TV ads placed by the Republicans. *AP* is the number of campaign appearances by a presidential candidate. *UN* is the percent undecided according to tracking polls. *PE* is Perot's support, also from tracking polls. (Ross Perot was a maverick candidate.) *RS* is the historical average Republican share of the vote. There is a dummy variable D_{1992}, which is 1 in 1992 and 0 in the other years. There is another dummy D_{1996} for 1996. A regression equation is fitted by OLS, and the Republican share of the vote is

$$- 0.326 - 2.324 \times D_{1992} - 5.001 \times D_{1996}$$
$$+ 0.430 \times (\text{Rep. } TV - \text{Dem. } TV) + 0.766 \times (\text{Rep. } AP - \text{Dem. } AP)$$
$$+ 0.066 \times (\text{Rep. } TV - \text{Dem. } TV) \times (\text{Rep. } AP - \text{Dem. } AP)$$
$$+ 0.032 \times (\text{Rep. } TV - \text{Dem. } TV) \times UN + 0.089 \times (\text{Rep. } AP - \text{Dem. } AP) \times UN$$
$$+ 0.006 \times (\text{Rep. } TV - \text{Dem. } TV) \times RS + 0.017 \times (\text{Rep. } AP - \text{Dem. } AP) \times RS$$
$$+ 0.009 \times UN + 0.002 \times PE + 0.014 \times RS + \text{error.}$$

(a) What are dummy variables, and why might D_{1992} be included in the equation?

(b) According to the model, if the Republicans buy another 500 GRPs in a state, other things being equal, will that increase their share of the vote in that state by $0.430 \times 5 \doteq 2.2$ percentage points? Answer yes or no, and discuss briefly. (The 0.430 is the coefficient of Rep. TV − Dem. TV in the second line of the equation.)

8. In the Nurses' Health Study of post-menopausal women, 6,224 subjects were on hormone replacement therapy (HRT) and 27,034 were not. The investigators want to show that HRT reduces the risk of heart attack. The investigators find out whether each subject was on HRT, and whether she experienced a heart attack during the study period. For each subject, baseline measurements are made on potential confounders: age, height, weight, cigarette smoking (yes or no), hypertension (yes or no), and high cholesterol level (yes or no).

(a) Should the investigators use OLS or logistic regression? Why?

(b) State the model explicitly. What is the design matrix X? n? p?
How will the yes/no variables be represented in the design matrix?
What is Y?

(c) Which parameter is the crucial one?

(d) Do the investigators hope to see a positive estimate or a negative estimate for the crucial parameter? How can they determine whether the estimate is statistically significant?

(e) What are the assumptions in the model?

(f) Why is a model needed in the first place?

(g) To what extent is the argument convincing? Discuss briefly.

Comment. Details of the study have been changed a little for purposes of this question; see chapter end notes.

9. People often use observational studies to demonstrate causation, but there's a big problem. What is an observational study, what's the problem, and how do people try to get around it? Discuss. If possible, give examples to illustrate your points.

10. There is a population of N subjects, indexed by $i = 1, \ldots, N$. Each subject will be assigned to treatment T or control C. Subject i has a response y_i^T if assigned to treatment and y_i^C if assigned to control. Each response is 0 ("failure") or 1 ("success"). For instance, in an experiment to see whether aspirin prevents death from heart attack, survival over the followup period would be coded as 1, death would be coded as 0. If you assign subject i to treatment, you observe y_i^T but not y_i^C. Conversely, if you assign subject i to control, you observe y_i^C but not y_i^T. These responses are fixed (not random).

Each subject i has a $1 \times p$ vector of personal characteristics w_i, unaffected by assignment. In the aspirin experiment, these characteristics might include weight and blood pressure just before the experiment starts. You can always observe w_i. Population parameters of interest are

$$\alpha^T = \frac{1}{N} \sum_{i=1}^{N} y_i^T, \quad \alpha^C = \frac{1}{N} \sum_{i=1}^{N} y_i^C, \quad \alpha^T - \alpha^C.$$

The first parameter is the fraction of successes we would see if all subjects were put into treatment. We could measure this directly—by putting all the subjects into treatment—but would then lose our chance to learn about the second parameter, which is the fraction of successes if all subjects were in the control condition. The third parameter is the difference between the first two parameters. It measures the effectiveness of treatment, on average across all the subjects. This parameter is the most interesting of the three. It cannot be measured directly, because we cannot put subjects both into treatment and into control.

Suppose $0 < n < N$. In a "randomized controlled experiment," n subjects are chosen at random without replacement and assigned to treatment; the remaining $N - n$ subjects are assigned to control. Can you estimate the three population parameters of interest? Explain. Hint: see discussion questions 7–8 in chapter 5.

11. (This continues question 10.) The assignment variable X_i is defined as follows: $X_i = 1$ if i is in treatment, else $X_i = 0$. The probit model says that given the assignments, subjects are independent, the probability of success for subject i being $\Phi(X_i\alpha + w_i\beta)$, where Φ is the standard normal distribution function and w_i is a vector of personal characteristics for subject i (question 10).

    (a) Would randomization justify the probit model?

    (b) The logit model replaces Φ by $\Lambda(x) = e^x/(1 + e^x)$. Would randomization justify the logit model?

    Explain briefly. Hint: see discussion questions 9–10 in chapter 5.

12. Malaria is endemic in parts of Africa. A vaccine is developed to protect children against this disease. A randomized controlled experiment is done in a small rural village: half the children are chosen at random to get the vaccine, and half get a placebo. Some epidemiologists want to analyze the data using the setup described in question 10. What is your advice?

13. As in question 12, but this time, the epidemiologists have 20 isolated rural villages. They choose 10 villages at random for treatment. In these villages, everybody will get the vaccine. The other 10 villages will serve as the control group: nobody gets the vaccine. Can the epidemiologists use the setup described in question 10?

14. Suppose we accept the model in question 10, but data are collected on X_i and Y_i in an observational study not a controlled experiment. Subjects assign themselves to treatment ($X_i = 1$) or control ($X_i = 0$), and we observe the response Y_i as well as the covariates w_i. One person suggests separating the subjects into several groups with similar w_i's. For each group on its own, we can compare the fraction of successes in treatment to the fraction of successes in control. Another person suggests fitting a probit model: conditional on the X's and covariates, the probability that

$Y_i = 1$ is $\Phi(X_i\alpha + w_i\beta)$. What are the advantages and disadvantages of the two suggestions?

15. Paula has observed values on four independent random variables with common density $f_{\alpha,\beta}(x) = c(\alpha,\beta)|\alpha x - \beta| \exp[-(\alpha x - \beta)^2]$, where $\alpha > 0$, $-\infty < \beta < \infty$, and $c(\alpha,\beta)$ is chosen so that $\int_{-\infty}^{\infty} f_{\alpha,\beta}(x)dx = 1$. She estimates α, β by maximum likelihood and computes the standard errors from the observed information. Before doing the t-test to see whether $\hat{\beta}$ is significantly different from 0, she decides to get some advice. What do you say?

16. Jacobs and Carmichael (2002) are comparing various sociological theories that explain why some states have the death penalty and some do not. The investigators have data for 50 states (indexed by i) in years $t = 1971, 1981, 1991$. The response variable Y_{it} is 1 if state i has the death penalty in year t, else 0. There is a vector of explanatory variables X_{it} and a parameter vector β, the latter being assumed constant across states and years. Given the explanatory variables, the investigators assume the response variables are independent and

$$\log[-\log P(Y_{it} = 0|X)] = X_{it}\beta.$$

(This is a "complementary log log" or "cloglog" model.) After fitting the equation to the data by maximum likelihood, the investigators determine that some coefficients are statistically significant and some are not. The results favor certain theories over others. The investigators say,

> "All standard errors are corrected for heteroscedasticity by White's method. . . . Estimators are robust to misspecification because the estimates are corrected for heteroscedasticity."

(The quote is slightly edited.) "Heteroscedasticity" means, unequal variances (section 4.4). White's method is discussed in the end notes to chapter 4: it estimates SEs for OLS when the ϵ's are heteroscedastic, using equation (4.20). "Robust to misspecification" means, works pretty well even if the model is wrong.

Discuss briefly, answering these questions. Are the parameter estimates robust, or the estimated standard errors? If the former, what do the estimates mean when the model is wrong? If the latter, according to the model, is $\text{var}(Y_{it}|X)$ different for different combinations of i and t? Are these differences taken into account by the asymptotic SEs? Do asymptotic SEs for the MLE need correction for heteroscedasticity?

17. Ludwig is working hard on a statistics project. He is overheard muttering to himself, "Ach! Schrecklich! So many Parameters! So little Data!" Is he worried about bias, endogeneity, or non-identifiability?

18. Garrett (1998) considers the impact of left-wing political power (LPP) and trade-union power (TUP) on economic growth. There are 25 years of data on 14 countries. Countries are indexed by $i = 1, \ldots, 14$; years are indexed by $t = 1, \ldots, 25$. The growth rate for country i in year t is modeled as

$$a \times \text{LPP}_{it} + b \times \text{TUP}_{it} + c \times \text{LPP}_{it} \times \text{TUP}_{it} + X_{it}\beta + \epsilon_{it},$$

where X_{it} is a vector of control variables. Estimates for a and b are negative, suggesting that right-wing countries grow faster. Garrett rejects this idea, because the estimated coefficient c of the interaction term is positive. This term is interpreted as the "combined impact" of left-wing political power and trade-union power, Garrett's conclusion being that the country needs both kinds of left-wing power in order to grow more rapidly. Assuming the model is right, does $c \times \text{LPP} \times \text{TUP}$ measure the combined impact of LPP and TUP? Answer yes or no, and explain.

19. This continues question 18; different notation is used: part (b) might be a little tricky. Garrett's model includes a dummy variable for each of the 14 countries. The growth rate for country i in year t is modeled as

$$\alpha_i + Z_{it}\gamma + \epsilon_{it},$$

where Z_{it} is a 1×10 vector of explanatory variables, including LPP, TUP, and the interaction. (In question 18, the country dummies didn't matter, and were folded into X.) Beck (2001) uses the same model—except that an intercept is included, and the dummy for country #1 is excluded. So, in this second model, the growth rate in country $i > 1$ and year t is

$$\alpha^* + \alpha_i^* + Z_{it}\gamma^* + \epsilon_{it};$$

whereas the growth rate in country #1 and year t is

$$\alpha^* + Z_{1t}\gamma^* + \epsilon_{1t}.$$

Assume both investigators are fitting by OLS and using the same data.

(a) Why can't you have an intercept, as well as a dummy variable for each of the 14 countries?

(b) Show that $\hat{\gamma} = \hat{\gamma}^*$, $\hat{\alpha}_1 = \hat{\alpha}^*$, and $\hat{\alpha}_i = \hat{\alpha}^* + \hat{\alpha}_i^*$ for $i > 1$.

Hints for (b). Let M be the design matrix for the first model; M^*, for the second. Find a lower triangular matrix L—which will have 1's on the diagonal and mainly be 0 elsewhere—such that $ML = M^*$. How does this relationship carry over to the parameters and the estimates?

20. Yule used a regression model to conclude that outrelief causes pauperism (section 1.4). He presented his paper at a meeting of the Royal Statistical Society on 21 March 1899. Sir Robert Giffen, Knight Commander of the Order of the Bath, was in the chair. There was a lively discussion, summarized in the *Journal of the Royal Statistical Society* (Vol. LXII, Part II, pp. 287–95).

    (a) According to Professor FY Edgeworth, if one diverged much from the law of normal errors, "one was on an ocean without rudder or compass"; this normal law of error "was perhaps more universal than the law of gravity." Do you agree? Discuss briefly.

    (b) According to Sir Robert, practical men who were concerned with poor law administration knew that "if the strings were drawn tightly in the matter of out-door relief, they could immediately observe a reduction of pauperism itself." Yule replied,

> "he was aware that the paper in general only bore out conclusions which had been reached before ... but he did not think that lessened the interest of getting an independent test of the theories of practical men, purely from statistics. It was an absolutely unbiassed test, and it was always an advantage in a method that it was unbiassed."

What do you think of this reply? Is Yule's test "purely from statistics"? Is it Yule's methods that are "unbiassed," or his estimates of the parameters given his model? Discuss briefly.

6.6 End notes for chapter 6

Who reads books? Data are available from the August supplement to the Current Population Survey of 2002. Also see table 1244 in *Statistical Abstract of the United States 2003*.

The MLE. For a more detailed discussion of the MLE, with the outline of an argument for theorem 1, see

               http://www.stat.berkeley.edu/users/census/mle.pdf

There are excellent graduate-level texts by Lehmann (1991ab) and Rao (1973), with careful statements of theorems and proofs. Lehmann (1999) might be the place to start: fewer details, more explanations. For exponential families, the calculus is easier; see, e.g., Barndorff-Nielsen (1980). In particular, there is (with minor conditions) a unique max.

The theory for logits is prettier than for probits, because the logit model defines an exponential family. However, the following example shows that even in a logit model, the likelihood may not have a maximum: theorems have regularity conditions to eliminate this sort of exceptional case. Suppose X_i is real and logit $P(Y_i = 1 \mid X_i = x) = \theta x$. We have two independent data points. At the first, $X_1 = -1$, $Y_1 = 0$. At the second, $X_2 = 1$, $Y_2 = 1$. The log likelihood function is $L(\theta) = -2\log(1 + e^{-\theta})$, which increases steadily with θ.

Deviance. In brief, there is a model with p parameters. The null hypothesis constrains p_0 of these parameters to be 0. Maximize the log likelihood over the full model. Denote the maximum by M. Then maximize the log likelihood subject to the constraint, getting a smaller maximum M_0. The deviance is $2(M - M_0)$. If the null hypothesis holds, n is large, and certain regularity conditions hold, the deviance is asympotically chi-squared, with p_0 degrees of freedom. Deviance is also called the "Neyman-Pearson statistic" or the "Wilks statistic." Deviance is the analog of F (section 4.8), although the scaling is a little different. Details are beyond our scope.

The score test. In many applications, the score test will be more robust. The score test uses the statistic

$$\frac{1}{n} L'(\hat{\theta}_0) I_{\hat{\theta}_0}^{-1} L'(\hat{\theta}_0),$$

where $\hat{\theta}_0$ is the MLE in the constrained model, and L' is the partial derivative of the log likelihood function: L' is viewed as a row vector on the left and a column vector on the right. The asymptotic distribution is still chi-squared with p_0 degrees of freedom. Rao (1973, pp. 415–20) discusses the various likelihood tests.

The information matrix. Suppose the X_i are IID with density f_θ. The jkth entry in the Fisher information matrix can be estimated as

$$\frac{1}{n} \sum_{i=1}^{n} \frac{\partial f_\theta(X_i)}{\partial \theta_j} \frac{\partial f_\theta(X_i)}{\partial \theta_k} \frac{1}{f_\theta(X_i)^2},$$

at $\theta = \hat{\theta}$, the MLE. In some circumstances, this is easier to compute than observed information, and more stable. With endpoint maxima, neither method is likely to work very well.

Identifiability. A constant function $f(\theta)$ is identifiable for trivial (and irritating) reasons: there are no θ_1, θ_2 with $f(\theta_1) \neq f(\theta_2)$. Although many texts blur the distinction between identifiability and estimability, it seemed better to separate them. The flaw in the terminology is this. A parameter may not be estimable (no estimator for it is exactly unbiased) but there could still exist a very accurate estimator (small bias, small variance).

A technical side issue. According to our definition, $f(\theta)$ is identifiable if $f(\theta_1) \neq f(\theta_2)$ implies $P_{\theta_1} \neq P_{\theta_2}$. The informal discussion may correspond better to a slightly sharper definition: there should exist a function ϕ with $\phi(P_\theta) = f(\theta)$; measurability conditions are elided.

The bigger picture. Many statisticians frown on under-identified models: if a parameter is not identifiable, two or more values are indistinguishable, no matter how much data you have. On the other hand, most applied problems *are* under-identified. Identification is achieved only by imposing somewhat arbitrary assumptions (independence, constant coefficients, etc.). That is one of the central tensions in the field. Efforts have been made to model this tension as a bias-variance tradeoff. Truncating the number of parameters introduces bias but reduces variance, and the optimal truncation can be considered. Generally, however, the analysis takes place in a context that is already highly stylized. For discussion, see Evans and Stark (2002).

Evans and Schwab. The focus is on tables 1–3 and table 6 in the paper. In table 6, we consider only the likelihood estimates for line (2); line (1) repeats estimates from the single-equation model. Data from High School and Beyond (HS&B) are available, under stringent confidentiality agreements, as part of NELS—the National Educational Longitudinal Surveys. The basic books on HS&B are Coleman et al (1982), Coleman and Hoffer (1987). Twenty years later, these books are still worth reading: the authors had real insight into the school system, and the data analysis is quite interesting. Coleman and Hoffer (1987) include several chapters on graduation rates, admission to college, success in college, and success in the labor force, although Evans and Schwab pay little attention to these data.

The total sample sizes for students in Catholic and public schools in table 1 of Evans and Schwab appear to have been interchanged. There may be other data issues too. See table 2.1 in Coleman and Hoffer (1987), which reports noticeably higher percentages of students with incomes above \$38,000. Moreover, table 2 in Evans and Schwab should be compared with Coleman and Hoffer (1987, table 5.3): graduation rates appear to be inconsistent.

Table 1.1 in Coleman et al (1982) shows a realized sample in 1980 of 26,448 students in public schools, and 2831 in Catholic schools. Evans and Schwab have 10,767 in public schools, and 2527 in Catholic schools (after switching entries in table 1). The difference in sample size for the Catholic schools probably reflects sample attrition from 1980 to 1984, but the difference for public schools seems too large to be explained that way. Some information on dropout rates can be gleaned from US Department of Education (1987). Compare also table 1.1 in Coleman et al (1982) with table 2.9 in Coleman and Hoffer (1987).

The discussion questions. Powers and Rock are using a version of Heckman's (1976, 1978, 1979) model, as are Evans and Schwab. The model is discussed with unusual care by Briggs (2004). Many experiments have been analyzed with logits and probits, for example, Pate and Hamilton (1992). In question 7, the model has been simplified a little. The Nurses' Health Study used a Cox model with additional covariates and body mass index (weight/height^2) rather than height and weight. The 6224 refers to women on combined estrogen and progestin; the 27,034 are never-users. See Grodstein et al (1996). The experimental evidence shows the observational studies to have been quite misleading: Petitti (2002), Writing Group for the Women's Health Initiative Investigators (2002).

Question 10 outlines the most basic of the response schedule models. A subject has a potential response at each level of treatment (T or C). One of these is observed, the other not. It is often thought that models are justified by randomization: but see question 11. Question 12 points to a weakness in response-schedule models: if a subject's response depends on treatments given to other subjects, the model does not apply. This is relevant to studies of school effects. Question 18 looks at the "baseline model" in Garrett (1998, table 5.3); some complications in the data analysis have been ignored.

Numerical methods. Suppose f is a smooth function on the line, and we want to find x near x_0 with $f(x) = 0$. "Newton's method," also called the "Newton-Raphson method," is simple—and often works. If $f(x_0) = 0$, stop. Otherwise, approximate f by the linear function $f_0(x) = a + b(x - x_0)$, where $a = f(x_0)$ and $b = f'(x_0)$. Solve the linear equation $f_0 = 0$ to get a new starting point. Iterate. There are many variations on this idea.

Quadrature. If f is a smooth function on the unit interval [0,1], we can approximate $\int_0^1 f(x)dx$ by $\frac{1}{n}\sum_{j=0}^{n-1} f(\frac{j}{n})$. This method approximates f by a step function with horizontal steps; the integral is approximated by the sum of the areas of rectangular blocks. The "trapezoid rule" approximates f on the interval $[\frac{j-1}{n}, \frac{j}{n}]$ by a line segment running from the point $(\frac{j-1}{n}, f(\frac{j-1}{n}))$ to the point $(\frac{j}{n}, f(\frac{j}{n}))$. The integral is approximated by the sum of trapezoidal areas. This is better, as the diagram illustrates. There are many variations (Simpson's rule, Newton-Cotes methods, etc.).

If you want to read more about numerical methods, try—

Acton FS (1997). *Numerical Methods That Work*. Mathematical Association of America.

Atkinson K (1993). *Elementary Numerical Analysis*. 2nd ed. Wiley. Newton's method for solving equations is discussed on pp. 68–77, and quadrature on pp. 161–95.

Epperson JF (2001). *An Introduction to Numerical Methods and Analysis*. Wiley. Newton's method for solving equations is discussed on pp. 87–94, and quadrature on pp. 60–67.

Lanczos C (1988). *Applied Analysis*. Dover Publications.

Strang G (1986). *Introduction to Applied Mathematics*. Wellesley-Cambridge. Newton's method and the method of steepest descent for finding minima are covered on pp. 373–79.

Acton and Lanczos are classics, written for the mathematically-inclined. Atkinson and Epperson are more conventional textbooks. Strang is clear and concise, with a personal style; might be the place to start.

Logistic regression: the brief history. The logistic curve was originally used to model population growth (Verhulst 1845, Yule 1925). If $p(t)$ is the population at time t, Malthusian population theory suggested an equation of the form

$$\frac{1}{p}\frac{dp}{dt} = a - bp.$$

The solution is

$$p(t) = \frac{a}{b}\Lambda(at + c),$$

where Λ is the logistic distribution function. (The first thing to check is that $\Lambda'/\Lambda = 1 - \Lambda$.) The linear function $a - bp$ on the right hand side of the differential equation might be viewed by some as a first approximation to a more realistic decreasing function.

In 1920, the population of the United States was 106 million, and models based on the logistic curve showed that the population would never exceed 200 million (Pearl and Reed 1920, Hotelling 1927). As the US population increased beyond that limit, enthusiasm for the logistic growth law waned, although papers keep appearing on the topic. For reviews of population models, including the logistic, see Dorn (1950) and Hajnal (1955). Feller (1940) shows that normal and Cauchy distributions fit growth data as well as the logistic.

An early biomedical application of logistic regression was Truett, Cornfield, and Kannel (1967). These authors fit a logistic regression to data from the Framingham study of coronary heart disease. The risk of death in the study period was related to a vector of covariates, including age, blood cholesterol level, systolic blood pressure, relative weight, blood hemoglobin level, smoking (at 3 levels), and abnormal electrocardiogram (a dummy variable). There were 2187 men and 2669 women, with 387 deaths and 271 subjects lost to followup (these were just censored). The analysis was stratified by sex and sometimes by age.

The authors argue that the relationship must be logistic. Their model seems to be like this, with death in the study period coded as $Y_i = 1$, survival as $Y_i = 0$, and X_i a row vector of covariates. Subjects are a random sample from a population. Given $Y_i = 1$, the distribution of X_i is multivariate normal with mean μ_1. Given $Y_i = 0$, the distribution is normal with the same covariance matrix G but a different mean μ_0. Then $P(Y_i = 1|X_i)$ would indeed be logistic. This is easily verified, using Bayes' rule and theorem 3.2.

The upshot of the calculation: $\operatorname{logit} P(Y_i = 1|X) = \alpha + X_i\beta$, where $\beta = G^{-1}(\mu_1' - \mu_0')$ is the interesting parameter vector. The intercept is a nuisance parameter, $\alpha = \operatorname{logit} P(Y_i = 1) + \frac{1}{2}(\mu_0 G^{-1}\mu_0' - \mu_1 G^{-1}\mu_1')$. If $P(X_i \in dx|Y_i = 1) = C_\beta \exp(\beta x)P(X_i \in dx|Y_i = 0)$, conclusions are similar; again, there will be a nuisance intercept.

According to Truett, Cornfield, and Kannel, the distribution of X_i has to be multivariate normal, by the central limit theorem. But why is the central limit theorem relevant? Indeed, the distribution of X_i clearly *wasn't* normal: (i) there were dummy variables in X_i, and (ii) data on the critical linear combinations are long-tailed. Furthermore, the subjects were a population, not a random sample. Finally, why should we think that parameters are invariant under interventions??

7

The Bootstrap

7.1 Introduction

The bootstrap is a powerful tool for approximating the bias and standard error of an estimator in a complex statistical model. However, results are dependable only if the sample is reasonably large. We begin with some toy examples, where the bootstrap is not needed but the algorithm is easy to understand. Then we go on to examples that are more interesting.

Example 1. The sample mean. Let X_i be IID for $i = 1, \ldots, n$, with mean μ and variance σ^2. We use the sample mean $\overline{X}$ to estimate μ. Is this estimator biased? What is its standard error? Of course, we know by statistical theory (chapter 4) that the estimator is unbiased. We know the SE is $\sigma/\sqrt{n}$. And we know that σ^2 can be estimated by the sample variance,

$$\hat{\sigma}^2 = \frac{1}{n} \sum_{i=1}^{n} (X_i - \overline{X})^2.$$

(With large samples, it is immaterial whether we divide by n or $n - 1$.)

For the sake of argument, suppose we've forgotten the theory but remember how to use the computer. What can we do to estimate the bias in

$\overline{X}$? to estimate the SE? Here comes the bootstrap idea at its simplest. Take the data—the observed values of the X_i's—as a little population. Simulate n draws, made at random with replacement, from this little population. This is a *bootstrap sample*. Figure 1 shows the procedure in box-model format.

Figure 1. Bootstrapping the sample mean.

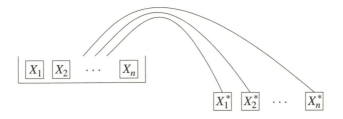

Let $X_1^*, \ldots, X_n^*$ be the bootstrap sample. Each X_i will come into the bootstrap sample some small random number of times, zero being a possible number, and in random order. From the bootstrap sample, we compute the bootstrap estimate for the average of the little population (the numbers in the box). The *bootstrap estimator* is just the average of the bootstrap sample:

$$\overline{X}^* = \frac{1}{n} \sum_{i=1}^{n} X_i^*.$$

One bootstrap sample may not tell us very much, but we can draw many bootstrap samples to get the sampling distribution of $\overline{X}^*$. Let's index these samples by k. There will be a lot of indices, so we'll put parens around the k. In this notation, the kth bootstrap estimator is $\overline{X}_{(k)}$: we don't need both a superscript $*$ and a subscript (k). Suppose we have N *bootstrap replicates*, indexed by $k = 1, \ldots, N$:

$$\overline{X}_{(1)}, \ldots, \overline{X}_{(k)}, \ldots, \overline{X}_{(N)}.$$

Please keep separate:

- N, the number of bootstrap replicates;
- n, the size of the real sample.

Usually, we can make N as large as we need, because computer time is cheap. Making n larger could be an expensive proposition.

What about bias? On the computer, we're resampling from the real sample, whose mean is $\overline{X}$. According to our rules of the moment, we're not allowed to compute $E\left(\overline{X}_{(k)}\right)$ using probability theory. But we can approximate the expectation by

$$\overline{X}_{\text{ave}} = \frac{1}{N} \sum_{k=1}^{N} \overline{X}_{(k)},$$

the mean of the N bootstrap replicates. The computer's bottom line:

$$\overline{X}_{\text{ave}} \doteq \overline{X}.$$

In our simulation, the expected value of the sample mean *is* the population mean. The bootstrap is telling us that the sample mean is unbiased.

Our next desire is the SE of the sample mean. Let

$$V = \frac{1}{N} \sum_{k=1}^{N} \left[\overline{X}_{(k)} - \overline{X}_{\text{ave}} \right]^2.$$

This is the variance of the N bootstrap replicates. The SD is $\sqrt{V}$, which tells us how close a typical $\overline{X}_{(k)}$ is to $\overline{X}$. That's what we're looking for.

> The bootstrap SE is the SD of the bootstrap replicates.

The bootstrap SE says how good the original $\overline{X}$ was, as an estimate for μ.

Why does this work? We've simulated $k = 1, \ldots, N$ replicates of $\overline{X}$, and used the sample variance to approximate the real variance. The only problem is this. We should be drawing from the distribution that the real sample came from. Instead, we're drawing from an approximation, namely, the *empirical distribution* of the sample $\{X_1, \ldots, X_n\}$. See figure 1. If n is reasonably large, this is a good approximation. If n is small, the approximation isn't good, and the bootstrap isn't likely to work.

> *Bootstrap principle for the sample mean.* Provided that the sample is reasonably large, the distribution of $\overline{X}^* - \overline{X}$ will be a good approximation to the distribution of $\overline{X} - \mu$. In particular, the SD of $\overline{X}^*$ will be a good approximation to the standard error of $\overline{X}$.

On the computer, we imitated the sampling model for the data. We assumed the data come from IID random variables, so we simulated IID data on the computer—drawing at random with replacement from a box. This is important. Otherwise, the bootstrap is doing the wrong thing. As a technical matter, we've been talking rather loosely about the bootstrap distribution of $\overline{X}^* - \overline{X}$, but the distribution is conditional on the data $X_1, \ldots, X_n$.

The notation is a little strange, and so is the terminology. For instance, $\overline{X}_{\text{ave}}$ looks imposing, but it's just something we use to check that the sample mean is unbiased. The "bootstrap estimator" $\overline{X}^*$ is *not* a new estimator for the parameter μ. It's something we generate on the computer to help us understand the behavior of the estimator we started with—the sample mean. The "empirical distribution of the sample" isn't a distribution for the sample. Instead, it's an approximation to the distribution that we sampled from. The approximation puts mass $1/n$ at each of the n sample points. Lacking other information, this is perhaps the best we can do.

Example 2. Regression. Suppose $Y = X\beta + \epsilon$, where the design matrix X is $n \times p$. Suppose that X is fixed (not random) and has full rank. The parameter vector β is $p \times 1$, unknown, to be estimated by OLS. The errors $\epsilon_1, \ldots, \epsilon_n$ are IID with mean 0 and variance σ^2, also unknown. What is the bias in the OLS estimator $\hat{\beta} = (X'X)^{-1}X'Y$? What is the covariance matrix of $\hat{\beta}$? The answers, of course, are 0 and $\sigma^2(X'X)^{-1}$; we would estimate σ^2 as the mean square of the residuals.

Again, we've forgotten the formulas but have computer time on our hands. We'll use the bootstrap to get at bias and variance. We don't want to resample the Y_i's, because they're not IID: $E(Y_i) = X_i\beta$ differs from one i to another, X_i being the ith row of the design matrix X. The ϵ_i are IID, but we can't get our hands on them. A puzzle.

Suppose there's an intercept in the model, so the first column of X is all 1's. Then $\bar{e} = 0$, where $e = Y - X\hat{\beta}$ is the vector of residuals. We can resample the residuals, and that's the thing to do. The residuals $e_1, \ldots, e_n$ are a new little population, whose mean is 0. We draw n times at random with replacement from this population to get bootstrap errors $\epsilon_1^*, \ldots, \epsilon_n^*$. These are IID and $E(\epsilon_i^*) = 0$. The ϵ_i^* behave like the ϵ_i. Figure 2 summarizes the procedure.

Figure 2. Bootstrapping a regression model.

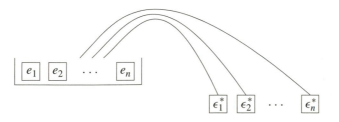

The next step is to regenerate the Y_i's:

$$Y^* = X\hat{\beta} + \epsilon^*.$$

Each e_i comes into ϵ^* some small random number of times (zero is a possible number) and in random order. So e_1 may get paired with X_7 and X_{19}. Or, e_1 may not come into the sample at all. The design matrix X doesn't change, because we assumed it was fixed. Notice that Y^* follows the regression model: errors are IID with expectation 0. We've imitated the original model on the computer. There is a difference, though. On the computer, we know the true parameter vector. It's $\hat{\beta}$. We also know the true distribution of the disturbances—IID draws from $\{e_1, \ldots, e_n\}$. So we can get our hands on the distribution of $\hat{\beta}^* - \hat{\beta}$, where $\hat{\beta}^*$ is the bootstrap estimator $\hat{\beta}^* = (X'X)^{-1}X'Y^*$.

> *Bootstrap principle for regression.* With a reasonably large n, the distribution of $\hat{\beta}^* - \hat{\beta}$ is a good approximation to the distribution of $\hat{\beta} - \beta$. In particular, the empirical covariance matrix of the $\hat{\beta}^*$ is a good approximation to the theoretical covariance matrix of $\hat{\beta}$.

What is an "empirical" covariance matrix? Suppose we generate N bootstrap data sets, indexed by $k = 1, \ldots, N$. For each one, we would have a bootstrap OLS estimator, $\hat{\beta}_{(k)}$. We have N bootstrap replicates, indexed by k:

$$\hat{\beta}_{(1)}, \ldots, \hat{\beta}_{(k)}, \ldots \hat{\beta}_{(N)}.$$

The empirical covariance matrix is

$$\frac{1}{N} \sum_{k=1}^{N} \left[\hat{\beta}_{(k)} - \hat{\beta}_{\text{ave}} \right] \left[\hat{\beta}_{(k)} - \hat{\beta}_{\text{ave}} \right]', \text{ where } \hat{\beta}_{\text{ave}} = \frac{1}{N} \sum_{k=1}^{N} \hat{\beta}_{(k)}.$$

This is something you can work out. By way of comparison, the theoretical covariance matrix depends on the unknown σ^2:

$$E\left\{ [\hat{\beta} - E(\hat{\beta})][\hat{\beta} - E(\hat{\beta})]' \right\} = \sigma^2 (X'X)^{-1}.$$

What about bias? As shown in chapter 4, there is no bias: $E(\hat{\beta}) = \beta$. In the simulation, $\hat{\beta}_{\text{ave}} = \hat{\beta}$, apart from a little bit of random error. After all, $\hat{\beta}$—the estimated β in the real data—is what we told the computer to take as the true parameter vector. And $\hat{\beta}_{\text{ave}}$ is the average of N bootstrap replicates $\hat{\beta}_{(k)}$, which is a good approximation to $E\left[\hat{\beta}_{(k)}\right]$.

On the computer, we imitated the sampling model for the data. By assumption, the real data came from a regression model with fixed X and IID errors having mean 0. That is what we had to simulate on the computer: otherwise, the bootstrap would have been doing the wrong thing.

We've been talking about the bootstrap distribution of $\hat{\beta}^* - \hat{\beta}$. This is conditional on the data $Y_1, \ldots, Y_n$. After conditioning, we can treat the residuals—which were computed from $Y_1, \ldots, Y_n$—as data rather than random variables. The randomness in the bootstrap comes from resampling the residuals. Again, the catch is this. We'd like to be drawing from the real distribution of the ϵ_i's. Instead, we're drawing from the empirical distribution of the e_i's. If n is reasonably large and the design matrix is not too crazy, this is a good approximation.

Example 3. Autoregression. There are parameters a, b. These are unknown. Somehow, we know that $|b| < 1$. For $i = 1, 2, \ldots, n$, we have $Y_i = a + bY_{i-1} + \epsilon_i$. Here, Y_0 is a fixed number. The ϵ_i are IID with mean 0 and variance σ^2, unknown. The equation has a *lag term*, Y_{i-1}: this is the Y for the previous i. We're going to estimate a and b by OLS, so let's put this into the format of a regression problem: $Y = X\beta + \epsilon$ with

$$Y = \begin{pmatrix} Y_1 \\ Y_2 \\ \vdots \\ Y_n \end{pmatrix}, \quad X = \begin{pmatrix} 1 & Y_0 \\ 1 & Y_1 \\ \vdots & \vdots \\ 1 & Y_{n-1} \end{pmatrix}, \quad \beta = \begin{pmatrix} a \\ b \end{pmatrix}, \quad \epsilon = \begin{pmatrix} \epsilon_1 \\ \epsilon_2 \\ \vdots \\ \epsilon_n \end{pmatrix}.$$

The algebra works out fine: the ith row in the matrix equation $Y = X\beta + \epsilon$ gives us $Y_i = a + bY_{i-1} + \epsilon_i$, which is where we started. The OLS estimator is $\hat{\beta} = (X'X)^{-1}X'Y$. We write $\hat{a}$ and $\hat{b}$ for the two components of $\hat{\beta}$.

But something is fishy. There is a correlation between X and ϵ. Look at the second column of X. It's full of ϵ's, tucked away inside the Y's. Maybe we shouldn't use $\hat{\sigma}^2(X'X)^{-1}$? And what about bias? Although the standard theory doesn't apply, the bootstrap works fine. We can use the bootstrap to estimate variance and bias, in this non-standard situation where explanatory variables are correlated with errors,

The bootstrap can be done following the pattern set by example 2, even though the design matrix is random. You fit the model, getting $\hat{\beta}$ and residuals $e = Y - X\hat{\beta}$. You freeze Y_0, as well as

$$\hat{\beta} = \begin{pmatrix} \hat{a} \\ \hat{b} \end{pmatrix}$$

and e. You resample the e's to get bootstrap disturbance terms $\epsilon_1^*, \ldots, \epsilon_n^*$. The new point is that you have to generate the Y_i^*'s one at a time, using $\hat{a}, \hat{b}$, and the ϵ_i^*'s:

$$Y_1^* = \hat{a} + \hat{b}Y_0 + \epsilon_1^*,$$
$$Y_2^* = \hat{a} + \hat{b}Y_1^* + \epsilon_2^*,$$
$$\vdots$$
$$Y_n^* = \hat{a} + \hat{b}Y_{n-1}^* + \epsilon_n^*.$$

The first line is OK because Y_0 is a constant. The second line is OK because when we need Y_1^*, we have it from the line before. And so forth. So, we have a bootstrap data set:

$$Y^* = \begin{pmatrix} Y_1^* \\ Y_2^* \\ \vdots \\ Y_n^* \end{pmatrix}, \quad X^* = \begin{pmatrix} 1 & Y_0 \\ 1 & Y_1^* \\ \vdots \\ 1 & Y_{n-1}^* \end{pmatrix}, \quad \epsilon^* = \begin{pmatrix} \epsilon_1^* \\ \epsilon_2^* \\ \vdots \\ \epsilon_n^* \end{pmatrix}.$$

Then we compute the bootstrap estimator, $\hat{\beta}^* = (X^{*\prime}X^*)^{-1}X^{*\prime}Y^*$. Notice that we had to regenerate the design matrix because of the second column. That is why there is a $*$ on X^*. This procedure could be repeated many times on the computer, to get N bootstrap replicates. The same residuals e are used throughout. But ϵ^* changes from one replicate to another. So do X^*, Y^*, and $\hat{\beta}^*$.

> *Bootstrap principle for autoregression.* With a reasonably large n, the distribution of $\hat{\beta}^* - \hat{\beta}$ is a good approximation to the distribution of $\hat{\beta} - \beta$. In particular, the SD of $\hat{b}^*$ is a good approximation to the standard error of $\hat{b}$. The average of $\hat{b}^* - \hat{b}$ is a good approximation to the bias in $\hat{b}$.

In example 3, there will be some bias: the average of the $\hat{b}^*$'s will differ from $\hat{b}$ by a significant amount. The lag terms—the Y's from the earlier i's—do create some bias in the OLS estimator.

Example 4. A model with pooled time-series and cross-sectional variation. We combine example 3 above with example 3 in section 4.5. For $t = 1, \ldots, m$ and $j = 1, 2$, we assume

$$Y_{t,j} = a_j + bY_{t-1,j} + cW_{t,j} + \epsilon_{t,j}.$$

Think of t as time, and j as an index for geographical areas. The $Y_{0,j}$ are fixed, as are the W's. The a_1, a_2, b, c are scalar parameters, to be estimated from the data, $W_{t,j}, Y_{t,j}$ for $t = 1, \ldots, m$ and $j = 1, 2$. (For each t and j, $W_{t,j}$ and $Y_{t,j}$ are scalars.) The pairs $(\epsilon_{t,1}, \epsilon_{t,2})$ are IID with mean 0 and a positive definite 2×2 covariance matrix K. This too is unknown and to be estimated. One-step GLS is used to estimate a_1, a_2, b, c—although the GLS model (4.19) doesn't hold, because of the lag term: see example 3. The bootstrap will help us evaluate bias in feasible GLS, and the quality of the plug-in estimators for SEs (section 4.4).

We have to get the model into the matrix framework. Let $n = 2m$. For Y, we just stack up the $Y_{t,j}$:

$$Y = \begin{pmatrix} Y_{1,1} \\ Y_{1,2} \\ Y_{2,1} \\ Y_{2,2} \\ \vdots \\ Y_{m,1} \\ Y_{m,2} \end{pmatrix}.$$

This is $n \times 1$. Ditto for the errors:

$$\epsilon = \begin{pmatrix} \epsilon_{1,1} \\ \epsilon_{1,2} \\ \epsilon_{2,1} \\ \epsilon_{2,2} \\ \vdots \\ \epsilon_{m,1} \\ \epsilon_{m,2} \end{pmatrix}.$$

For the design matrix, we'll need a little trick, so let's do β next:

$$\beta = \begin{pmatrix} a_1 \\ a_2 \\ b \\ c \end{pmatrix}.$$

Since Y is $n \times 1$ and β is 4×1, the design matrix has to be $n \times 4$. The last column is the easiest: you just stack the W's. Column 3 is also pretty easy: stack the Y's, with a lag. Columns 1 and 2 have the dummies for the two geographical areas. These have to be organized so that a_1 goes with $Y_{t,1}$ and a_2 goes with $Y_{t,2}$:

$$X = \begin{pmatrix} 1 & 0 & Y_{0,1} & W_{1,1} \\ 0 & 1 & Y_{0,2} & W_{1,2} \\ 1 & 0 & Y_{1,1} & W_{2,1} \\ 0 & 1 & Y_{1,2} & W_{2,2} \\ \vdots & \vdots & \vdots & \vdots \\ 1 & 0 & Y_{m-1,1} & W_{m,1} \\ 0 & 1 & Y_{m-1,2} & W_{m,2} \end{pmatrix}.$$

Let's check it out. The matrix equation is $Y = X\beta + \epsilon$. The first line of this equation says

$$Y_{1,1} = a_1 + bY_{0,1} + cW_{1,1} + \epsilon_{1,1}.$$

Just what we need. The next line is

$$Y_{1,2} = a_2 + bY_{0,2} + cW_{1,2} + \epsilon_{1,2}.$$

This is good. And then we get

$$Y_{2,1} = a_1 + bY_{1,1} + cW_{2,1} + \epsilon_{2,1},$$

$$Y_{2,2} = a_2 + bY_{1,2} + cW_{2,2} + \epsilon_{2,2}.$$

These are all fine, and so are the rest.

Now, what about the covariance matrix for the errors? It's pretty easy to check that $\text{cov}(\epsilon) = G$, where the $n \times n$ matrix G has K repeated along the main diagonal:

(1)
$$G = \begin{pmatrix} K & 0_{2\times2} & \cdots & 0_{2\times2} \\ 0_{2\times2} & K & \cdots & 0_{2\times2} \\ \vdots & \vdots & \ddots & \vdots \\ 0_{2\times2} & 0_{2\times2} & \cdots & K \end{pmatrix}.$$

Before going on to bootstrap the model in example 4, let's pause here to review one-step GLS—equation (4.25). You make a first pass at the data, estimating β by OLS. This gives $\hat{\beta}_{\text{OLS}}$ with a residual vector $e = Y - X\hat{\beta}_{\text{OLS}}$. We use e to compute an estimate $\hat{K}$ for K. (We'll also use the residuals for another purpose, in the bootstrap.) Then we use $\hat{K}$ to estimate G. Notice that the residuals naturally come in pairs. There is one pair for each time period, because there are two geographical areas. Rather than a single subscript on e it will be better to have two, t and j, with $t = 1, \ldots, m$ for time and $j = 1, 2$ for geography. Let

$$e_{t,1} = e_{2t-1} \quad \text{and} \quad e_{t,2} = e_{2t}.$$

This notation makes the pairing explicit.

Now $\hat{K}$ is the empirical covariance matrix of the pairs:

$$(2) \qquad \hat{K} = \frac{1}{m} \sum_{t=1}^{m} \begin{pmatrix} e_{t,1} \\ e_{t,2} \end{pmatrix} (e_{t,1} \ \ e_{t,2}).$$

Plug $\hat{K}$ into the formula (1) for G to get $\hat{G}$, and then $\hat{G}$ into (4.22) to get

$$(3) \qquad \hat{\beta}_{\text{FGLS}} = (X'\hat{G}^{-1}X)^{-1}X'\hat{G}^{-1}Y.$$

This is one-step GLS. The "F" in $\hat{\beta}_{\text{FGLS}}$ is for "feasible." Plug $\hat{G}$ into the right hand side of (4.24) to get an estimated covariance matrix for $\hat{\beta}_{\text{FGLS}}$, namely,

$$(4) \qquad (X'\hat{G}^{-1}X)^{-1}.$$

Feasible GLS may be biased, especially with a lag term. And (4) is only an "asymptotic" formula: under some regularity conditions, it gives essentially the right answers with large samples. What happens with small samples? What about the sample size that we happen to have? And what about the bias?? The bootstrap should give us a handle on these questions.

Resampling the Y's is not a good idea: see example 2 for the reasoning. Instead, we bootstrap the model following the pattern in examples 2 and 3. We freeze $Y_{0,j}$ and the W's, as well as

$$\hat{\beta}_{\text{FGLS}} = \begin{pmatrix} \hat{a}_1 \\ \hat{a}_2 \\ \hat{b} \\ \hat{c} \end{pmatrix}$$

and the residuals e from the OLS fit. To regenerate the data, we start by resampling the e's. As noted above, the residuals come in pairs. The pairing has to be preserved in order to capture the covariance between $\epsilon_{t,1}$ and $\epsilon_{t,2}$. Therefore, we resample pairs of residuals (figure 3, next page).

Figure 3. Bootstrapping a model with pooled time-series and cross-sectional variation.

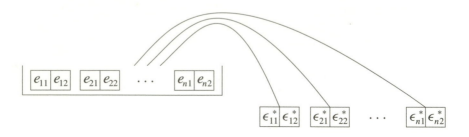

More formally, we generate IID pairs $(\epsilon_{i,1}^*, \epsilon_{i,2}^*)$, choosing at random with replacement from the paired residuals. The chance that $(\epsilon_{t,1}^*, \epsilon_{t,2}^*) = (e_{7,1}, e_{7,2})$ is $1/m$. Ditto if 7 is replaced by 19. Or any other number. Since we have a_1 and a_2 in the model, $\sum_{s=1}^m e_{s,1} = \sum_{s=1}^m e_{s,2} = 0$. (For the proof, e is orthogonal to the columns of X: the first two columns are the relevant ones.) In other words, $E(\epsilon_{t,1}^*) = E(\epsilon_{t,2}^*) = 0$. We have to generate the $Y_{t,j}^*$'s, as in example 3, one t at a time, using $\hat{a}_1, \hat{a}_2, \hat{b}$, and the $\epsilon_{t,j}^*$'s:

$$Y_{1,1}^* = \hat{a}_1 + \hat{b}Y_{0,1} + \hat{c}W_{1,1} + \epsilon_{1,1}^*,$$
$$Y_{1,2}^* = \hat{a}_2 + \hat{b}Y_{0,2} + \hat{c}W_{1,2} + \epsilon_{1,2}^*,$$
$$Y_{2,1}^* = \hat{a}_1 + \hat{b}Y_{1,1}^* + \hat{c}W_{2,1} + \epsilon_{2,1}^*,$$
$$Y_{2,2}^* = \hat{a}_2 + \hat{b}Y_{1,2}^* + \hat{c}W_{2,2} + \epsilon_{2,2}^*,$$

and so forth. No need to regenerate $Y_{0,j}$ or the W's: these are fixed. Now we bootstrap the estimator, getting $\hat{\beta}_{\text{FGLS}}^*$. This means doing OLS, getting residuals, then $\hat{K}^*$ as in (2), $\hat{G}^*$ as in (1), and finally

(5) $$\hat{\beta}_{\text{FGLS}}^* = (X^{*\prime}\hat{G}^{*-1}X^*)^{-1}X^{*\prime}\hat{G}^{*-1}Y^*.$$

We have to do this many times on the computer, to get some decent approximation to the distribution of $\hat{\beta}_{\text{FGLS}}^* - \hat{\beta}$. Notice the stars on the design matrix in (5). When a bootstrap design matrix is generated on the computer, the column with the Y's changes every time.

Bootstrap principle for feasible GLS. With a reasonably large n, the distribution of $\hat{\beta}_{\text{FGLS}}^* - \hat{\beta}_{\text{FGLS}}$ is a good approximation to the distribution of $\hat{\beta}_{FGLS} - \beta$. In particular, the empirical covariance matrix of $\hat{\beta}^*_{FGLS}$ is a good approximation to the theoretical covariance matrix of $\hat{\beta}_{FGLS}$. The average of $\hat{\beta}^*_{FGLS} - \hat{\beta}_{FGLS}$ is a good approximation to the bias in $\hat{\beta}_{FGLS}$.

More specifically, we would simulate N data sets, indexed by $k = 1, \ldots, N$. Each data set would consist of simulated design matrix $X_{(k)}$ and a simulated response vector $Y_{(k)}$. For each data set, we would compute $\hat{G}_{(k)}$ and a bootstrap replicate of the one-step GLS estimator,

$$\hat{\beta}_{FGLS,(k)} = \left[X'_{(k)} \, \hat{G}_{(k)}^{-1} X_{(k)} \right]^{-1} X'_{(k)} \, \hat{G}_{(k)}^{-1} Y_{(k)}.$$

Some things don't depend on k: for instance, $Y_{0,j}$ and $W_{t,j}$. We keep $\hat{\beta}_{FGLS}$—the one-step GLS estimate from the real data—fixed throughout, as the true parameter vector in the simulation. We keep the error distribution fixed too: the box in figure 3 stays the same through all the bootstrap replications.

This is a complicated example, but it is in this sort of example that you might want to use the bootstrap. The standard theory doesn't apply. There will be some bias, which can be detected by the bootstrap. There probably won't be any useful finite-sample results, although there may be some asymptotic formula like (4). The bootstrap is also asymptotic, but it often gets there faster than the competition. The next section has a real example, with a model for energy demand. Work the exercises, in preparation for the example.

Exercise set A

1. Let $X_1, \ldots, X_{50}$ be IID $N(\mu, \sigma^2)$. The sample mean is $\overline{X}$. True or false: $\overline{X}$ is an unbiased estimate of μ, but is likely to be off μ by something like $\sigma/\sqrt{50}$, just due to random error.

2. Let $X_{i(k)}$ be IID $N(\mu, \sigma^2)$, for $i = 1, \ldots, 50$ and $k = 1, \ldots, 100$. Let

$$\overline{X}_{(k)} = \frac{1}{50} \sum_{i=1}^{50} X_{i(k)}, \qquad s_{(k)}^2 = \frac{1}{50} \sum_{i=1}^{50} \left[X_{i(k)} - \overline{X}_{(k)} \right]^2,$$

$$\overline{X}_{ave} = \frac{1}{100} \sum_{k=1}^{100} \overline{X}_{(k)}, \qquad V = \frac{1}{100} \sum_{k=1}^{100} \left[\overline{X}_{(k)} - \overline{X}_{ave} \right]^2.$$

True or false, and explain:

(a) $\{\overline{X}_{(k)} : k = 1, \ldots, 100\}$ is a sample of size 100 from $N(\mu, \sigma^2/50)$.

(b) V is around $\sigma^2/50$.

(c) $|\overline{X}_{(k)} - \overline{X}_{\text{ave}}| < 2\sqrt{V}$ for about 95 of the k's.

(d) $\sqrt{V}$ is a good approximation to the SE of $\overline{X}$, where $\overline{X}$ was defined in exercise 1.

(e) The sample SD of the $\overline{X}_{(k)}$'s is a good approximation to the SE of $\overline{X}$.

(f) $\overline{X}_{\text{ave}}$ is $N(\mu, \sigma^2/5000)$.

3. (This continues exercise 2.) Fill in the blanks, and explain.

(a) $\overline{X}_{\text{ave}}$ is nearly μ, but is off by something like _____. Options:

$$\sigma \qquad \sigma/\sqrt{50} \qquad \sigma/\sqrt{100} \qquad \sigma/\sqrt{5000}$$

(b) $\overline{X}_{\text{ave}}$ is nearly μ, but is off by something like _____. Options:

$$\sqrt{V} \qquad \sqrt{V}/\sqrt{50} \qquad \sqrt{V}/\sqrt{100} \qquad \sqrt{V}/\sqrt{5000}$$

(c) The SD of the $\overline{X}_{(k)}$'s is around _____. Options:

$$\sqrt{V} \qquad \sqrt{V}/\sqrt{50} \qquad \sqrt{V}/\sqrt{100} \qquad \sqrt{V}/\sqrt{5000}$$

Exercises 1–3 illustrate the *parametric bootstrap*: we're resampling from a given parametric distribution, the normal. The notation looks awkward, but will be handy later.

7.2 Bootstrapping a model for energy demand

In the 1970s, long before the days of the SUV, we had an energy crisis in the United States. An insatiable demand for Arab oil, coupled with an oligopoly, led to price controls and gas lines. The crisis generated another insatiable demand, for energy forecasts. The Department of Energy tried to handle both problems. This section will discuss RDFOR, the Department's Regional Demand Forecasting model for energy demand.

We consider only the industrial sector. (The other sectors are residential, commercial, transportation.) The chief equation was this:

(6) $\qquad Q_{t,j} = a_j + bC_{t,j} + cH_{t,j} + dP_{t,j} + eQ_{t-1,j} + fV_{t,j} + \delta_{t,j}$.

Here, t is time in years: $t = 1961, 1962, \ldots, 1978$. The index j ranges over geographical regions, 1 through 10. Maine is in region 1 and California in region 10. On the left hand side, $Q_{t,j}$ is the log of energy consumption by the industrial sector in year t and region j.

On the right hand side of the equation, Q appears again, lagged by a year: $Q_{t-1,j}$. The coefficient e of the lag term was of policy interest, because e

was thought to measure the speed with which the economy would respond to energy shocks. Other terms can be defined as follows.

- $C_{t,j}$ is the log of cooling degree days in year t and region j. Every day that the temperature is one degree above 65° is a cooling degree day: energy must be supplied to cool the factories down. If we have 15 days with a temperature of 72°, that makes $15 \times (72 - 65) = 105$ cooling degree days. It's conventional to choose 65° as the baseline temperature. Temperatures are in Fahrenheit: this is the US Department of Energy.

- $H_{t,j}$ is the log of heating degree days in year t and region j. Every day that the temperature is one degree below 65° is a heating degree day: energy must be supplied to heat the factories up. If we have 15 days with a temperature of 54°, that makes $15 \times (65 - 54) = 165$ heating degree days.

- $P_{t,j}$ is the log of the energy price for the industrial sector in year t and region j.

- $V_{t,j}$ is the log of value added in the industrial sector in year t and region j. "Value added" means receipts from sales less costs of production; the latter include capital, labor, and materials. (This is a quick sketch of a complicated national-accounts concept.)

- There are 10 region-specific intercepts, a_j. There are 5 coefficients (b, c, d, e, f) that are constant across regions, making $10 + 5 = 15$ parameters so far. Watch it: e is a parameter here, not a residual vector.

- δ is an error term. The $(\delta_{t,j} : j = 1, \ldots, 10)$ are IID 10-vectors for $t = 1961, \ldots, 1978$, with mean 0 and a 10×10 covariance matrix K that expresses inter-regional dependence.

- The δ's are independent of all the right hand side variables, except the lag term.

Are the assumptions sensible? For now, don't ask, don't tell: it won't matter in the rest of this section. (The end notes comment on assumptions.)

The model is like example 4, with 18 years of data. There are 10 regions rather than 2. Analysts at the Department of Energy estimated the model by one-step GLS, equation (3). Results are shown in column A of table 1 (next page). For instance, the lag coefficient e is estimated as 0.684. Furthermore, standard errors were computed by the "plug-in" method, equation (4). Results are shown in column B. The standard error on the 0.684 is 0.025. The quality of these plug-in standard errors is an issue. Bias is also an issue, for two reasons. (i) There is a lag term. (ii) With a known covariance structure, GLS is unbiased; but one-step GLS is another story.

One-step GLS is working hard in this example. Besides the 10 intercepts and 5 slopes, there is a 10×10 covariance matrix that has to be estimated from the data. The matrix has 10 variances on the diagonal and 45 covariances above the diagonal. We only have 18 years of data on 10 regions—at best, 180 data points. The bootstrap will show there is bias in feasible GLS. It will also show that the plug-in SEs are seriously in error.

We bootstrap the model just as in the previous section. This involves generating 100 simulated data sets on the computer. We tell the computer to take $\hat{\beta}_{FGLS}$, column A, as ground truth for the parameters. (This is a truth about the computer code, not a truth about the economy.) What do we use for the errors? Answer: we resample the residuals from the OLS fit. This is like example 4, with 18 giant tickets in the box, each ticket being a 10-vector of residuals. For instance, 1961 contributes a 10-vector with a component for each region. So does 1962, and so forth, up to 1978.

When we resample, each ticket comes out a small random number of times (perhaps zero). The tickets come out in random order too. For example, the 1961 ticket might get used to simulate 1964 and again to simulate 1973; the 1962 ticket might not get used at all. What about the explanatory

Table 1. Bootstrapping RDFOR.

	GLS		Bootstrap			
	(A)	(B)	(C)	(D)	(E) RMS plug-in SE	(F) RMS bootstrap SE
	Estimate	Plug-in SE	Mean	SD		
a_1	−.95	.31	−.94	.54	.19	.43
a_2	−1.00	.31	−.99	.55	.19	.43
a_3	−.97	.31	−.95	.55	.19	.43
a_4	−.92	.30	−.90	.53	.18	.41
a_5	−.98	.32	−.96	.55	.19	.44
a_6	−.88	.30	−.87	.53	.18	.41
a_7	−.95	.32	−.94	.55	.19	.44
a_8	−.97	.32	−.96	.55	.19	.44
a_9	−.89	.29	−.87	.51	.18	.40
a_{10}	−.96	.31	−.94	.54	.19	.42
cdd b	.022	.013	.021	.025	.0084	.020
hdd c	.10	.031	.099	.052	.019	.043
price d	−.056	.019	−.050	.028	.011	.022
lag e	.684	.025	.647	.042	.017	.034
va f	.281	.021	.310	.039	.014	.029

variables on the right hand side of (6), like cooling degree days? We just leave them as we found them; they were assumed exogenous. Similarly, we leave $Q_{1960,j}$ alone. The lag terms for $t = 1961, 1962, \ldots$ have to be regenerated as we go.

For each simulated data set, we compute a one-step GLS estimate, $\hat{\beta}^*_{\text{FGLS}}$. This is a 15×1 vector (10 regional intercepts, 5 coefficients). The mean of these vectors is shown in column C. For example, the coefficient of the lag term is 14th in order, so $\hat{e}^*$ is the 14th entry in $\hat{\beta}^*_{\text{FGLS}}$. The mean of the 100 $\hat{e}^*$'s is 0.647. The SD of the 100 bootstrap estimates is shown in column D. For instance, the SD of the 100 $\hat{e}^*$'s is 0.042. The bootstrap has now delivered its output, in columns C and D. We will use the output to analyze bias and variance in feasible GLS.

Bias. As noted above, the mean of the 100 $\hat{e}^*$'s is 0.647. This is somewhat lower than the assumed true value of 0.684 in column A. The difference may look insignificant. Look again. We have a sample of size 100. The sample average is 0.647. The sample SD is 0.042. The SE for the sample average is $0.042/\sqrt{100} = 0.0042$. Bias is highly significant, and larger in size than the plug-in SE (column B). The bootstrap has shown that FGLS is biased. This completes our discussion of bias, and we turn to variance.

Variance. We use column D again: the bootstrap SEs are just the SDs in column D. To review the logic, the 100 $\hat{e}^*$'s are a sample from the true distribution—true within the confines of the computer simulation. The mean of the sample is a good estimate for the mean of the population, i.e., the expected value of $\hat{e}^*$. The SD of the sample is a good estimate for the SD of $\hat{e}^*$. This tells you how far the FGLS estimator is likely to get from its expected value. (If in doubt, go back to the previous section.)

Plug-in SEs vs the bootstrap. Column B reports the plug-in SEs. Column D reports the bootstrap SEs. Comparing columns B and D, you see that the plug-in method and the bootstrap are very different. The plug-in SEs are a lot smaller. But, maybe the plug-in method is right and the bootstrap is wrong? That is where column E comes in. Column E will show that the plug-in SEs are a lot too small. (Column E is special; the usual bootstrap stops with columns C and D.)

For each simulated data set, we compute not only the one-step GLS estimator but also the plug-in covariance matrix. The square root of the mean of the diagonal is shown in column E. Within the confines of the computer simulation—where the modeling assumptions are true by virtue of the computer code—column D gives the true SEs for one-step GLS, up to a little random error. Column E tells you what the plug-in method is doing, on average. The plug-in method is too small, by a factor of 2 or 3. Estimating all those covariances is making the data work too hard. That is what the bootstrap has shown us.

Some details. The bootstrap is a bit complicated. Explicit notation may make the story easier to follow. We're going to have 100 simulated data sets. Let's index these by a subscript $k = 1, \ldots, 100$. We put parens around k to distinguish it from other subscripts. Thus, $Q_{t,j,(k)}$ is the log quantity of energy demand in year t and region j, in the kth simulated data set. The response vector $Y_{(k)}$ in the kth data set is obtained by stacking up the $Q_{t,j,(k)}$. First we have $Q_{1961,1,(k)}$, then $Q_{1961,2,(k)}$, and so on, down to $Q_{1961,10,(k)}$. Next comes $Q_{1962,1,(k)}$, and so forth, all the way down to $Q_{1978,10,(k)}$. In terms of a formula, $Q_{t,j,(k)}$ is the $[10(t - 1961) + j]^{th}$ entry in $Y_{(k)}$, for $t = 1961, 1962, \ldots$ and $j = 1, \ldots, 10$.

There's no need to have a subscript (k) on the other variables, like cooling degree days or value added: these don't change. The design matrix in the kth simulated data set is $X_{(k)}$. There are 10 columns for the regional dummies (example 4 had two regional dummies), followed by one column each for cooling degree days, heating degree days, price, lagged quantity, value added. These are all stacked in the same order as $Y_{(k)}$. Most of the columns stay the same throughout the simulation, but the column with the lags keeps changing. That is why a subscript k is needed on the design matrix.

For the kth simulated data set, we compute the one-step GLS estimator as

(7) $$\hat{\beta}_{\text{FGLS},(k)} = \left[X'_{(k)} \hat{G}^{-1}_{(k)} X_{(k)}\right]^{-1} X'_{(k)} \hat{G}^{-1}_{(k)} Y_{(k)},$$

where $\hat{G}_{(k)}$ is estimated from OLS residuals in a preliminary pass through the kth simulated data set. Here is a little more detail. The formula for the OLS residuals is

(8) $$Y_{(k)} - X_{(k)}\left[X'_{(k)} X_{(k)}\right]^{-1} X'_{(k)} Y_{(k)}.$$

The OLS residual $r_{t,j,(k)}$ for year t and region j is the $[10(t - 1961) + j]^{th}$ entry in (8). (Why r? Because e is a parameter.) For each year from 1961 through 1978, we have a 10-vector of residuals, whose empirical covariance matrix is

$$\hat{K}_{(k)} = \frac{1}{18} \sum_{t=1961}^{1978} \begin{pmatrix} r_{t,1,(k)} \\ r_{t,2,(k)} \\ \vdots \\ r_{t,10,(k)} \end{pmatrix} \begin{pmatrix} r_{t,1,(k)} & r_{t,2,(k)} & \cdots & r_{t,10,(k)} \end{pmatrix}.$$

If in doubt, look back at example 4. String 18 copies of $\hat{K}_{(k)}$ down the diagonal of a 180×180 matrix to get the $\hat{G}_{(k)}$ in (7):

$$\hat{G}_{(k)} = \begin{pmatrix} \hat{K} & 0_{10 \times 10} & \cdots & 0_{10 \times 10} \\ 0_{10 \times 10} & \hat{K} & \cdots & 0_{10 \times 10} \\ \vdots & \vdots & \ddots & \vdots \\ 0_{10 \times 10} & 0_{10 \times 10} & \cdots & \hat{K} \end{pmatrix}.$$

The kth replicate bootstrap estimator $\hat{\beta}_{FGLS,(k)}$ in (7) is a 15-vector, with estimates for the 10 regional intercepts followed by $\hat{b}_{(k)}, \hat{c}_{(k)}, \hat{d}_{(k)}, \hat{e}_{(k)}, \hat{f}_{(k)}$. The simulated estimate for the lag coefficient $\hat{e}_{(k)}$ is therefore the 14th entry in $\hat{\beta}_{FGLS,(k)}$. The 0.647 under column C in the table was obtained as

$$\hat{e}_{ave} = \frac{1}{100} \sum_{k=1}^{100} \hat{e}_{(k)}.$$

Up to a little random error, this is $E[\hat{e}_{(k)}]$, i.e., the expected value of the one-step GLS estimator in the simulation. The 0.042 was obtained as

$$\sqrt{\frac{1}{100} \sum_{k=1}^{100} \left(\hat{e}_{(k)} - \hat{e}_{ave}\right)^2}.$$

Up to another little random error, this is the SE of the one-step GLS estimator in the simulation. (Remember, e is a parameter not a residual vector.)

For each simulated data set, we compute not only the one-step GLS estimator but also the plug-in covariance matrix

(9) $$\left[X'_{(k)} \hat{G}_{(k)}^{-1} X_{(k)}\right]^{-1}.$$

We take the mean over k of each of the 15 diagonal elements in (9). The square root of the means goes into column E. That column tells the truth about the plug-in SEs: they're much too small.

The squaring and unsquaring may be a little hard to follow, so let's try a general formula. We generate a sequence of variances on the computer. The square root of each variance is an SE. Then

$$\text{RMS SE} = \sqrt{\text{mean (SE}^2)} = \sqrt{\text{mean variance}}.$$

Bootstrapping the bootstrap. Finally, what about the bootstrap? Does it do any better than the asymptotics? It turns out we can calibrate the bootstrap by doing an even larger simulation (column F). For each of our 100 simulated data sets $[X_{(k)}, Y_{(k)}]$, we compute the analog of column D. For this purpose, each simulated data set spawns 100 simulated data sets of its own. All in all, there are $100^2 = 10{,}000$ data sets to keep track of, but with current technology, not a problem. For each simulated data set, we get simulated bootstrap SEs on each of the 15 parameter estimates. The RMS of the simulated bootstrap SEs is shown in column F. The bootstrap runs out of gas too, but it comes a lot closer to truth (column D) than the plug-in SEs (column E).

As noted before, usual applications of the bootstrap stop with columns C and D. Columns E and F are special. Column E uses the bootstrap to check on the plug-in SEs. Column F uses the bootstrap to check on itself.

What is truth? For the simulation, column C gives expectations and D gives SEs (up to a little random error). For the real data, these are only approximations, because (i) the real world may not follow the model, and (ii) even if it did, we're sampling from the empirical distribution of the residuals, not the theoretical distribution of the errors. If the model is wrong, the estimates and their SEs are meaningless statistics. If the model is right, the estimates in column A of table 1 are biased, and the SEs in column B are far too small: this requires an extrapolation from the computer model to the real world.

Exercise set B

1. There is a statistical model with a parameter θ. You need to estimate θ. Which is a better description of the bootstrap? Explain briefly.
   (i) The bootstrap will help you find an estimator for θ.
   (ii) Given an estimator $\hat{\theta}$ for θ, the bootstrap will help you find the bias and SE of $\hat{\theta}$.

2. Which terms in equation (6) are observable, and which are unobservable? Which are parameters?

3. Does the model reflect the idea that energy consumption in 1975 might have been different from what it was? If so, how?

4. In table 1, at the end of column A, you will find the number 0.281. How is this number related to equation (6)?

5. To what extent are the one-step GLS estimates biased in this application? Which numbers in the table prove your point? How?

6. Are plug-in SEs biased in this application? Which numbers in the table prove your point? How?

7. Are bootstrap standard errors biased in this application? Which numbers in the table prove your point? How?

8. Paula has observed values on four independent random variables with common density $f_{\alpha,\beta}(x) = c(\alpha, \beta)|\alpha x - \beta| \exp[-(\alpha x - \beta)^2]$, where $\alpha > 0$, $-\infty < \beta < \infty$, and $c(\alpha, \beta)$ is chosen so that $\int_{-\infty}^{\infty} f_{\alpha,\beta}(x)dx = 1$. She estimates α, β by maximum likelihood and computes the standard errors from the observed information. Before doing the t-test to see whether β is significantly different from 0, she consults a statistician, who tells her to use the bootstrap because observed information is only useful with large samples. What is your advice? (See discussion question 6.15.)

9. (Hard.) In example 3, if $1 \leq i < n$, show that $E(\epsilon_i | X) = \epsilon_i$.

7.3 End notes for chapter 7

Terminology. In the olden days, boots had straps so you could pull them on. The term "bootstrap" comes from the expression, to lift yourself up by your own bootstraps.

Theory. Freedman (1981, 1984) describes the theoretical basis for applying the bootstrap to different kinds of regression models.

Centering. In example 2, without an intercept, you would have to center the residuals. Likewise, in example 4, you need the two regional intercepts a_1, a_2. With RDFOR, it is the 10 regional intercepts that center the residuals. Without centering, the bootstrap may be way off (Freedman 1981).

Which set of residuals? We could resample FGLS residuals. However, $\hat{G}$ in (4) is computed from the OLS residuals. A comparison between asymptotics and the bootstrap seemed fairer if OLS residuals were resampled in the latter, so that is what we did.

Autoregression. A regression of Y_t on "lagged" values (e.g., Y_{t-1}) and control variables is called an "autoregression," with "auto" meaning self: Y is explained in part by its own previous values. With the autoregression in example 3, if $|b| < 1$ the conventional theory is a good approximation when the sample size is large; however, if $|b| \geq 1$, the theory gets more complicated (Anderson 1959). Bias in coefficient estimates due to lags is a well-known phenomenon (Hurwicz 1950). Bias in asymptotic standard errors is a less familiar topic.

RDFOR. The big problem with the bootstrap is that the residuals are too small. For OLS, there is an easy fix: divide by $n - p$ not n. In a complicated model like RDFOR, what would you use for p? The right answer turns out to depend on unknown parameters: feasible GLS isn't real GLS. Using the bootstrap to remove bias is tempting, but the reduction in bias is generally offset by an increase in variance. Doss and Sethuraman (1989) have a theorem which captures this idea.

Section 2 is based on Freedman and Peters (1984abc, 1985). Technically, $P_{t,j}$ is a price index and $Q_{t,j}$ is a quantity index. ("Divisia" indices were used in constructing the data.) Further simulation studies show the bias in FGLS is mainly due to the presence of the lag term.

RDFOR, developed by the Department of Energy, is somewhat unrealistic as a model for energy demand (Freedman-Rothenberg-Sutch 1983). Among other things, P and δ can scarcely be independent (chapter 8). However, failures in the model do not explain bias in FGLS, or the poor behavior of the plug-in SEs. Differences between columns A and C in table 1, or differences among columns D-E-F, are not the results of specification error. In the computer simulation, the model holds true by virtue of the coding.

Plug-in SEs. These are more politely referred to as *nominal* or *asymptotic* SEs: "nominal" contrasts with "actual," and asymptotics work when the sample is large enough (see below).

The role of asymptotics. Statistical procedures are often defended on the basis of their "asymptotic" properties—the way they behave when the sample is large. See, for instance, Beck (2001, p. 273): "methods can be theoretically justified based on their large-[sample] behavior." This is an oversimplification. If we have a sample of size 100, what would happen with a sample of size 100,000 is not a decisive consideration. Asymptotics are useful because they give clues to behavior for samples like the one you actually have. Furthermore, asymptotics set a threshhold. Procedures that do badly with large samples are unlikely to do well with small samples.

With the central theorem, the asymptotics take hold rather quickly: when the sample size is 25, the normal curve is a often a good approximation to the probability histogram for the sample average; when the sample size is 100, the approximation is often excellent. With feasible GLS, if there are a lot of covariances to estimate, the asymptotics take hold rather slowly.

Other papers. Bias in plug-in SEs for feasible GLS is rediscovered from time to time. See, e.g., Beck and Katz (1995). Their use of White's method for estimating the SEs in OLS may have the same sort of problems as the plug-in SEs, because their estimated covariance matrices can be quite unstable. As a result, t-statistics will show unexpected behavior. Moreover, in the application of interest, feasible GLS is likely to give more accurate estimates of the parameters than OLS. Also see Beck (2001).

8

Simultaneous Equations

8.1 Introduction

This chapter explains *simultaneous-equation* models, and how to estimate them using *instrumental variables* (or *two-stage least squares*). These techniques are needed to avoid *simultaneity bias* (aka *endogeneity bias*). The lead example will be hypothetical supply and demand equations for butter in the state of Wisconsin. The source of endogeneity bias will be explained, and so will methods for working around this problem.

Then we discuss two real examples—(i) the way education and fertility influence each other, and (ii) the effect of school choice on social capital. These examples indicate how social scientists use two-stage least squares to handle (i) reciprocal causation and (ii) self-selection of subjects into the sample. (In the social sciences, two-stage least squares is often seen as *the* solution to problems of statistical inference.) At the end of the chapter there is a literature review, which puts modeling issues into a broader perspective.

We turn now to butter. Supply and demand need some preliminary discussion. For an economist, butter supply is not a single quantity but a relationship between quantity and price. The supply curve shows the quantity of butter that farmers would bring to market at different prices. In the left

Figure 1. Supply and demand. The vertical axis shows quantity; the horizontal axis, price.

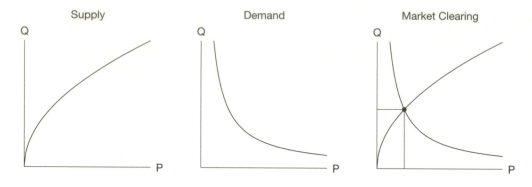

hand panel of figure 1, price is on the horizontal axis and quantity on the vertical. (Economists usually do it the other way around.)

Notice that the supply curve slopes up. Other things being equal—ceteris paribus, as they say—if the price goes up so does the quantity offered for sale. Farmers will divert their efforts from making cheese or delivering milk to churning butter. If the price gets high enough, farmers will start buying suburbs and converting them back to pasture. As you can see from the figure, the curve is concave: each extra dollar brings in less butter than the dollar before it. (For instance, suburban land is expensive land.)

Demand is also a relationship between quantity and price. The demand curve in the middle panel of figure 1 shows the total amount of butter that consumers would buy at different prices. This curve slopes down. Other things being equal, as price goes up the quantity demanded goes down. This curve is convex—one expression of "the law of diminishing marginal utility." (The second piece of cake is never as good as the first; if you will pay $10 for the first pound of butter, you might only pay $8 for the second, and so forth: that is convexity of P as a function of Q.)

According to economic theory, the free market price is determined by the crossing point of the two curves. This "law of supply and demand" is illustrated in the right hand panel of figure 1. At the free market price, the market clears: supply equals demand. If the price were set lower, the quantity demanded would exceed the quantity supplied, and disappointed buyers would bid the price up. If the price were set higher, the quantity supplied would exceed the quantity demanded, and frustrated suppliers would lower their prices. With price control, you just sell the butter to the government. That is why price controls lead to butter mountains. With rent control, overt bidding is illegal;

there is excess demand for housing, as well as under-the-counter payments of one kind or another. Relative to free markets, politicians set rents too low and butter prices too high.

Supply curves and demand curves are response schedules (section 5.4). The supply curve shows the response of farmers to different prices. The demand curve shows the response of consumers. These curves are somewhat hypothetical, because at any given time, we only get to see one price and one quantity. The extent to which supply curves and demand curves exist, in the sense that (say) planetary orbits exist, may be debatable. For now, let us set such questions aside and proceed with the usual theory.

Other things affect supply and demand besides price. Supply is affected by the costs of factors of production, e.g., the agricultural wage rate and the price of hay (labor and materials). These are "determinants of supply." Demand is affected by prices for complements (things that go with butter, like toast) and substitutes (like margarine). These are "determinants of demand." The list could be extended.

Suppose the supply curve is stable while the demand curve moves around (left hand panel, figure 2). Then the observations—the market clearing prices and quantities—would trace out the supply curve. Conversely, if the supply curve shifts while the demand curve remains stable, the observations would trace out the demand curve (middle panel). In reality, as economists see things, both curves are changing, so we get the right hand panel of figure 2. To estimate the curves, more assumptions must be introduced. Economists call this "specifying the model." We need to specify the determinants of supply and demand, as well as the functional form of the curves.

Figure 2. Tracing out supply and demand curves. The vertical axis shows quantity; the horizontal axis, price.

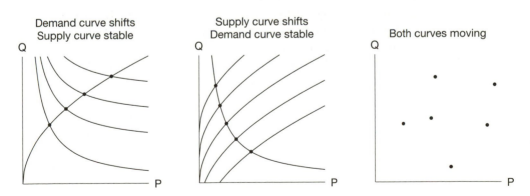

Demand curve shifts
Supply curve stable

Supply curve shifts
Demand curve stable

Both curves moving

Our model has two "endogenous variables," the quantity and price of butter, denoted Q and P. The specification will say how these endogenous variables are determined by "exogenous variables." The exogenous variables in the supply equation are the agricultural wage rate W and the price H of hay (labor and materials). These are the determinants of supply. The exogenous variables in the demand equation are the prices T of toast and M of margarine (complements and substitutes). These are the determinants of demand. For the moment, "exogeneity" just means "externally determined" and "endogeneity" means "determined within the model." Technical definitions will come shortly.

We consider a linear specification. The model has two linear equations in two unknowns, Q and P. For each time period t,

(1a) Supply $Q = a_0 + a_1 P + a_2 W + a_3 H + \delta_t$,

(1b) Demand $Q = b_0 + b_1 P + b_2 T + b_3 M + \epsilon_t$.

On the right hand side, there are parameters, the a's and b's. There is price P. There are the determinants of supply in (1a) and the determinants of demand in (1b). There are random disturbance terms δ_t and ϵ_t: otherwise, the data would never fit the equations. Everything is linear and additive. (Linearity makes things simple; however, economists might transform the variables in order to get curves like those sketched in figures 1 and 2.)

To complete the specification, we need to make some assumptions about (δ_t, ϵ_t). Error terms have expectation 0. As pairs, (δ_t, ϵ_t) are independent and identically distributed for $t = 1, \ldots, n$, but δ_t is allowed to be correlated with ϵ_t. The variance of δ_t and the variance of ϵ_t may be different. Equation (1a) is a linear supply schedule; (1b) is a linear demand schedule. We should write $Q_{t,P,W,H,T,M}$ instead of Q—after all, these are response schedules—but inconsistency seems a better choice.

Each equation describes a hypothetical experiment. In (1a), we set P, W, H, T, M, and observe how much butter the farmers bring to market. By assumption, T and M have no effect on supply: they're not in the equation. On the other hand, P, W, H should have additive linear effects. In (1b), we set P, W, H, T, M and observe how much butter the consumers will buy: W and H should have no effect on demand, while P, T, M should have additive linear effects. The disturbance terms are invariant under all interventions. So are the parameters, which remain the same for all combinations of W, H, T, M.

There is a third hypothetical experiment, which could be described by taking equations (1a) and (1b) together. The exogenous variables W, H, T, M

can be set to any particular values of interest, perhaps within certain ranges, and the two equations solved together for the two unknowns Q and P, giving us the quantity and price we would see in a free market—with the prescribed values for the exogenous variables.

So far, we have three hypothetical experiments, where we can set the exogenous variables. In the social sciences, experiments are unusual. More often, equations are estimated using observational data. Another assumption is needed: that Nature runs experiments for us.

Suppose, for instance, that we have 20 years of data in Wisconsin. Economists would assume that Nature generated the data for us as if by choosing W_t, H_t, T_t, M_t for $t = 1, \ldots, 20$ from some joint distribution, independently of the δ's and ϵ's. Thus, by assumption, W_t, H_t, T_t, M_t are independent of the error terms. This is "exogeneity" in its technical sense.

Nature substitutes her values for W_t, H_t, T_t, M_t into the right hand side of (1a) and (1b). She gets the supply and demand equations that are operative in year t:

(2a) Supply $\qquad Q = a_0 + a_1 P + a_2 W_t + a_3 H_t + \delta_t$,

(2b) Demand $\qquad Q = b_0 + b_1 P + b_2 T_t + b_3 M_t + \epsilon_t$.

According to the model—here comes the law of supply and demand—the market price P_t and the quantity sold Q_t in period t are determined as if by solving (2a) and (2b) for the two unknowns Q and P:

(3a) $\qquad Q_t = \dfrac{a_1(b_0 + b_2 T_t + b_3 M_t + \epsilon_t) - b_1(a_0 + a_2 W_t + a_3 H_t + \delta_t)}{a_1 - b_1}$,

(3b) $\qquad P_t = \dfrac{(b_0 + b_2 T_t + b_3 M_t + \epsilon_t) - (a_0 + a_2 W_t + a_3 H_t + \delta_t)}{a_1 - b_1}$.

We do not get to see the parameters or the disturbance terms. All we get to see are Q_t, P_t, and the exogenous variables W_t, H_t, T_t, M_t. Our objective is to estimate the parameters in (2a)-(2b), from these observational data. That will tell us, for example, how farmers and consumers would respond to price controls. Notice that the model would allow us to make causal inferences from observational data—if the underlying assumptions are right.

A regression of Q_t on P_t and the exogenous variables leads to *simultaneity bias*, also called *endogeneity bias*, because there are disturbance terms in the formula (3b) for P_t. Apart from rare parameter combinations, P_t is correlated with δ_t and ϵ_t. In other words, P_t is endogenous. That is the new statistical problem. Of course, Q_t is endogenous too: there are disturbance terms in (3a).

This section presented a simple econometric model with a supply equation and a demand equation—equations (2a) and (2b). The source of endogeneity bias was identified: disturbance terms turn up in the formula (3b) for P_t. (This "reduced form" equation is of no further interest here, although it may be helpful in other contexts.) The way to get around endogeneity bias is to estimate equations (2a) and (2b) by instrumental variables rather than OLS. This new technique will be explained in sections 2 and 3. Section 6.4 discussed endogeneity bias in a different kind of model, with a binary response variable.

Exercise set A

1. In equation (1a), should a_1 be positive or negative? What about a_2, a_3?
2. In equation (1b), should b_1 be positive or negative? What about b_2, b_3?
3. In the butter model of this section:
   (a) Does the law of supply and demand hold true?
   (b) Is the supply curve concave? strictly concave?
   (c) Is the demand curve convex? strictly convex?
   (Economists prefer log linear specifications. . . .)
4. An economist wants to use the butter model to determine how farmers will respond to price controls. Which of the following equations is the most relevant—(2a), (2b), (3a), (3b)? Explain briefly.

8.2 Instrumental variables

We begin with a slightly abstract linear model

$$(4) \qquad\qquad\qquad Y = X\beta + \delta,$$

where Y is an observable $n \times 1$ random vector, X is an observable $n \times p$ random matrix, and β is an unobservable $p \times 1$ parameter vector. The δ_i are IID with mean 0 and finite variance σ^2; they are unobservable random errors. This is the standard regression model, except that X is endogenous, i.e., X and δ are dependent. Conditional on X, the OLS estimates are biased by $(X'X)^{-1}X'E(\delta|X)$: see (4.9). This is *simultaneity bias*.

To handle simultaneity bias, economists and other social scientists would estimate (4) using *instrumental-variables regression*, also called *two-stage least squares*: the acronyms are IVLS and IISLS (or 2SLS, if you prefer Arabic numerals). The method requires an $n \times q$ matrix of *instrumental* or *exogenous* variables, with $n > q \geq p$. The matrix will be denoted Z. The matrices $Z'X$ and $Z'Z$ need to be of full rank, p and q respectively. If $q > p$, the system is *over-identified*. If $q = p$, the system is *just-identified*. If $q < p$, the case which is excluded by assuming $q \geq p$, the system is *under-identified*—parameters will not be identifiable (section 6.2). Let's make a cold list of the assumptions.

    (i) X is $n \times p$ and Z is $n \times q$ with $n > q \geq p$.
    (ii) $Z'X$ and $Z'Z$ have full rank, p and q respectively.
    (iii) $Y = X\beta + \delta$.
    (iv) The δ_i are IID, with mean 0 and variance σ^2.
    (v) Z is exogenous, i.e., $Z \perp\!\!\!\perp \delta$.

Assumptions (i) and (ii) are easy to check from the data. The others are substantially more mysterious.

    The idea behind IVLS is to multiply both sides of (4) by Z', getting

$$(5) \qquad Z'Y = Z'X\beta + Z'\delta.$$

This is a least squares problem. The response variable is $Z'Y$. The design matrix is $Z'X$ and the error term is $Z'\delta$. The parameter vector is still β. (The matrix equation unpacks to q ordinary equations in p unknowns—the components of β: if $q < p$, we have too many parameters and too little data.)

    Econometricians use feasible GLS (section 4.4) to estimate (5), rather than OLS. This is because $\mathrm{cov}(Z'\delta|Z) = \sigma^2 Z'Z \neq \sigma^2 I_{q \times q}$ (exercise 3C4). By assumptions (i)-(ii), $Z'Z$ has an inverse; and the inverse has a square root (exercise B1 below). We multiply both sides of (5) by $(Z'Z)^{-1/2}$ to get

$$(6) \quad \left[(Z'Z)^{-1/2}Z'Y\right] = \left[(Z'Z)^{-1/2}Z'X\right]\beta + \eta, \text{ where } \eta = (Z'Z)^{-1/2}Z'\delta.$$

    Apart from a little wrinkle to be discussed below, equation (6) is the usual regression model. As far as the errors are concerned,

$$(7) \qquad\qquad\qquad E(\eta|Z) = 0$$

because Z was assumed exogenous: see (iv)-(v). Moreover,

$$
\begin{aligned}
(8) \qquad \mathrm{cov}(\eta\,|\,Z) &= E\left[(Z'Z)^{-1/2}Z'\delta\delta'Z(Z'Z)^{-1/2}\,\big|\,Z\right] \\
&= (Z'Z)^{-1/2}Z'E\left[\delta\delta'\,|\,Z\right]Z(Z'Z)^{-1/2} \\
&= (Z'Z)^{-1/2}Z'\sigma^2 I_{n\times n}Z(Z'Z)^{-1/2} \\
&= \sigma^2(Z'Z)^{-1/2}(Z'Z)(Z'Z)^{-1/2} \\
&= \sigma^2 I_{q\times q}.
\end{aligned}
$$

The big move is in the third line: $E[\delta\delta'|Z] = \sigma^2 I_{n\times n}$, because Z was assumed to be exogenous, and the δ_i were assumed to be IID with mean 0 and variance σ^2: see (iv)-(v). Otherwise, we're just factoring constants out of the expectation and juggling matrices.

The OLS estimate for β in (6) is

(9) $$\tilde{\beta} = (M'M)^{-1}M'L,$$

where $M = (Z'Z)^{-1/2}Z'X$ is the design matrix and $L = (Z'Z)^{-1/2}Z'Y$ is the response variable. (Exercise B1 shows that all the inverses exist.)

The IVLS estimator in the original system (4) is usually given as

(10) $$\hat{\beta}_{IVLS} = \left[X'Z(Z'Z)^{-1}Z'X\right]^{-1}X'Z(Z'Z)^{-1}Z'Y.$$

We will show that $\hat{\beta}_{IVLS} = \tilde{\beta}$, completing the derivation of the IVLS estimator. This takes a bit of algebra. For starters, because $Z'Z$ is symmetric,

(11a) $$M'M = X'Z(Z'Z)^{-1/2}(Z'Z)^{-1/2}Z'X = X'Z(Z'Z)^{-1}Z'X,$$

and

(11b) $$M'L = X'Z(Z'Z)^{-1/2}(Z'Z)^{-1/2}Z'Y = X'Z(Z'Z)^{-1}Z'Y.$$

Substituting (11a) and (11b) into (9) proves that $\hat{\beta}_{IVLS} = \tilde{\beta}$.

Standard errors can be obtained from equation (12), whose derivation is left to exercise E1 below:

(12) $$\widehat{\text{cov}}(\hat{\beta}_{IVLS}|Z) = \hat{\sigma}^2\left[X'Z(Z'Z)^{-1}Z'X\right]^{-1},$$

where

(13) $$\hat{\sigma}^2 = \|Y - X\hat{\beta}_{IVLS}\|^2/(n-p).$$

It is conventional to divide by $n - p$ in (13), but theorem 4.4 does not apply— we're not in the OLS model.

Equation (10) is pretty dense. For some people, it helps to check that all the multiplications make sense. For instance, Z is $n \times q$, so Z' is $q \times n$. Then $Z'Z$ and $(Z'Z)^{-1}$ are $q \times q$. Next, X is $n \times p$, so X' is $p \times n$. Thus, $X'Z$ is $p \times q$ and $Z'X$ is $q \times p$, which makes $X'Z(Z'Z)^{-1}Z'X$ a $p \times p$ matrix. What about $X'Z(Z'Z)^{-1}Z'Y$? Well, $X'Z$ is $p \times q$, $(Z'Z)^{-1}$ is $q \times q$, and $Z'Y$ is $q \times 1$. So $X'Z(Z'Z)^{-1}Z'Y$ is $p \times 1$. This is pretty dense too, but there is a simple bottom line: $\hat{\beta}_{IVLS}$ is $p \times 1$, like it should be.

The little wrinkle in (6). Given Z, the design matrix $M = (Z'Z)^{-1/2}Z'X$ is still related to the errors $\eta = (Z'Z)^{-1/2}Z'\delta$, because of the endogeneity of X. This leads to *small-sample bias*. However, with luck, M will be practically constant, and a little bit of correlated randomness can't hurt much. See theorem 1 in section 8.

Exercise set B

1. By assumptions (i)-(ii), $Z'X$ is $q \times p$ of rank p, and $Z'Z$ is $q \times q$ of rank q. Show that:

   (a) $Z'Z$ is positive definite and invertible; the inverse has a square root.

   (b) $X'Z(Z'Z)^{-1}Z'X$ is positive definite, hence invertible. Hint. Suppose c is $p \times 1$. Can $c'X'Z(Z'Z)^{-1}Z'Xc \leq 0$?

   Note. Without assumptions (i)-(ii), (10) and (12) wouldn't make sense.

2. Let U_i be IID random variables. Let $\overline{U} = \frac{1}{n}\sum_{i=1}^{n} U_i$. True or false, and explain:

   (a) $E(U_i)$ is the same for all i.

   (b) $\text{var}(U_i)$ is the same for all i.

   (c) $E(U_i) = \overline{U}$.

   (d) $\text{var}(U_i) = \frac{1}{n}\sum_{i=1}^{n}(U_i - \overline{U})^2$.

   (e) $\text{var}(U_i) = \frac{1}{n-1}\sum_{i=1}^{n}(U_i - \overline{U})^2$.

8.3 Estimating the butter model

Our next project is to estimate the butter model using IVLS. We'll start with the supply equation (2a). The equation is often written this way:

$$(14) \qquad Q_t = a_0 + a_1 P_t + a_2 W_t + a_3 H_t + \delta_t \quad \text{for} \quad t = 1, \ldots, 20.$$

The actual price and quantity in year t are substituted for the free variables Q and P that define the supply schedule. Reminder: according to the law of supply and demand in the model, Q_t and P_t were obtained by solving the pair of equations (2a)-(2b) for the two unknowns Q and P.

Let's get (14) into the format of (4). The response variable Y is the 20×1 column vector of Q_t's, and δ is just the column of δ_t's. To get β, we stack up a_0, a_1, a_2, a_3. The design matrix X is 20×4. Column 1 is all 1's, to accommodate the intercept. Then we get a column of P_t's, a column of W_t's, and a column of H_t's. (Keep these in the right order, as indexed by time t.) Column 1 is constant, and totally exogenous. Columns 3 and 4 are exogenous, by assumption. But column 2 is endogenous. That's the problem.

To get the matrix Z of exogenous variables, we start with columns 1, 3, and 4 in X. But we need at least one more instrument, to make up for the column of prices. Where to look? The answer is, in the demand equation. Just add a column of T_t's and a column of M_t's. Both of these are exogenous, by assumption. Now $q = 5$, and we're good to go. The demand equation is handled the same way: the extra instruments come from the supply equation.

Our model is a hypothetical, but one of the first applications of IVLS was to estimate supply and demand equations for butter (Wright 1928, p. 316). See Angrist and Krueger (2001) for discussion.

Exercise set C

1. An economist is specifying a model for the butter market in Illinois. She likes the model that we used for Wisconsin. She is willing to assume that the determinants of supply (wage rates and hay prices) are exogenous; also that the determinants of demand (toast prices and margarine prices) are exogenous. After reading sections 1–2 and looking at equation (10), she wants to use OLS not IVLS, and is therefore willing to assume that P_t is exogenous. What is your advice?

2. Let $e = Y - X\hat{\beta}_{IVLS}$ be the residuals from IVLS. True or false, and explain:

    (a) $\sum_i e_i = 0$.
    (b) $e \perp X$.
    (c) $\|Y\|^2 = \|X\hat{\beta}_{IVLS}\|^2 + \|e\|^2$.
    (d) $\hat{\sigma}^2 = \|e\|^2/(n-p)$.

3. Which is smaller, $\|Y - X\hat{\beta}_{IVLS}\|^2$ or $\|Y - X\hat{\beta}_{OLS}\|^2$? Discuss briefly.

4. Is $\hat{\beta}_{IVLS}$ biased or unbiased? What about $\hat{\sigma}^2 = \|Y - X\hat{\beta}_{IVLS}\|^2/(n-p)$ as an estimator for σ^2?

8.4 What are the two stages?

In the olden days, the model (4) was estimated in two stages.

Stage I. Regress X on Z. (This first-stage regression can be done one column at a time.) The fitted values are $\hat{X} = Z\hat{\gamma}$, where $\hat{\gamma} = (Z'Z)^{-1}Z'X$.

Stage II. Regress Y on $\hat{X}$.

In short,

(15) $$\hat{\beta}_{IISLS} = (\hat{X}'\hat{X})^{-1}\hat{X}'Y.$$

The idea: $\hat{X}$ is almost a function of Z, and has been "purged" of endogeneity. If you just sit down at the computer and run regressions, however, you may get the wrong SEs. The computer estimates σ^2 as $\|Y - \hat{X}\hat{\beta}_{\mathrm{IISLS}}\|^2/(n-p)$, but you want $\|Y - X\hat{\beta}_{\mathrm{IISLS}}\|^2/(n-p)$, without the hat on the X. The fix is easy, once you know the problem: get the residuals from (4).

By slightly tedious algebra, $\hat{\beta}_{\mathrm{IISLS}} = \hat{\beta}_{\mathrm{IVLS}}$. To begin the argument, let $H_Z = Z(Z'Z)^{-1}Z'$. The IVLS estimator in (10) can be rewritten as

$$(16) \quad \hat{\beta}_{\mathrm{IVLS}} = (X'H_Z X)^{-1}X'H_Z Y = \left[(H_Z X)'(H_Z X)\right]^{-1}(H_Z X)'(H_Z Y).$$

Indeed, H_Z is the hat matrix which projects onto the column space of Z (section 4.2). So H_Z is a symmetric idempotent matrix:

$$(H_Z X)'(H_Z X) = X'H_Z X \quad \text{and} \quad (H_Z X)'(H_Z Y) = X'H_Z,$$

proving (16). Next, we claim that

$$\hat{\beta}_{\mathrm{IVLS}} = \left[(H_Z X)'(H_Z X)\right]^{-1}(H_Z X)'Y.$$

This follows from (16), because $Y = H_Z Y + (I_{n \times n} - H_Z)Y$ and

$$(H_Z X)'(I_{n \times n} - H_Z)Y = X'(H_Z - H_Z)Y = 0_{p \times 1}.$$

In short, regressing Y on $H_Z X$ gives $\hat{\beta}_{\mathrm{IVLS}}$. But that is also the recipe for $\hat{\beta}_{\mathrm{IISLS}}$: the fitted values in Stage I are $H_Z X = \hat{X}$. The proof that $\hat{\beta}_{\mathrm{IISLS}} = \hat{\beta}_{\mathrm{IVLS}}$ is complete. The message of this section: old-fashioned IISLS coincides with new-fangled IVLS.

Invariance assumptions

Invariance assumptions need to be made in order to draw causal conclusions from non-experimental data: parameters are invariant—unchanging—under interventions, and so are errors or their distributions (sections 5.4–5). Exogeneity is another concern. In a real example, as opposed to a hypothetical about butter, real questions would have to be asked about these assumptions. Why are the equations "structural," in the sense that the required invariance assumptions hold true? Applied papers seldom address such assumptions, or the narrower statistical assumptions: for instance, why are errors IID?

The tension here is worth considering. We want to use regression to draw causal inferences from non-experimental data. To do that, we need to know that certain parameters and certain distributions would remain invariant if we were to intervene. Invariance can seldom be demonstrated experimentally. If it could, we probably wouldn't be discussing invariance assumptions, at least in that application. What then is the source of the knowledge?

"Economic theory" seems like a natural answer, but an incomplete one. Theory has to be anchored in reality. Sooner or later, invariance needs empirical demonstration, which is easier said than done. Outside of economics, the situation is perhaps even less satisfactory, because theory is less well developed, interventions are harder to define, and the hypothetical experiments are murkier.

8.5 A social-science example: education and fertility

Simultaneous equations are often used to model reciprocal causation—U influences V, and V influences U. Here is an example. Rindfuss et al (1980) propose a simultaneous-equations model to explain the process by which a woman decides how much education to get, and when to have children. The authors' explanation is as follows.

> "The interplay between education and fertility has a significant influence on the roles women occupy, when in their life cycle they occupy these roles, and the length of time spent in these roles. . . . This paper explores the theoretical linkages between education and fertility. . . . It is found that the reciprocal relationship between education and age at first birth is dominated by the effect from education to age at first birth with only a trivial effect in the other direction.
>
> "No factor has a greater impact on the roles women occupy than maternity. Whether a woman becomes a mother, the age at which she does so, and the timing and number of subsequent births set the conditions under which other roles are assumed. . . . Education is another prime factor conditioning female roles. . . .
>
> "The overall relationship between education and fertility has its roots at some unspecified point in adolescence, or perhaps even earlier. At this point aspirations for educational attainment as a goal in itself and for adult roles that have implications for educational attainment first emerge. The desire for education as a measure of status and ability in academic work may encourage women to select occupational goals that require a high level of educational attainment. Conversely, particular occupational or role aspirations may set standards of education that must be achieved. The obverse is true for those with either low educational or occupational goals. Also, occupational and educational aspirations are affected by a number of prior factors, such as mother's education, father's education, family income, intellectual ability, prior educational experience, race, and number of siblings. . . ."

The model used by Rindfuss et al (the paper is reprinted at the back of the book) consists of two linear equations in two unknowns, ED and AGE:

(17) $$ED = a\,AGE + A_i + \delta_i,$$

(18) $$AGE = a'ED + A'_i + \epsilon_i.$$

According to the model, a woman—indexed by the subscript i—chooses her educational level ED_i and age at first birth AGE_i as if by solving the two equations for the two unknowns.

These equations are response schedules (section 4.5). The coefficients a and a' are parameters, to be estimated from the data. (Here, the prime doesn't denote a transpose: a is scalar.) The terms A_i and A_i' take background factors into account:

$$(19) \qquad A_i = a_0 + b\,OCC_i + c_1 RACE_i + \cdots + c_7 YCIG_i,$$
$$(20) \qquad A_i' = a_0' + b'\,FEC_i + c_1' RACE_i + \cdots + c_7' YCIG_i.$$

Variables are defined in table 1, and the notes to the table describe the sample survey that collected the data.

The parameters $a_0, b, c_1, \ldots$ are to be estimated from the data. The random errors (δ_i, ϵ_i) are assumed to have mean 0, and (as pairs) to be indepen-

Table 1. Variables in the model (Rindfuss et al 1980).

The endogenous variables	
ED	Respondent's education
	(Years of schooling completed at first marriage)
AGE	Respondent's age at first birth
The exogenous variables	
OCC	Respondent's father's occupation
RACE	Race of respondent (Black = 1, other = 0)
NOSIB	Respondent's number of siblings
FARM	Farm background (coded 1 if respondent grew up
	on a farm, else coded 0)
REGN	Region where respondent grew up (South = 1, other = 0)
ADOLF	Broken family (coded 0 if both parents present
	when respondent was 14, else coded 1)
REL	Religion (Catholic = 1, other = 0)
YCIG	Smoking (coded 1 if respondent smoked before age 16,
	else coded 0)
FEC	Fecundability (coded 1 if respondent had
	a miscarriage before first birth; else coded 0)

Notes: The data are from a probability sample of 1766 women 35–44 years of age residing in the continental United States. The sample was restricted to ever-married women with at least one child. OCC was measured on Duncan's scale (section 5.1), combining information on education and income. Notation differs from Rindfuss et al.

dent and identically distributed from woman to woman. The model allows δ_i and ϵ_i to be correlated; δ_i may have a different distribution from ϵ_i.

Rindfuss et al use two-stage least squares to fit the equations. Notice that they have excluded FEC from equation (19), and OCC from equation (20). Without these identifying restrictions, the system would be under-identified (section 2 above).

The main empirical finding is this. The estimated coefficient of AGE in (17) is not statistically significant, i.e., a could be zero. The woman who dropped out of school because she got pregnant at age 16 would have dropped out anyway: the causal arrow points from ED to AGE, not the other way.

This finding depends on the model. When looked at coldly, the argument may seem implausible. A critique can be given along the following lines.

(i) Assumptions about the errors. Why are the errors independent and identically distributed across the women? Independence may be reasonable, but heterogeneity is more plausible than homogeneity.

(ii) Omitted variables. Important variables have been omitted from the model, including two that were identified by Rindfuss et al themselves—aspirations and intellectual ability. (See the quotes at the beginning of the section.) Since Malthus (1798), it has been considered that wealth is an important factor in determining education and marriage. Wealth is not in the model. Social class matters, and OCC measures only one of its aspects.

(iii) Why additive linear effects?

(iv) Constant coefficients. Rindfuss et al are assuming that the same parameters apply to all women alike, from poor blacks in the cities of the Northeast to rich whites in the suburbs of the West. Why?

(v) Are FEC and OCC exogenous?

(vi) What about the identifying restrictions?

(vii) Are the equations structural?

It is easier to think about questions (v–vii) in the context of a model that restricts attention to a more homogeneous group of women, where the only relevant background factors are OCC and FEC. The equations—response schedules—are as follows.

(21) $$ED = a_0 + a\,AGE + b\,OCC + \delta_i,$$

(22) $$AGE = a_0' + a'\,ED + b'\,FEC + \epsilon_i.$$

What do these equations tell us? Two hypothetical experiments help answer this question. In both experiments, fathers are assigned to jobs; and daughters are assigned to have a miscarriage before giving birth to their first child (FEC = 1), or not to have a miscarriage (FEC = 0).

> Experiment #1. Daughters are assigned to the various levels of AGE. ED is observed as the response. In other words, the hypothetical experimenter chooses when the woman has her first child, but allows her to decide when to leave school.

> Experiment #2. Daughters are assigned to the various levels of ED. Then AGE is observed as the response. The hypothetical experimenter decides when the woman has had enough education, but lets her have a baby when she wants to.

The statistical terminology is rather dry. The experimenter makes fathers do one job rather than another: surgeons cut pastrami sandwiches and taxi drivers run the central banks. Women are made to miscarry at one time and have their first child at another.

The equations can now be translated. According to (21), in the first experiment, ED does not depend on FEC. (That is one of the identifying restrictions assumed by Rindfuss et al.) Moreover, ED depends linearly on AGE and OCC, plus an additive random error. According to (22), in the second experiment, AGE does not depend on OCC. (That is the other identifying restriction assumed by Rindfuss et al.) Moreover, AGE depends linearly on ED and FEC, plus an additive random error. Even for thought experiments, this is a little fanciful.

We return now to the full model, equations (17–20). The data were collected in a sample survey, not an experiment (notes to table 1). Rindfuss et al must be assuming that Nature assigned OCC, RACE, . . . , FEC independently of the disturbance terms δ and ϵ in (17) and (18). That assumption is what makes OCC, RACE, . . . , FEC exogenous. Rindfuss et al must further be assuming that women chose ED and AGE as if by solving the two equations (17) and (18) for the two unknowns, ED and AGE. Without this assumption, simultaneous-equation modeling seems irrelevant. (The comparable element in the butter model is the law of supply and demand.)

The equations estimated from the survey data must apply as well to experimental situations where ED and AGE are manipulated. For instance, women who freely choose their educational levels and times to have children do so using the same pair of equations—with the same parameter values and error terms—as women made to give birth at certain ages. The data analysis in the paper doesn't justify such assumptions: how could it? But these constancy assumptions are the basis for causal inference from non-experimental data.

Without the response schedules, it is hard to see what "effects" might mean, apart from slopes of some plane that has been fitted to survey data. It would remain unclear why the plane should be fitted by two-stage least squares, or what role the significance tests are playing. Rindfuss et al have an interesting question, and there is much wisdom in their paper. But they have not demonstrated a connection between the social problem they are studying and the statistical technique they are using.

Simultaneous equations that derive from response schedules are structural. Structural equations describe real observational studies, and the hypothetical experiments that usually remain behind the scenes. Unless equations are structural, they have no causal implications (section 5.5).

More on Rindfuss et al

Rindfuss et al have arguments to support their position, but these are not convincing. For instance, exogeneity is discussed in the paper, and in Rindfuss and St. John (1983). However, the discussion misses the critical point: variables labelled as "instrumental" or "exogenous," like OCC, RACE, ..., FEC, need to be independent of the error terms. Why would that be so? Moreover, justifications given for the identifying restrictions seem artificial.

Hofferth and Moore (1979, 1980) obtain different results using different instruments, as noted by Hofferth (1984). Rindfuss et al (1984) say that

> "instrumental variables.... require strong theoretical assumptions.... and can give quite different results when alternative assumptions are made.... it is usually difficult to argue that behavioral variables are truly exogenous and that they affect only one of the endogenous variables but not the other." [pp. 981–82]

Thus, results depend quite strongly on assumptions about identifying restrictions and exogeneity, and there is no good way to justify one set of assumptions rather than another. Also see Bartels (1991), who comments on the impact of exogeneity assumptions and the difficulty of verification. Rindfuss and St. John (1983) give useful detail on the model. There is an interesting exchange between Geronimus and Korenman (1993) and Hoffman et al (1993) on the costs of teenage pregnancy.

8.6 Covariates

In the butter hypothetical, we could take the exogenous variables as non-manipulable covariates. The assumption would be that Nature chooses

$$(W_t, H_t, T_t, M_t) : t = 1, \ldots, 20$$

independently of the random error terms

$$(\delta_t, \epsilon_t) : t = 1, \ldots, 20.$$

The error terms would still be assumed IID (as pairs) with mean 0, and a 2×2 covariance matrix. We still have two hypothetical experiments: (i) set the price P to farmers, and see how much butter comes to market; (ii) set the price P to consumers and see how much butter is bought. By assumption, the answer to (i) is still

(23) $Q = a_0 + a_1 P + a_2 W_t + a_3 H_t + \delta_t,$

while the answer to (ii) is

(24) $Q = b_0 + b_1 P + b_2 T_t + b_3 M_t + \epsilon_t.$

For the observational data, we would still need to assume that Q_t and P_t in year t are determined as if by solving (23) and (24) for the two unknowns, Q and P, which gets us back to (2a) and (2b).

With Rindfuss et al, OCC, RACE, ..., FEC could be taken as non-manipulable covariates, eliminating some of the difficulty in the hypothetical experiments. The identifying restrictions—FEC is excluded from (19) and OCC from (20)—remain mysterious, as does the assumed linearity. How could you verify such assumptions?

Terminology. Often, "covariate" just means a right hand side variable in a regression equation—especially if that variable is only included to control for a possible confounder. Sometimes, "covariate" signifies a non-manipulable characteristic, like age or sex. Non-manipulable variables are occasionally called "concomitants."

8.7 Linear probability models

Schneider et al (1997) use two-stage least squares—with lots of bells and whistles—to study the effects of school choice on social capital. (The paper is reprinted at the back of the book.) "Linear probability models" are used to control for confounders and self-selection. The estimation strategy is quite intricate. Let's set the details aside, and think about the logic. First, here is what Schneider et al say they're doing, and what they found:

> "While the possible decline in the level of social capital in the United States has received considerable attention by scholars such as Putnam and Fukuyama, less attention has been paid to the local activities of citizens that help define a nation's stock of social capital . . . . giving parents greater choice over the public schools their children attend creates incentives for parents as

'citizen/consumers' to engage in activities that build social capital. Our empirical analysis employs a quasi-experimental approach .... the design of governmental institutions can create incentives for individuals to engage in activities that increase social capital .... active participation in school choice increases levels of involvement with voluntary organizations.... School choice can help build social capital."

Social capital is a very complicated concept, and quantification is more of a challenge than Schneider et al are willing to recognize. PTA membership—one measure of social capital, according to Schneider et al—is closer to ground level. (PTA means Parent-Teachers Association.) Schneider et al suggest that school choice promotes PTA membership. They want to prove this by running regressions on observational data. We'll look at results in their tables 1–2.

The analysis involves about 600 families with children in school in New York school districts 1 and 4. Schneider et al find that "active choosers" are more likely to be PTA members, other things being equal. Is this causation, or self-selection? The sort of parents who exercise choice might be the sort of parents who go to PTA meetings. The investigators use a two-stage model to correct for self-selection—like Evans and Schwab, but with a linear specification instead of probits.

There are statistical controls for universal choice, dissatisfaction, school size, black, Hispanic, Asian, length of residence, education, employed, female, church attendance (table 2 in the paper). School size, length of residence, and education are continuous variables. So is church attendance: frequency of attendance is scaled from 1 to 7. The other variables are all dummies. "Universal choice" is 1 for families in district 4, and 0 in district 1. "Dissatisfaction" is 1 if the parents often think about moving the child to another school, and 0 otherwise: note 12 in the paper. The statistical controls for family i are denoted W_i. (The paper uses different notation.)

The dummy variable Y_i is 1 if family i exercises school choice. There is another dummy Z_i for PTA membership. The object is to show by instrumental variables that in some sense, Y_i influences Z_i. There are two instruments, both dummy variables: when choosing a school, did the parents think its values mattered? did they think the diversity of the student body mattered? The instrumental variables for family i are denoted X_i.

The assumptions

Each family (indexed by i) has a pair of latent variables (U_i, V_i), with $E(U_i) = E(V_i) = 0$. The (U_i, V_i) are IID across families i, but U_i and V_i may be correlated. The (U_i, V_i) are independent of the (X_i, W_i). The social physics is like this (figure 3):

Figure 3. PTA membership explained.

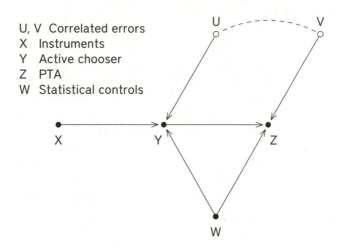

U, V Correlated errors
X   Instruments
Y   Active chooser
Z   PTA
W   Statistical controls

$$(25) \qquad P\{Y_i = 1 \mid X, W, U, V\} = X_i a + W_i b + U_i,$$
$$(26) \qquad P\{Z_i = 1 \mid y, X, W, U, V\} = cy + W_i d + V_i.$$

Here, X is the $n \times 2$ matrix whose ith row is X_i, and so forth. Given X, W, U, V, the response variables (Y_i, Z_i) are independent in i.

Equation (25) is an "assignment equation." The assignment equation says how likely it is for family i to exercise school choice. Equation (26) is a response schedule. The response schedule says what happens to family i if y is set to 1 or to 0—be an active chooser or not. (Remember, $Y_i = 1$ if family i exercises school choice, and $Z_i = 1$ if the parents are PTA members; it might be better to write $Z_{i,y}$, but we won't.) The crucial parameter in (26) is c, the effect of active choice on PTA membership. This c is scalar; a, b, d are vectors because X_i, W_i are vectors. Equations (25) and (26) are "linear probability models:" probabilities are expressed as linear combinations of control variables, plus latent variables that are meant to capture unmeasured personal characteristics. In the bivariate probit model for Catholic schools, the assignment equation is (6.9) and the response schedule is (6.8).

Equations (1) and (2) in the paper look different from (25) and (26). They are different. In (1), well, Schneider et al aren't distinguishing between $Y = 1$ and $P(Y = 1)$. Equation (2) has the same defect. Furthermore, the equation is part of a fitting algorithm rather than a model. The algorithm involves two-stage least squares. That is why "predicted active chooser" appears on the right hand side of the equation. ("Active choosers" are parents who exercise school choice: their children don't go to the default local public school.)

The vision behind (25) and (26) is this. Families have X_i's and W_i's. Whatever these may be—here comes the exogeneity assumption—Nature chooses (U_i, V_i) as IID pairs from a certain fixed probability distribution, which is unknown to us. Having chosen all these variables, Nature then flips a coin to see if $Y_i = 1$ or 0. According to (25), the probability that $Y_i = 1$ is $X_i a + W_i b + U_i$. Nature is supposed to take the Y_i she just generated, and plug it in for y in (26). Then she flips a coin to see if $Z_i = 1$. According to (26), the probability that $Z_i = 1$ is $cY_i + W_i d + V_i$.

We do not get to see the parameters a, b, c, d or the latent variables U_i, V_i. All we get to see is X_i, W_i, Y_i, Z_i. Still, Schneider et al can estimate c by some complicated version of two-stage least squares: $\hat{c} = 0.128$ and $\widehat{SE} = 0.064$, so $t = 0.128/0.064 = 2$ and $P = 0.05$. (See table 2 in the paper.) School choice matters. QED.

The questions

This paper leaves too many loose ends to be convincing. Why are the variables used as instruments independent of the latent variables? For that matter, what makes the control variables independent of the latent variables? Why are the latent variables IID across subjects? Where does linearity come from? Why are the parameters a, b, c, d the same for all subjects? What justifies the identifying restriction—no X on the right hand side of (26)?

The questions keep coming. Table B1 indicates that the dummies for dissatisfaction and district 4 were excluded from the assignment equation; so was school size. Why? There are 580 subjects in the PTA model (table 1 in Schneider et al). What about the other $400 + 401 - 580 = 221$ respondents (table A1)? Or the $113 + 522 + 225 + 1642 = 2502$ non-respondents? At a more basic level, what intervention are Schneider et al talking about, and what suggests stability under interventions? As with previous examples— Evans and Schwab, Rindfuss et al—there is a disconnect between the research questions and the data processing.

Exercise set D

Schneider et al is reprinted at the back of the book. The estimated coefficient for school size reported in table 2 is -0.000; i.e., the estimate was somewhere between 0 and -0.0005. When doing the numbers in exercises 1 and 2, you may assume the estimate is -0.0003.

1. Using the data in table 2 of Schneider et al, estimate the probability that a respondent with the following characteristics will be a PTA member: (i) active chooser, (ii) lives in district 1, (iii) dissatisfied, (iv) child attends a school which has 300 students, (v) black, (vi) lived in district 1 for 11 years before survey, (vii) completed 12 years of schooling, (viii) employed, (ix) female, (x) atheist—never goes to church—never!!

2. Repeat, for a respondent who is not an active chooser but has otherwise the same characteristics as the respondent in exercise 1.

3. What is the difference between the numbers for the two respondents in exercises 1 and 2? How do Schneider et al interpret the difference?

4. Given the model, the numbers you have computed for the two respondents in exercises 1 and 2 are best interpreted as _____. Options:

   probabilities   estimated probabilities   estimated expected probabilities

5. What is it in the data that makes the coefficient of school size so close to 0? (For instance, would -0.3 be feasible?)

6. Do equations (1) and (2) in the paper state the model?

7. (a) Does table 1 in Schneider et al show the sample is representative or unrepresentative?
   (b) What percentage of the sample had incomes below $20,000?
   (c) Why isn't there an income variable in table 2? table B1?
   (d) To what extent have Schneider et al stated the model? the statistical assumptions?
   (e) Are Schneider et al trying to estimate the effect of an intervention? If so, what is that intervention?

8.8 More on IVLS

This section looks at some fine points in the theory of IVLS. Exercise set E is hard, but depends only on the material in sections 2–4. After the exercises, there are some computer simulations to illustrate the twists and turns; IVLS is described in the multivariate normal case. There are suggestions for further reading.

Some technical issues

(i) Initially, more instruments may be better; but if q is too close to n, then $\hat{X} \doteq X$ and IISLS may not do much purging.

(ii) The OLS estimator has smaller variance than IVLS, sometimes to the extent that OLS winds up with smaller mean squared error than IVLS:

(simultaneity bias)^2 + OLS variance $<$ (small-sample bias)^2 + IVLS variance.

There is a mathematical inequality for the asymptotic variance-covariance matrices:

$$\widehat{\text{cov}}\,(\hat{\beta}_{\text{OLS}}|X) \leq \widehat{\text{cov}}\,(\hat{\beta}_{\text{IVLS}}|Z)$$

where $A \leq B$ means that $B - A$ is non-negative definite. As noted in exercise C3, OLS has the smaller $\hat{\sigma}^2$. Next, $Z(Z'Z)^{-1}Z'$ is the projection matrix onto the columns of Z, so

$$Z(Z'Z)^{-1}Z' \leq I_{n \times n},$$
$$X'Z(Z'Z)^{-1}Z'X \leq X'I_{n \times n}X = X'X,$$
$$[X'Z(Z'Z)^{-1}Z'X]^{-1} \geq (X'X)^{-1}.$$

Equation (12) completes the argument.

(iii) If the instruments are only weakly related to the endogenous variables, the randomness in $Z'X$ can be similar in size to the randomness in X. Then small-sample bias can be quite large—even when the sample is large (Bound et al 1995).

(iv) If $Z'Z$ is nearly singular, that can also make trouble.

(v) Even after conditioning on Z, the means and variances of matrices like

$$[X'Z(Z'Z)^{-1}Z'X]^{-1}$$

can be infinite—due to the inverses. That is one reason for talking about "asymptotic" means and variances.

(vi) Theoretical treatments of IVLS usually assume that n is large, while p and q are relatively small; furthermore, $Z'Z \doteq nA$ and $Z'X \doteq nB$, where A is $q \times q$ positive definite and B is $q \times p$ of rank p. Difficulties listed above are then precluded. For example, if Z_i is the ith row of Z, and X_i is the ith row of X, it might be assumed that the triplets (Z_i, X_i, δ_i) are IID with several moments; that $Z_i \perp\!\!\!\perp \delta_i$; that $E(\delta_i) = 0$; that $Y_i = X_i \beta + \delta_i$; that $E(Z_i'Z_i)$ is non-singular and $E(Z_i'X_i)$ has rank p. Under these circumstances, the IV estimator given by (10) is asymptotically normal; the asymptotic mean is β and the asymptotic covariance is given by (12). We write $N(0_{p \times 1}, I_{p \times p})$ for the joint distribution of p independent $N(0, 1)$ variables.

THEOREM 1. Under regularity conditions like those sketched above,

$$\hat{\sigma}^{-1}[X'Z(Z'Z)^{-1}Z'X]^{1/2}(\hat{\beta}_{\text{IVLS}} - \beta)$$

is asymptotically $N(0_{p \times 1}, I_{p \times p})$ as n gets large.

Exercise set E

This a set of hard exercises; the first three use the notation of section 2.

1. Pretend $Z'X$ is constant. Complete the argument for (12) by showing that

$$\text{cov}(\hat{\beta}_{\text{IVLS}}|Z) = \sigma^2 [X'Z(Z'Z)^{-1}Z'X]^{-1}.$$

2. Verify that $\hat{\beta}_{\text{IVLS}} = (Z'X)^{-1}Z'Y$ in the just-identified case ($q = p$). In particular, OLS is a special case of IISLS, with $Z = X$.

3. A chance for bonus points. Three investigators are studying the following model: $Y_i = X_i\beta + \epsilon_i$ for $i = 1, \ldots, n$. The random variables are all scalar, as is the unknown parameter β. The unobservable ϵ_i are IID with mean 0 and finite variance, but X is endogenous. Fortunately, the investigators also have an $n \times 1$ vector Z, which is exogenous and not orthogonal to X. Investigator #1 wishes to fit the model by OLS. Investigator #2 wants to regress Y on X and Z; the coefficient of X in this multiple regression would be the estimator for β. Investigator #3 suggests $\hat{\beta} = Z'Y/Z'X$. Which of the three estimators would you recommend? To focus the discussion, let's assume that $(X_i, Y_i, Z_i, \epsilon_i)$ are IID four-tuples, jointly normal, mean 0, and $\text{var}(X_i) = \text{var}(Z_i) = 1$. Let's also assume that n is large.

4. Another chance for bonus points. Suppose that $(X_i, Y_i, Z_i, \epsilon_i)$ are independent four-tuples of scalar random variables for $i = 1, \ldots, n$, with a common jointly normal distribution. All means are 0 and n is large. Suppose further that $Y_i = X_i\beta + \epsilon_i$. The variables X_i, Y_i, Z_i are observable, and every pair of them has a positive correlation which is less than 1. However, ϵ_i is not observable, and β is an unknown constant. Is the correlation between Z_i and ϵ_i identifiable? Can Z be used as an instrument for estimating β? Explain briefly.

5. Last chance for bonus points. In the over-identified case, what's wrong with estimating σ^2 from the residuals in (6), normalizing by $q - p$?

Simulations to illustrate IVLS

Let $(Z_i, \delta_i, \epsilon_i)$ be IID jointly normal with mean 0. Here, δ_i and ϵ_i are scalars, but Z_i is $1 \times q$, with $q \geq 1$. Assume $Z_i \perp\!\!\!\perp (\delta_i, \epsilon_i)$, the components of Z_i are independent with variance 1, but $\text{cov}(\delta_i, \epsilon_i)$ may not vanish. Let C be a fixed $q \times 1$ matrix, with $\|C\| > 0$. Let $X_i = Z_iC + \delta_i$, a scalar random variable: in the notation of section 2, $p = 1$. The model is

$$Y_i = X_i\beta + \epsilon_i \text{ for } i = 1, \ldots, n.$$

We stack in the usual way: Y_i is the ith component of the vector Y and ϵ_i is the ith component of the vector ϵ, while X_i is the ith row of the matrix X and Z_i is the ith row of the matrix Z. Thus, Z is exogenous ($Z \perp\!\!\!\perp \epsilon$) and X is endogenous unless $\mathrm{cov}(\delta_i, \epsilon_i) = 0$. We can estimate the scalar parameter β by OLS or IVLS and compare the MSEs. Generally, OLS will be inconsistent, due to simultaneity bias; IVLS will be consistent. If n is small or $\|C\|$ is small, then small-sample bias will be an issue. We can also compare methods for estimating $\mathrm{var}(\epsilon_i)$.

Ideally, IISLS would replace X_i by $Z_i C$. However, C isn't known. The actual estimator replaces X_i by $Z_i \hat{C}$, where $\hat{C}$ is obtained by regressing X on Z. Since X is endogenous, so is $\hat{C}$, and this is the source of small-sample bias. When n is large, $\hat{C} \doteq C$, and the problem goes away. If $p > 1$, then X_i and δ_i should be $1 \times p$, β should be $p \times 1$, C should be $q \times p$. We would require $q \geq p$ and $\mathrm{rank}(C) = p$.

Terminology. As the sample size gets large, a *consistent* estimator converges to the truth; an *inconsistent* estimator does not. This differs from ordinary English usage.

Further reading on econometric technique

Greene WH (2003). *Econometric Analysis.* 5th ed. Prentice Hall. A standard graduate-level textbook. Broad coverage. Theoretical.

Kennedy P (1998). *A Guide to Econometrics.* 4th ed. MIT Press. Informal, clear, useful.

Maddala GS (1992). *Introduction to Econometrics.* 2nd ed. Macmillan. Chatty and clear. 3rd ed. Wiley (2001).

Theil H (1971). *Principles of Econometrics.* Wiley. This is a formal treatment, but it is clear and accurate. On pp. 444, 451, Theil writes X for the matrix of instrumental variables, Z for the endogenous variables, which is opposite to the convention adopted here.

Wooldridge JM (2003). *Introductory Econometrics.* 2nd ed. Thomson. A standard undergraduate textbook. Applied focus.

8.9 Issues in statistical modeling

It is an article of faith in much applied work that disturbance terms are IID—Independent and Identically Distributed—across observations. Sometimes, this assumption is replaced by other assumptions that are more complicated but equally artificial. For example, when observations are ordered in time, the disturbance terms ϵ_t are sometimes assumed to follow an "autoregression," e.g., $\epsilon_t = \lambda \epsilon_{t-1} + \delta_t$, where now λ is a parameter to be estimated, and it is the δ_t that are IID. However, there is an alternative that should always be kept in mind. Disturbances are DDD—Dependent and Differently Distributed—across subjects. In the autoregression, for example, the δ_t could easily be DDD, and introducing yet another model would only postpone the moment of truth.

A second article of faith for many applied workers is that functions are linear with coefficients that are constant across subjects. The alternative is that functions are non-linear, with coefficients (or parameters more generally) that vary across subjects. The dueling acronyms would be LCC (Linear with Constant Coefficients) and NLNC (Non-Linear with Non-constant Coefficients). Some models have "random coefficients," which only delays the inevitable. For every subject, coefficients are assumed to be drawn at random from the same distribution. Why would that be so?

When reading a statistical study, try to spot the questions it is asking, and the answers it gives to those questions. What data are used? What kind of statistical analysis gets the authors from the data to the answers? What are the assumptions behind the analysis? What is allowed to vary and what is taken as constant? Are the constancy assumptions plausible? If causal inferences are made from observational data, why are parameters invariant under interventions? Where are the response schedules? Do the response schedules describe reasonable thought experiments?

For applied workers who are going to publish research based on statistical models, the recommendation is to archive the data, the equations, and the programs. This would allow replication, at least in the narrowest sense of the term (Dewald et al 1986, Hubbard et al 1998). Assumptions should be made explicit. It should be made clear which assumptions were checked, and how the checking was done. It should also be made clear which assumptions were not checked.

From the modeling perspective, a variety of relevant techniques have not been considered in this book, including regression diagnostics, specification tests, and model selection procedures. These techniques might be helpful. For instance, diagnostics are seldom reported in applied papers, and should probably be used more often.

In the end, however, such things work only if there is some relatively localized breakdown in the modeling assumptions—a technical problem which has a technical fix. There is no way to infer the "right" model from the data unless there is strong prior theory to limit the universe of possible models. (More technically, diagnostics and specification tests usually have good power only against restricted classes of alternatives.) That kind of strong theory is rarely available in the social sciences.

Model selection procedures like AIC (Akaike's Information Criterion) only work—under suitable regularity conditions—"in the limit," as sample size goes to infinity. Even then, AIC overfits. Therefore, behavior in finite samples needs be assessed. Such assessments are unusual—especially by contrast with asymptotics for regression or the MLE. Moreover, AIC and the like are commonly used in cases where the regularity conditions do not hold, so operating characteristics of the procedures are unknown, even with very large samples. Specification tests are open to similar objections.

Bayesian methods are sometimes thought to solve the model selection problem (and other problems too). However, in non-parametric settings, even a strictly Bayesian approach can lead to inconsistency, often because of overfitting. "Priors" that have infinite mass or depend on the data merely cloud the issue. For reviews, see Diaconis and Freedman (1998), Eaton and Freedman (2004), Freedman (1995).

How does the bootstrap fit into this picture? The bootstrap is in many cases a helpful way to compute standard errors—*given* the model. Although the bootstrap usually cannot answer basic questions about validity of the model, it can sometimes be used to assess impacts of relatively minor failures in assumptions. The bootstrap has been used to create chance models from data sets, and some observers will find this pleasing. One take-home message from chapter 7: with feasible GLS, if many covariances are estimated from the data, plug-in standard errors need to be taken with correspondingly many grains of salt.

The difficulties in modeling are not unknown. For example, Hendry (1980, p. 390) writes that "Econometricians have found their Philosophers' Stone; it is called regression analysis and is used for transforming data into 'significant' results!" This seriously under-estimates the number of philosophers' stones. Hendry's position is more complicated than the quote might suggest. Other responses from the modeling perspective are quite predictable.

Philosophers' Stones: Late Twentieth Century

Correlation, partial correlation, Cross lagged correlation, Principal components, Factor analysis, OLS, GLS, PLS, IISLS, IIISLS, IVLS, FIML, LIML, SEM, GLM, HLM, HMM, GMM, ANOVA, MANOVA, Meta-analysis, Logits, Probits, Ridits, Tobits, RESET, DFITS, AIC, BIC, MAXENT, MDL, VAR, AR, ARIMA, ARFIMA, ARCH, GARCH, LISREL, Partial likelihood, Proportional hazards, Hinges, Froots, Flogs with median polish, MARS, LARS, LASSO, Neural nets, Expert systems, Bayesian expert systems, Ignorance priors, BUGS, EM, LM, MCMC, DAGs, TETRAD, TETRAD II. . . .

The Modelers' Responses

*We know all that. Nothing is perfect. Linearity has to be a good
first approximation. Log linearity has to be a good first approximation. The assumptions are reasonable. The assumptions don't
matter. The assumptions are conservative. You can't prove the assumptions are wrong. The biases will cancel. We can model the
biases. We're only doing what everybody else does. Now we use
more sophisticated techniques. If we don't do it, someone else will.
What would you do? The decision-maker has to be better off with
us than without us. We all have mental models, not using a model
is still a model. The models aren't totally useless. You have to do
the best you can with the data. You have to make assumptions in
order to make progress. You have to give the models the benefit of
the doubt. Where's the harm?*

8.10 Critical literature

For the better part of a century, many scholars in many different disci-
plines have expressed considerable skepticism about the possibility of disentangling complex causal processes by means of statistical modeling. We
review some of this critical literature here, and a bit of the positive literature.
Our starting-point is the exchange between Keynes (1939, 1940) and Tinbergen (1940). Tinbergen was one of the pioneers of econometric modeling.
Keynes expressed blank disbelief about this line of research:

> "No one could be more frank, more painstaking, more free from sub-
> jective bias or *parti pris* than Professor Tinbergen. There is no one,
> therefore, so far as human qualities go, whom it would be safer to trust
> with black magic. That there is anyone I would trust with it at the present
> stage, or that this brand of statistical alchemy is ripe to become a branch
> of science, I am not yet persuaded. But Newton, Boyle and Locke all
> played with alchemy. So let him continue." (Keynes 1940, p. 156)

Other familiar citations in the economics literature include Liu (1960),
Lucas (1976), and Sims (1980). Lucas was concerned about parameters
that changed under intervention. Manski (1995) returns to the problem of
under-identification that was posed so sharply by Liu and Sims. In brief, a
priori exclusion of variables from causal equations can seldom be justified, so
there will typically be more parameters than data. Manski suggests methods
for bounding quantities that cannot be estimated. Sims' idea was to use low-
dimensional models for policy analysis, instead of complex high-dimensional

ones. Leamer (1978) discusses the issues created by inferring specifications from the data, as does Hendry (1980). Engle, Hendry, and Richard (1983) distinguish several kinds of exogeneity assumptions, with different implications for causal inference.

Heckman (2000) traces the development of econometric thought from Haavelmo and Frisch onwards. Potential outcomes and structural parameters play a central role, but "the empirical track record of the structural [modeling] approach is, at best, mixed" [p. 49]. Instead, the fundamental contributions of econometrics are the insights

> "that causality is a property of a model, that many models may explain the same data and that assumptions must be made to identify causal or structural models...." [p. 89]

Moreover, econometricians have clarified "the possibility of interrelationships among causes," as well as "the conditional nature of causal knowledge and the impossibility of a purely empirical approach to analyzing causal questions" [pp. 89–90]. Heckman concludes that

> "The information in any body of data is usually too weak to eliminate competing causal explanations of the same phenomenon. There is no mechanical algorithm for producing a set of 'assumption free' facts or causal estimates based on those facts." [p. 91]

Some econometricians have turned to natural experiments for the evaluation of causal theories. These investigators stress the value of strong research designs, with careful data collection and thorough, context-specific, data analysis. Angrist and Krueger (2001) have a useful survey.

Rational choice theory is a frequently-offered justification for statistical modeling in economics and cognate fields. Therefore, any discussion of empirical foundations must take into account a remarkable series of papers, initiated by Kahneman and Tversky (1974), that explores the limits of rational choice theory. These papers are collected in Kahneman, Slovic, and Tversky (1982), Kahneman and Tversky (2000). The heuristics-and-biases program of Kahneman and Tversky has attracted its own critics (Gigerenzer 1996). The critique is interesting, and has some merit. But in the end, the experimental evidence demonstrates severe limits to the power of rational choice theory (Kahneman and Tversky 1996).

The experimental evidence shows that if people are trying to maximize expected utility, they don't do it very well. Errors are large and repetitive, go in predictable directions, and fall into recognizable categories. Rather than making decisions by optimization—or bounded rationality, or satisficing—people seem to use plausible heuristics that can be classified and analyzed.

Rational choice theory is generally not a good basis for justifying empirical models of behavior, because it does not describe the way real people make real choices.

Sen (2002), drawing in part on the work of Kahneman and Tversky, gives a far-reaching critique of rational choice theory, with many counter-examples to the assumptions. The theory has its place, according to Sen, but also leads to "serious descriptive and predictive problems" [p. 23]. Nelson and Winter (1982) reached similar conclusions in their study of firms and industries. The axioms of orthodox economic theorizing, profit maximization and equilibrium, create a "flagrant distortion of reality" [p. 21].

Almost from the beginning, there were critiques of modeling in other social sciences too. Bernert (1983) and Platt (1996) review the historical development in sociology. Abbott (1997) finds that variables like income and education are too abstract to have much explanatory power; so do models built on those variables. There is a broader examination of causal modeling in Abbott (1998). He finds that "an unthinking causalism today pervades our journals and limits our research" [p. 150]. He recommends more emphasis on descriptive work and on smaller-scale theories more tightly linked to observable facts—middle-range theories, in Robert Merton's useful phrase. Clogg and Haritou (1997) consider difficulties with regression, noting that you can all too easily include endogenous variables as regressors. Hedström and Swedberg (1998) present a lively collection of essays by a number of sociologists who are quite skeptical about regression models. Rational choice theory also takes its share of criticism.

Goldthorpe (1999, 2000, 2001) describes several ideas of causation and corresponding methods of statistical proof, which have different strengths and weaknesses. He is skeptical of regression, but finds rational choice theory to be promising—unlike other scholars cited above. He favors use of descriptive statistics to infer social regularities, and statistical models that reflect generative processes. He finds the manipulationist account of causation to be generally inadequate for the social sciences. Ní Bhrolcháin (2001) has some particularly forceful examples to illustrate the limits of modeling.

Lieberson (1985) finds that in social science, non-experimental data are routinely analyzed as if they had been generated experimentally, the typical mode of analysis being a regression model with some control variables. This enterprise has "no more merit than a quest for a perpetual-motion machine" [p. ix]. Finer-grain analytic methods are needed for causal inference, more closely adapted to the details of the problem at hand. The role of counter-factuals is explained (pp. 45–48). Lieberson and Lynn (2002) are equally skeptical about mimicking experimental control through complex

statistical models: simple analysis of natural experiments would be prefer-
able. Sobel (1998) reviews the literature on social stratification, concluding
that "the usual modeling strategies are in need of serious change" [p. 345].
Also see Sobel (2000). Like Lieberson, Berk (2004) doubts the possibility of
inferring causation by statistical modeling, absent a strong theoretical basis
for the models—which rarely is to be found.

Paul Meehl was a leading empirical psychologist. His 1954 book has
data showing the advantage of using regression, rather than experts, to make
predictions. On the other hand, his 1978 paper, "Theoretical risks and tabular
asterisks: Sir Karl, Sir Ronald, and the slow progress of soft psychology," saw
hypothesis tests—and cognate black arts—as stumbling blocks that slowed
the progress of psychology. Meehl and Waller (2002) is closer in spirit to
Meehl (1954). The 2002 paper discusses the choice between two similar
path models, viewed as reasonable approximations to some underlying causal
structure, but does not reach the critical question—how to assess the adequacy
of the approximations.

Steiger (2001) provides a critical review of structural equation models.
Larzalere et al (2004) offer a more general discussion of difficulties with
causal inference by purely statistical methods. Abelson (1995) has a distinctive viewpoint on statistics in psychology. There is a well-known book on the
logic of causal inference, by Cook and Campbell (1979). Also see Shadish,
Cook, and Campbell (2002), who have among other things a useful discussion
of manipulationist versus non-manipulationist ideas of causation.

In political science, Achen (1982, 1986) provides an interesting defense
of statistical models; Achen (2002) is substantially more skeptical. King,
Keohane, and Verba (1994) are well-known enthusiasts of regression. After
a careful review of the evidence, Green and Shapiro (1994) conclude "despite its enormous and growing prestige in the discipline, rational choice
theory has yet to deliver on its promise to advance the empirical study of politics" [p. 7]. Fearon (1991) discusses the role of counter-factuals. Brady and
Collier (2004) edited a volume of essays comparing regression methods with
case studies; invariance is discussed under the rubric of causal homogeneity.
The introductory chapter (Brady, Collier, and Seawright 2004) finds that

> "it is difficult to make causal inferences from observational data, espe-
> cially when research focuses on complex political processes. Behind the
> apparent precision of quantitative findings lie many potential problems
> concerning equivalence of cases, conceptualization and measurement,
> assumptions about the data, and choices about model specification. . . .
> The interpretability of quantitative findings is strongly constrained by
> the skill with which these problems are addressed." [pp. 9–10]

One of the difficulties with regression models is accounting for the ϵ's. Where do they come from, what do they mean, and why do they have the required statistical properties? Error terms are often said to represent the overall effects of factors omitted from the equation. But this characterization has problems of its own, as shown by Pratt and Schlaifer (1984, 1988).

In Holland (1986, 1988), there is a super-population model—rather than individualized error terms—to account for the randomness in causal models. However, justifying the super-population model is no easier than justifying assumptions about error terms. Stone (1993) presents a super-population model with some observed covariates and some unobserved; this paper is remarkable for its clarity.

Recently, strong claims have been made for non-linear methods that elicit the model from the data, and control for unobserved confounders, with little need for substantive knowledge (Spirtes-Glymour-Scheines 1993, Pearl 2000). However, the track record is not encouraging (Freedman 1997, 2004; Humphreys and Freedman 1996, 1999). There is a free-ranging discussion of such issues in McKim and Turner (1997). Other cites to the critical literature include Oakes (1986), Diaconis (1998), Freedman (1985, 1987, 1991, 1995, 1999, 2005).

Response schedules

The response-schedule model is the bridge between regression and causation, as discussed in section 5.4. This model was proposed by Neyman (1923). The paper is in Polish, but there is an English translation by Dabrowska and Speed in *Statistical Science* (1990), with discussion. The model has been rediscovered many times: see, for instance, Roy (1951) or Hodges and Lehmann (1964, section 9.4). The setup is often called "Rubin's model:" see for instance Holland (1986), who cites Rubin (1974). That simply mistakes the history.

Neyman's model covers observational studies—in effect, assuming these studies are experiments after suitable controls have been introduced. Indeed, Neyman does not require random assignment of treatments, assuming instead an urn model. The model is non-parametric, with a finite number of treatment levels. Response schedules were developed further by Holland and Rubin among others, with extensions to real-valued treatment variables and parametric models, including linear causal relationships.

As demonstrated in chapters 5–8, response schedules help clarify the process by which causation can be, under some circumstances, inferred by running regressions on observational data. The mathematical elegance of response schedules should not be permitted to obscure the basic issue. To what extent are the assumptions valid, for the applications of interest?

8.11 Evaluating the models in chapters 6–8

Chapter 6 discussed a probit model for the effect of Catholic schools (Evans and Schwab 1995). Chapter 8 considered a simultaneous-equation model for education and fertility (Rindfuss et al 1980), and a linear probability model for social capital (Schneider et al 1997). In each case, we found serious difficulties. The studies under review are at the high end of the social science literature. They were chosen for their strengths, not their weaknesses. The problems are not in the studies, but in the modeling technology. More precisely, bad things happen when the technology is applied to real problems—without validating the assumptions behind the models. Taking assumptions for granted is what makes statistical techniques into philosophers' stones.

8.12 Summing up

In the social and behavioral sciences, far-reaching claims are often made for the superiority of advanced quantitative methods—by those who manage to ignore the far-reaching assumptions behind the models. In section 8.10, we saw there was considerable skepticism about disentangling causal processes by statistical modeling. Earlier in the book, we examined several high-end modeling exercises, and discovered good reasons for skepticism. Some kinds of problems may yield to sophisticated statistical technique; others will not. The goal of empirical research is—or should be—to increase our understanding of the phenomena, rather than displaying our mastery of technique.

References

Abbott A (1997). Of time and space: The contemporary relevance of the Chicago school. *Social Forces* 75: 1149–82.

Abbott A (1998). The causal devolution. *Sociological Methods and Research* 27: 148–81.

Abelson RP (1995). *Statistics as Principled Argument*. Lawrence Erlbaum Associates, Hillsdale, NJ.

Achen CH (1977). Measuring representation: Perils of the correlation coefficient. *American Journal of Political Science* 21: 805–15.

Achen CH (1982). *Interpreting and Using Regression*. Sage Publications.

Achen CH (1986). *The Statistical Analysis of Quasi-Experiments*. University of California Press, Berkeley.

Achen C (2002). Toward a new political methodology: Microfoundations and ART. *Annual Review of Political Science* 5: 423–50.

Adcock R, Collier D (2001). Measurement validity: A shared standard for qualitative and quantitative research. *American Political Science Review* 95: 529–46.

Alba RD, Logan JR (1993). Minority proximity to whites in suburbs: An individual-level analysis of segregation. *American Journal of Sociology* 98: 1388–1427.

Alberts B, Bray D, Lewis J, Raff M, Roberts K and Watson JD (1994). *Molecular Biology of the Cell*, 3rd. ed., Garland Publishing, New York. 4th ed. (2002).

Anderson TW (1959). On asymptotic distributions of estimates of parameters of stochastic difference equations. *Annals of Mathematical Statistics* 30: 676–87.

Angrist JD, Krueger AB (2001). Instrumental variables and the search for identification: From supply and demand to natural experiments. *Journal of Economic Persepctives* 19: 2–16.

Banerjee KS (1975). *Weighing Designs for Chemistry, Medicine, Economics, Operations Research, Statistics*. Dekker, New York.

Barndorff-Nielsen O (1980). Exponential families. Memoir #5, Department of Theoretical Statistics, Aarhus. Reprinted in S Kotz, NL Johnson, CB Read, eds (1982). *Encyclopedia of Statistical Sciences*. Wiley 2: 587–96.

Bartels LM (1991). Instrumental and "quasi-instrumental" variables. *American Journal of Political Science* 35: 777–800.

Beck N (2001). Time-series cross-section data: What have we learned in the past few years? *Annual Review of Political Science* 4: 271–93.

Beck N, Katz JN (1995). What to do (and not to do) with time-series cross-section data. *American Political Science Review* 89: 634–47.

Berk RA (2004). *Regression Analysis: A Constructive Critique*. Sage Publications.

Berk RA, Freedman DA (2003). Statistical assumptions as empirical commitments. In TG Blomberg and S Cohen, eds. *Law, Punishment, and Social Control: Essays in Honor of Sheldon Messinger*, 2nd ed. Aldine de Gruyter, New York, pp. 235–54.

Berkson J (1955). The statistical study of association between smoking and lung cancer. *Proceedings of the Mayo Clinic* 30: 319–48.

Bernert C (1983). The career of causal analysis in American sociology. *British Journal of Sociology* 34: 230–54.

Bernheim B, Shleifer A, Summers LH (1985). The strategic bequest motive. *Journal of Political Economy* 93: 1045–1076.

Blalock HM (1989). The real and unrealized contributions of quantitative sociology. *American Sociological Review* 54: 447–60.

Blau PM, Duncan OD (1967). *The American Occupational Structure*. Wiley. Reissued by the Free Press (1978). Data collection described on page 13, coding of status on pages 115–27, coding of education on pages 165–66, correlations and path diagram on pages 169–70.

Bound J, Jaeger DA, Baker RM (1995). Problems with instrumental variables estimation when the correlation between the instruments and endogenous variables is weak. *Journal of the American Statistical Association* 90: 443–50.

Brady HE, Collier D, eds. (2004). *Rethinking Social Inquiry: Diverse Tools, Shared Standards*. Rowman & Littlefield Publishers, Inc., Lanham, Maryland.

Brady HE, Collier D, Seawright J (2004). Refocusing the discussion of methodology. In HE Brady and D Collier, eds. *Rethinking Social Inquiry: Diverse Tools, Shared Standards*. Rowman & Littlefield Publishers, Inc., Lanham, Maryland, chapter 1.

Briggs DC (2004). Causal inference and the Heckman model. *Journal of Educational and Behavioral Statistics* 29: 397–420.

Cameron JM, Croarkin MC, Raybold RC (1977). *Designs for the Calibration of Standards of Mass*. National Bureau of Standards Technical Note 952.

Carpenter KJ (1981). *Pellagra*. Academic Press.

Clarke R, Armitage J (2002). Antioxidant vitamins and risk of cardiovascular disease. Review of large-scale randomised trials. *Cardiovascular Drugs and Therapy* 16: 411–15.

Clogg CC, Haritou A (1997). The regression method of causal inference and a dilemma confronting this method. In VR McKim and SP Turner, eds. *Causality in Crisis?* University of Notre Dame Press, pp. 83–112.

Coleman JS, Hoffer T (1987). *Public and Private High Schools: The Impact of Communities*. Basic Books, New York.

Coleman JS, Hoffer T, Kilgore S (1982). *High School Achievement: Public, Catholic, and Private Schools Compared*. Basic Books, New York.

Colwell RR (1996). Global climate and infectious disease: The cholera paradigm. *Science* 274: 2025–31.

Cook TD, Campbell DT (1979). *Quasi-Experimentation: Design & Analysis Issues for Field Settings*. Rand McNally, Chicago.

Cornfield J, Haenszel W, Hammond EC, Lilienfeld AM, Shimkin MB, Wynder EL (1959). Smoking and lung cancer: Recent evidence and a discussion of some questions. *Journal of the National Cancer Institute* 22: 173–203.

Davidson R, MacKinnon JG (2004). *Econometric Theory and Methods*. Oxford University Press.

Desrosières A (1993). *La politique des grands nombres: Histoire de la raison statistique*. Editions La Découverte, Paris. English translation by C Naish (1998), Harvard University Press.

Dewald WG, Thursby JG, Anderson RG (1986). Replication in empirical economics: The Journal of Money, Credit and Banking Project. *American Economic Review* 76: 587–603.

Diaconis P (1998). A place for philosophy? The rise of modeling in statistics. *Quarterly Journal of Applied Mathematics* 56: 797–805.

Diaconis P, Freedman DA (1998). Consistency of Bayes estimates for non-parametric regression: Normal theory. *Bernoulli Journal*, 4: 411–44.

Dijkstra TK, ed. (1988). *On Model Uncertainty and its Statistical Implications.* Lecture Notes in Economics and Mathematical Systems, No. 307, Springer.

Dorn HF (1950). Pitfalls in population forecasts and projections. *Journal of the American Statistical Association* 45: 311–34.

Doss H, Sethuraman J (1989). The price of bias reduction when there is no unbiased estimate. *Annals of Statistics* 17: 440–42.

Dubos R (1988). *Pasteur and Modern Science.* Springer.

Duncan OD (1984). *Notes on Social Measurement.* Russell Sage, New York.

Eaton ML, Freedman DA (2004). Dutch book against some "objective" priors. *Bernoulli Journal* 10: 861–72.

EC/IC Bypass Study Group (1985). Failure of extracranial-intracranial arterial bypass to reduce the risk of ischemic stroke: Results of international randomized trial. *New England Journal of Medicine* 313: 1191–1200. For commentary, see Sundt (1987).

Ehrenberg ASC, Bound JA (1993). Predictability and prediction. *Journal of the Royal Statistical Society* Series A 156: 167–206 (with discussion).

Engle RF, Hendry DF, Richard JF (1983). Exogeneity. *Econometrica* 51: 277–304.

Evans AS (1993). *Causation and Disease: A Chronological Journey.* Plenum, New York.

Evans RJ (1987). *Death in Hamburg: Society and Politics in the Cholera Years, 1830–1910.* Oxford University Press.

Evans WN, Schwab RM (1995). Finishing high school and starting college: Do Catholic schools make a difference? *Quarterly Journal of Economics* 110: 941–74.

Evans SN, Stark PB (2002). Inverse problems as statistics. *Inverse Problems* 18: R1–43.

Fearon J (1991). Counterfactuals and hypothesis testing in political science. *World Politics* 43: 169–95.

Feller W (1940). On the logistic law of growth and its empirical verifications in biology. *Acta Biotheoretica* 5: 51–66.

Feller W (1971). *An Introduction to Probability Theory and its Applications*. Vol. II, 2nd ed., Wiley, New York.

Fisher RA (1959). *Smoking: The Cancer Controversy*. Oliver and Boyd, Edinburgh.

Fleming A ed. (1946). *Penicillin: Its Practical Application*. Butterworth & Co., London.

Freedman DA (1981). Bootstrapping regression models. *Annals of Statistics* 9: 1218–28.

Freedman DA (1983). A note on screening regression equations. *American Statistician* 37: 152–55.

Freedman DA (1984). On bootstrapping two-stage least squares estimates in stationary linear models. *Annals of Statistics* 12: 827–42.

Freedman DA (1985). Statistics and the scientific method. In WM Mason and SE Fienberg, eds. *Cohort Analysis in Social Research: Beyond the Identification Problem*. Springer, pp. 343–90 (with discussion).

Freedman DA (1987). As others see us: A case study in path analysis. *Journal of Educational Statistics* 12: 101–223 (with discussion). Reprinted in J Shaffer, ed. (1992). *The Role of Models in Nonexperimental Social Science*, American Educational Research Association and American Statistical Association, Washington, DC, pp. 3–125.

Freedman DA (1991). Statistical models and shoe leather. In P Marsden, ed., *Sociological Methodology 1991*. American Sociological Association, Washington, DC, chapter 10 (with discussion).

Freedman DA (1995). Some issues in the foundation of statistics. *Foundations of Science* 1: 19–83 (with discussion). Reprinted in BC van Fraassen, ed. (1997). *Topics in the Foundation of Statistics*, Kluwer, Dordrecht, pp. 19–83.

Freedman DA (1997). From association to causation via regression. In VR McKim and SP Turner, eds. *Causality in Crisis?* University of Notre Dame Press, pp. 113–82 (with discussion). Reprinted in *Advances in Applied Mathematics* (1997) 18: 59–110.

Freedman DA (1999). From association to causation: Some remarks on the history of statistics. *Statistical Science* 14: 243–58. Reprinted in *Journal de la Société Française de Statistique* (1999) 140: 5–32 and in J Panaretos, ed (2003). *Stochastic Musings: Perspectives from the Pioneers of the Late 20th Century*. Lawrence Erlbaum Associates, Hillsdale, NJ, pp. 45–71.

Freedman DA (2004). Graphical models for causation, and the identification problem. *Evaluation Review* 28: 267–93. Reprinted in J Stock and D Wright, eds. (2005). *The Rothenberg Festschrift*. Cambridge University Press.

Freedman DA (2005). Linear statistical models for causation: A critical review. In B Everitt and D Howell, eds. *Wiley Encyclopedia of Statistics in Behavioral Science*.

Freedman DA, Lane D (1981). *Mathematical Methods in Statistics*. WW Norton, New York.

Freedman DA, Peters SC (1984a). Bootstrapping a regression equation: Some empirical results. *Journal of the American Statistical Association* 79: 97–106.

Freedman DA, Peters SC (1984b). Bootstrapping an econometric model: Some empirical results. *Journal of Business and Economic Statistics* 2: 150–58.

Freedman DA, Peters SC (1984c). Some notes on the bootstrap in regression problems. *Journal of Business and Economic Statistics* 2: 406–9.

Freedman DA, Peters SC (1985). Using the bootstrap to evaluate forecasting equations. *Journal of Forecasting* 4: 251–62.

Freedman DA, Petitti DB, Robins JM (2004). On the efficacy of screening for breast cancer. *International Journal of Epidemiology* 33: 43–73 (with discussion). Correspondence, pp. 1404–6.

Freedman DA, Pisani R, Purves RA (1998). *Statistics*. 3rd ed. WW Norton, New York.

Freedman DA, Rothenberg T, Sutch R (1983). On energy policy models. *Journal of Business and Economic Statistics* 1: 24–36 (with discussion).

Freedman DA, Stark PB (2003). What is the probability of an earthquake? In F Mulargia and RJ Geller, eds. *Earthquake Science and Seismic Risk Reduction*. NATO Science Series IV: Earth and Environmental Sciences, vol. 32, Kluwer, Dordrecht, The Netherlands, pp. 201–13.

Garrett G (1998). *Partisan Politics in the Global Economy*. Cambridge University Press.

Gauss CF (1809). *Theoria Motus Corporum Coelestium*. Perthes et Besser, Hamburg. Reprinted by Dover, New York (1963).

Geronimus AT, Korenman S (1993). The socioeconomic costs of teenage childbearing: Evidence and interpretation. *Demography* 30: 281–90.

Gibson JL (1988). Political intolerance and political repression during the McCarthy red scare. *American Political Science Review* 82: 511–29. Heinz Eulau Award from the American Political Science Association, as best paper published in 1988 in the *American Political Science Review*.

Gigerenzer G (1996). On narrow norms and vague heuristics. *Psychological Review* 103: 592–96.

Goldsmith ML (1946). *The Road to Penicillin: A History of Chemotherapy.* Drummond, London.

Goldthorpe JH (1999). *Causation, Statistics and Sociology.* Twenty-ninth Geary Lecture, Nuffield College, Oxford. Published by the Economic and Social Research Institute, Dublin, Ireland.

Goldthorpe JH (2000). *On Sociology: Numbers, Narratives, and Integration of Research and Theory.* Oxford University Press.

Goldthorpe JH (2001). Causation, statistics, and sociology. *European Sociological Review* 17: 1–20.

Gordis L (2004). *Epidemiology.* 3rd ed. Elsevier Saunders, Philadelphia.

Green DP, Shapiro I (1994). *Pathologies of Rational Choice Theory: A Critique of Applications in Political Science.* Yale University Press.

Grodstein F, Stampfer MJ, Manson JE et al (1996). Post menopausal estrogen and progestin use and the risk of cardiovascular disease. *New England Journal of Medicine* 335: 453–61.

Grogger J (1995). The effect of arrests on the employment and earnings of young men. *Quarterly Journal of Economics* 110: 51–71.

Hajnal J (1955). The prospects for population forecasts. *Journal of the American Statistical Association* 50: 309–22.

Hare R (1970). *The Birth of Penicillin and the Disarming of Microbes.* Allen & Unwin, London.

Hart HLA, Honoré AM (1985). *Causation in the Law.* 2nd ed. Oxford University Press.

Heckman JJ (1976). The common structure of statistical models of truncation, sample selection and limited dependent variables and a simple estimator for such models. *Annals of Economic and Social Measurement* 5: 475–92.

Heckman JJ (1978). Dummy endogenous variables in a simultaneous equation system. *Econometrica* 46: 931–59.

Heckman JJ (1979). Sample selection bias as a specification error. *Econometrica* 47: 153–61.

Heckman JJ (2000). Causal parameters and policy analysis in economics: A twentieth century retrospective. *The Quarterly Journal of Economics* 115: 45–97.

Hedström P, Swedberg R, eds. (1998). *Social Mechanisms.* Cambridge University Press.

Hendry DF (1980). Econometrics—alchemy or science? *Economica* 47: 387–406. Reprinted as chapter 1 in DF Hendry (2000). *Econometrics—Alchemy or Science?* Blackwell, Oxford.

Hercberg S et al (2004). The SU.VI.MAX study: A randomized, placebo-controlled trial of the health effects of antioxidant vitamins and minerals. *Arch Intern Med* 164: 2335–42.

Hodges JL Jr, Lehmann E (1964). *Basic Concepts of Probability and Statistics*. Holden-Day, San Francisco. 2nd ed., reprinted by SIAM, Philadelphia (2005).

Hofferth SL (1984). A comment on "social determinants of age at first birth." *Journal of Marriage and the Family* 46: 7–8.

Hofferth SL, Moore KA (1979). Early childbearing and later economic well-being. *American Sociology Review* 44: 784–815.

Hofferth SL, Moore KA (1980). Factors affecting early family formation: A path model. *Population and Environment* 3: 73–98.

Hoffman SD, Foster EM, Furstenberg FF Jr (1993). Reevaluating the costs of teenage childbearing. *Demography* 30: 1–13. Discussion, 281–96.

Holland PW (1986). Statistics and causal inference. *Journal of the American Statistical Association* 8: 945–70 (with discussion).

Holland PW (1988). Causal inference, path analysis, and recursive structural equation models. In C Clogg, ed. *Sociological Methodology 1988*, American Sociological Association, Washington, DC, chapter 13.

Hosmer DW, Lemeshow S (2000). *Applied Logistic Regression*. 2nd ed. Wiley.

Hotelling H (1927). Differential equations subject to error, and population estimates. *Journal of the American Statistical Association* 22: 283–314.

Howard-Jones N (1975). *The Scientific Background of the International Sanitary Conferences 1851–1938*. World Health Organization, Geneva.

Hubbard R, Vetter DE, Little EL (1998). Replication in strategic management: Scientific testing for validity, generalizability, and usefulness. *Strategic Management Journal* 19: 243–54.

Humphreys P, Freedman DA (1996). The grand leap. *British Journal for the Philosophy of Science* 47: 113–23.

Humphreys P, Freedman DA (1999). Are there algorithms that discover causal structure? *Synthese* 121: 29–54.

Hurwicz L (1950). Least-squares bias in time series. In TC Koopmans, ed. *Statistical Inference in Dynamical Economic Models*. Wiley, New York, pp. 365–83; also see p. 272.

IARC (1986). *Tobacco Smoking*. International Agency for Research on Cancer, Monograph 38, Lyon. Distributed by Oxford University Press.

Jacobs D, Carmichael JT (2002). The political sociology of the death penalty. *American Sociological Review* 67: 109–31. The quote, slightly edited, is from note 7, p. 117. The model is given on p. 116, although some of the details are unclear.

Jencks C, Phillips M, eds. (1998). *The Black-White Test Score Gap*. Brookings Institution Press, Washington, D.C.

Kahneman D, Slovic P, Tversky A, eds. (1982). *Judgment under Uncertainty: Heuristics and Biases*. Cambridge University Press.

Kahneman D, Tversky A (1974). Judgment under uncertainty: Heuristics and bias. *Science* 185: 1124–31.

Kahneman D, Tversky A (1996). On the reality of cognitive illusions. *Psychological Review* 103: 582–91.

Kahneman D, Tversky A, eds. (2000). *Choices, Values, and Frames*. Cambridge University Press.

Keefe FJ, Affleck G, Lefebvre J, Underwood L, Caldwell DS, Drew J, Egert J, Gibson J, Pargament K (2001). Living with rheumatoid arthritis: The role of daily spirituality and daily religious and spiritual coping. *The Journal of Pain* 2: 101–10.

Keynes JM (1939). Professor Tinbergen's method. *The Economic Journal* 49: 558–68.

Keynes JM (1940). Comment [on Tinbergen's reply]. *The Economic Journal* 50: 154–56.

King G, Keohane RO, Verba S (1994). *Designing Social Inquiry: Scientific Inference in Qualitative Research*. Princeton University Press.

Klein LR (1951). Estimating patterns of savings behavior from sample survey data. *Econometrica* 19: 438–54.

Labrie F et al (2004). Screening decreases prostate cancer mortality: 11-year follow-up of the 1988 Quebec prospective randomized controlled trial. *Prostate* 59: 311–18.

Larzalere RE, Kuhn BR, Johnson B (2004). The intervention selection bias: An underrecognized confound in intervention research. *Psychological Bulletin* 130: 289–303.

Last JM (2001). *A Dictionary of Epidemiology*. 4th ed. Oxford University Press.

Lawless JF (2003). *Statistical Models and Methods for Lifetime Data*. 2nd ed. Wiley-Interscience.

Leamer EE (1978). *Specification Searches*. Wiley.

Legendre AM (1805). *Nouvelles méthodes pour la détermination des orbites des comètes*. Courcier, Paris. Reprinted by Dover, New York (1959).

Lehmann EL (1998). *Nonparametrics*. Prentice Hall.

Lehmann EL (1999). *Elements of Large-Sample Theory*. Springer.

Lehmann EL (1991a). *Testing Statistical Hypotheses*. 2nd ed. Wadsworth & Brooks/Cole. 3rd ed., with JP Romano, Springer (2005).

Lehmann EL (1991b). *Theory of Point Estimation*. Wadsworth & Brooks/Cole. 2nd ed., with G Casella, Springer (1998).

Lieberson S (1985). *Making it Count*. University of California Press, Berkeley.

Lieberson S, Lynn FB (2002). Barking up the wrong branch: Alternatives to the current model of sociological science. *Annual Review of Sociology* 28: 1–19.

Liu TC (1960). Under-identification, structural estimation, and forecasting. *Econometrica* 28: 855–65.

Loudon I. (2000). *The Tragedy of Childbed Fever*. Oxford University Press.

Lucas RE Jr. (1976). Econometric policy evaluation: A critique. In K Brunner and A Meltzer, eds. *The Phillips Curve and Labor Markets*, vol. 1 of the Carnegie-Rochester Conferences on Public Policy, supplementary series to the *Journal of Monetary Economics*, North-Holland, Amsterdam, pp. 19–64 (with discussion).

Malthus TR (1798). *An Essay on the Principle of Population*. London, printed for J. Johnson, in St. Paul's Church-Yard.

Manski CF (1995). *Identification Problems in the Social Sciences*. Harvard University Press.

McKim VR, Turner SP, eds. (1997). *Causality in Crisis? Proceedings of the Notre Dame Conference on Causality*. University of Notre Dame Press.

Meehl PE (1954). *Clinical versus Statistical Prediction: A Theoretical Analysis and a Review of the Evidence*. University of Minnesota Press, Minneapolis.

Meehl PE (1978). Theoretical risks and tabular asterisks: Sir Karl, Sir Ronald, and the slow progress of soft psychology. *Journal of Consulting and Clinical Psychology* 46: 806–34.

Meehl PE, Waller NG (2002). The path analysis controversy: A new statistical approach to strong appraisal of verisimilitude. *Psychological Methods* 7: 283–337 (with discussion).

Mill JS (1843). *A System of Logic, Ratiocinative and Inductive*. Longmans, Green, Reader, and Dyer, London. 7th ed (1868).

Miller ER III et al (2005). Meta-analysis: High-dosage vitamin E supplementation may increase all-cause mortality. *Annals of Internal Medicine* 142: 37–46.

Nelson RR, Winter SG (1982). *An Evolutionary Theory of Economic Change*. Harvard University Press.

Neyman J (1923). Sur les applications de la théorie des probabilités aux experiences agricoles: Essai des principes. *Roczniki Nauk Rolniczych* 10: 1–51, in Polish. English translation by DM Dabrowska and TP Speed (1990), *Statistical Science* 5: 465–80 (with discussion).

Ní Bhrolcháin M (2001). "Divorce effects" and causality in the social sciences. *European Sociological Review* 17: 33–57.

Oakes MW (1990). *Statistical Inference*. Epidemiology Resources, Inc. Chestnut Hill, MA.

Oakland GB (1950). An application of sequential analysis to whitefish sampling. *Biometrics* 6: 59–67.

Pate AM, Hamilton EE (1992). Formal and informal deterrents to domestic violence: The Dade county spouse assault experiment. *American Sociological Review* 57: 691–97.

Pearl J (1995). Causal diagrams for empirical research. *Biometrika* 82: 669–710 (with discussion).

Pearl J (2000). *Causality: Models, Reasoning, and Inference*. Cambridge University Press.

Pearl R, Reed LJ (1920). On the rate of growth of the population of the United States since 1790 and its mathematical representation. *Proceedings of the National Academy of Science* 6: 275–88.

Pearson K (1911). *The Grammar of Science*. 3rd ed. Adam and Charles Black, London.

Pearson K, Lee A (1903). On the laws of inheritance in man: Inheritance of physical characters. *Biometrika* 2: 357–462. They give the joint distribution, with heights rounded to the nearest inch. We added uniform noise to get continuous data.

Petitti DB (1994). Coronary heart disease and estrogen replacement therapy. Can compliance bias explain the results of observational studies? *Annals of Epidemiology* 4: 115–18.

Petitti DB (1998). Hormone replacement therapy and heart disease prevention: Experimentation trumps observation. *Journal of the American Medical Association* 280: 650–52.

Petitti DB (2002). Hormone replacement therapy for prevention. *Journal of the American Medical Association* 288: 99–101.

Platt J (1996). *A History of Sociological Research Methods in America*. Cambridge University Press.

Powers DE, Rock DA (1999). Effect of coaching on SAT I: Reasoning test scores. *Journal of Educational Measurement* 36: 93–118.

Pratt JW, Schlaifer R (1984). On the nature and discovery of structure. *Journal of the American Statistical Association* 79: 9–33 (with discussion).

Pratt JW, Schlaifer R (1988). On the interpretation and observation of laws. *Journal of Econometrics* 39: 23–52.

Quetelet A (1835). *Sur l'homme et le développement de ses facultés, ou Essai de physique sociale*. Bachelier, Paris. English translation by R Knox (1842), Chambers, Edinburgh. Reprinted by Burt Franklin, New York (1968).

Rao CR (1973). *Linear Statistical Inference and its Applications*. 2nd ed. Wiley. Chapter 6 discusses likelihood techniques.

Raufman JP (1998). Cholera. *American Journal of Medicine* 104: 386–94.

Rindfuss RR, Bumpass L, St. John C (1980). Education and fertility: Implications for the roles women occupy. *American Sociological Review* 45: 431–47.

Rindfuss RR, Bumpass L, St. John C (1984). Education and the timing of motherhood: Disentangling causation. *Journal of Marriage and the Family* 46: 981–84.

Rindfuss RR, St. John C (1983). Social determinants of age at first birth. *Journal of Marriage and the Family* 45: 553–65.

Rodgers RC, Maranto CL (1989). Causal models of publishing productivity in psychology. *Journal of Applied Psychology* 74: 636–49.

Rosenberg CE (1962). *The Cholera Years*. Chicago University Press.

Roy AD (1951). Some thoughts on the distribution of earnings. *Oxford Economic Papers* 3: 135–46.

Rubin D (1974). Estimating causal effects of treatments in randomized and nonrandomized studies. *Journal of Educational Psychology* 66: 688–701.

Ruffin MT (1999). Screening for prostate cancer. *Journal of Family Practice* 48: 581–82.

Schneider M, Teske P, Marschall M (1997). Institutional arrangements and the creation of social capital: The effects of public school choice. *American Political Science Review* 91: 82–93.

Semmelweis I (1860). *The Etiology, Concept, and Prophylaxis of Childbed Fever*. English translation by KC Carter (1983), University of Wisconsin Press.

Sen AK (2002). *Rationality and Freedom*. Harvard University Press.

Shadish WR, Cook TD, Campbell DT (2002). *Experimental and Quasi-Experimental Designs for Generalized Causal Inference*. Houghton Mifflin, Boston.

Shapiro S, Venet W, Strax P, Venet L (1988). *Periodic Screening for Breast Cancer: The Health Insurance Plan Project and its Sequelae, 1963–1986*. Johns Hopkins University Press, Baltimore.

Shaw DR (1999). The effect of TV ads and candidate appearances on statewide presidential votes, 1988–96. *American Political Science Review* 93: 345–61.

Sims CA (1980). Macroeconomics and reality. *Econometrica* 48: 1–47.

Smith GCS, Pell JP (2004). Parachute use to prevent death and major trauma related to gravitational challenge: Systematic review of randomised controlled trials. *British Medical Journal* 327: 1459–61.

Smith RA (2003). Ideology masquerading as evidence-based medicine: The Cochrane review on screening for breast cancer with mammography. *Breast Diseases: A Yearbook Quarterly* 13: 298–307.

Snow J (1855). *On the Mode of Communication of Cholera*. Churchill, London. Reprinted by Hafner, New York (1965).

Sobel ME (1998). Causal inference in statistical models of the process of socioeconomic achievement—A case study. *Sociological Methods & Research* 27: 318–48.

Sobel ME (2000). Causal inference in the social sciences. *Journal of the American Statistical Association* 95: 647–51.

Spirtes P, Glymour C, Scheines R (1993). *Causation, Prediction, and Search*. Springer Lecture Notes in Statistics, no. 81. 2nd ed., MIT Press (2000).

Steiger JH (2001). Driving fast in reverse. *Journal of the American Statistical Association* 96: 331–38.

Stigler SM (1986). *The History of Statistics*. Harvard University Press.

Stone R (1993). The assumptions on which causal inferences rest. *Journal of the Royal Statistical Society* Series B 55: 455–66.

Sundt TM (1987). Was the international randomized trial of extracranial-intracranial arterial bypass representative of the population at risk? With editorial and comments by Goldring et al and reply by Barnett et al. *New England Journal of Medicine* 316: 809–10, 814–24. See EC/IC Bypass Study Group (1985).

Terris M, ed. (1964). *Goldberger on Pellagra*. Louisiana State University Press.

Timberlake M, Williams KR (1984). Dependence, political exclusion, and government repression: Some cross-national evidence. *American Sociological Review* 49: 141–46.

Tinbergen J (1940). On a method of statistical business-cycle research. A reply [to Keynes]. *The Economic Journal* 50: 141–54.

Truett J, Cornfield J, Kannel W (1967). A multivariate analysis of the risk of coronary heart disease in Framingham. *Journal of Chronic Diseases* 20: 511–24.

US Department of Education (1987). *High School and Beyond Third Follow-Up (1986) Sample Design Report*. Office of Educational Research and Improvement, Center for Education Statistics. Washington, DC.
http://nces.ed.gov/pubsearch/pubsinfo.asp?pubid=88402

US Preventive Services Task Force (2002). Screening for prostate cancer: Recommendation and rationale. *Annals of Internal Medicine* 137: 915–16.

Verhulst PF (1845). Recherches mathématiques sur la loi d'accroissement de la population. *Nouveaux mémoires de l'Académie Royale des Sciences et Belles-Lettres de Bruxelles* 18: 1–38.

Victora CG, Habicht JP, Bryce J (2004). Evidence-based public health: Moving beyond randomized trials. *American Journal of Public Health* 94: 400–405.

Virtamo J et al (2003). Incidence of cancer and mortality following alpha-tocopherol and beta-carotene supplementation: a postintervention follow-up. *Journal of the American Medical Association* 290: 476–85.

Waxman DJ, Strominger JL (1983). Penicillin-binding proteins and the mechanism of action of beta-lactam antibiotics. *Annual Review of Biochemistry* 52: 825–69.

White H (1980). A heteroskedasticity-consistent covariance matrix estimator and a direct test for heteroskedasticity. *Econometrica* 48: 817–38.

Widder DV (1946). *The Laplace Transform*. Princeton University Press.

Winkelstein W Jr (1995). A new perspective on John Snow's communicable disease theory. *American Journal of Epidemiology* 142 (9 Suppl) S3–9.

Wright PG (1928). *The Tariff on Animal and Vegetable Oils*. Macmillan, New York.

Writing Group for the Women's Health Initiative Investigators (2002). Risks and benefits of estrogen plus progestin in healthy postmenopausal women: Principal results from the women's health initiative randomized controlled trial. *Journal of the American Medical Association* 288: 321–33.

Yule GU (1899). An investigation into the causes of changes in pauperism in England, chiefly during the last two intercensal decades. *Journal of the Royal Statistical Society* 62: 249–95 (with discussion).

Yule GU (1925). The growth of population and the factors which control it. *Journal of the Royal Statistical Society* 88: 1–62 (with discussion).

Yule GU (1926). Why do we sometimes get nonsense-correlations between time series? *Journal of the Royal Statistical Society* 89: 1–69 (with discussion). This paper and the previous one were Yule's presidential addresses.

Prior publication

Plainly, this book draws on many sources. Work of others is acknowledged (I hope) in the text, and in chapter end notes. Furthermore, I've cannibalized many of my own publications—

Chapter 1. Freedman-Pisani-Purves (1988), Freedman-Petitti-Robins (2004), Freedman (1991, 1997, 1999, 2004, 2005). Table 1 is from FPP, copyright by WW Norton.

Chapter 2. Freedman-Pisani-Purves (1988), especially for figures 1 and 2, table 1, and exercise B8. These are copyright by WW Norton.

Chapter 4. Freedman (1983) and Freedman-Pisani-Purves (1988), for the discussion of data snooping.

Chapter 5. Freedman (1987, 1991, 1995, 1997, 1999, 2004, 2005).

Chapter 7. Freedman and Peters (1984abc, 1985), especially for table 1; Freedman, Rothenberg, and Sutch (1983).

Chapter 8. Freedman (1997, 1999, 2004, 2005).

Permission to reproduce copyright material is gratefully acknowledged.

Answers to Selected Exercises

Chapter 1 Observational Studies and Experiments

Exercise Set A

1. In table 1, there were 837 deaths from other causes in the total treatment group (screened plus refused) and 879 in the control group. Not much different.

Comments. (i) Groups are the same size, so we can look at numbers or rates. (ii) The difference in number of deaths is relatively small, and not statistically significant.

2. This comparison is biased. The control group includes women who would have accepted screening if they had been asked, and are therefore comparable to women in the screening group. But the control group also includes women who would have refused screening. The latter are poorer, less well educated, less at risk from breast cancer. (A comparison that includes only the subjects who follow the investigators' treatment plans is called "per protocol analysis," and is generally biased.)

3. Natural experiment. The fact that the Lambeth Company moved its pipe (i) sets up the comparison with Southwark & Vauxhall (table 2) and (ii) makes it harder to explain the difference in death rates between the Lambeth customers and the Southwark & Vauxhall customers on the basis of some difference between the two groups—other than the water. For instance, people were generally not choosing between the two water companies on the basis of how the water tasted. If they had been, self-selection and confounding would be bigger issues. The change in water intake point is one basis for the view that the data could be analyzed as if they were from a randomized controlled experiment.

4. Observational study. Hence the need for adjustment by regression.

5. (i) If -0.755, outrelief prevents poverty.
   (ii) If $+0.005$, outrelief has no real effect on poverty.

6. (i) $E(S_n) = n\mu$ and $\mathrm{var}(S_n) = n\sigma^2$.

(ii) $E(S_n/n) = \mu$ and $\text{var}(S_n/n) = \sigma^2/n$.

7.   (i) $E(S_n) = np$ and $\text{var}(S_n) = np(1 - p)$.

(ii) $E(S_n/n) = p$ and $\text{var}(S_n/n) = p(1 - p)/n$.

NB. For many purposes, variance has the wrong size and the wrong units. Take the square root of the variance to get the standard error.

8.   The law of large numbers says that with a big sample, the sample average will be close to the population average. More technically, let $X_1, X_2, \ldots$ be independent and identically distributed with $E(X_i) = \mu$. Then

$$(X_1 + X_2 + \cdots + X_n)/n \to \mu$$

with probability 1.

9.   Reverse causation is plausible: on the days when the joints don't hurt, subjects feel that religious coping worked.

10.   Association is not the same as causation. The big issue is confounding, and it is easy to get fooled. On the other hand, association is often a good clue. Sometimes, you can make a very tight argument for causation based on observational data. See text for discussion and examples.

Comment. If the material on experiments and observational studies is unfamiliar, you might want to read chapters 1, 2, and 9 in Freedman-Pisani-Purves (1998).

Chapter 2   The Regression Line

Exercise Set A

1.   (a)   False. The son is likely to be shorter: the 50–50 point is

$$68.7 - 0.501 \times \frac{2.81}{2.74} \times 67.7 + 0.501 \times \frac{2.81}{2.74} \times 72 \doteq 70.9 \text{ inches.}$$

The regression line is computed in part (b).

(b)   The slope is $0.501 \times 2.81/2.74 \doteq 0.514$. The intercept is

$$68.7 - 0.514 \times 67.7 \doteq 33.9 \text{ inches.}$$

The RMS error is $\sqrt{1 - 0.501^2} \times 2.81 \doteq 2.43$ inches.

Comment. The SD line says that sons are 1 inch taller than their fathers. However, it is the regression line that picks off the centers of the vertical strips, not the SD line, and the regression line is flatter than the SD line—the "regression effect." If the material on correlation and regression is unfamiliar, you might want to read chapters 8–12 in Freedman-Pisani-Purves (1998).

2. According to the model, if the weight $x_i = 0$, the measured length is $Y_i = a + \epsilon_i \neq a$. In short, a cannot be observed directly, due to measurement error. With ten measurements, the average is a, plus the average of the ten ϵ_i's. This still isn't a, but it's closer.

3. If we take $\hat{a} = 439.01$ and $\hat{b} = .05$, the residuals are $-0.01, 0.01, 0.00,$ $0.00, -0.01, -0.01$. The RMS error is the better statistic. It is about 0.008 cm. The MSE is 0.00007 cm^2. Wrong size, wrong units. (Residuals don't add to 0 because $\hat{a}$ and $\hat{b}$ were rounded.)

4. Use r for the left hand scatter plot. The middle one is U-shaped, and the right hand one has two clouds of points stuck together: r doesn't reveal these features of the data. If in doubt, read chapter 8 in Freedman-Pisani-Purves (1998).

Exercise Set B, Chapter 2

1. In equation (1), variance applies to data. So does correlation in (4).

2. These are estimates.

3. The regression line is $y = 439.0100 + 0.0495x$.

4. Data.

5. $35/12$ starts life as the variance of the list $\{1, 2, 3, 4, 5, 6\}$, which could be viewed as data. If you pick a number at random from the list, that's a random variable, whose variance is $35/12$.

6. The expected value is $180 \times 1/6 = 30$, which goes into the first blank. The variance is $180 \times (1/6) \times (5/6) = 25$. But it is $\sqrt{25} = 5$ that goes into the second blank.

7. The expected value is $1/6 = 0.167$. The variance is $(1/6 \times 5/6)/250 = 0.000556$. The SE is $\sqrt{0.000556} = 0.024$, The expected value goes into the first blank. The SE—not the variance—goes into the second blank.

8. (a) The observed value for the number of 1's is 17. The expected value is $100 \times 1/4 = 25$. The SE is $\sqrt{100 \times (1/4) \times (3/4)} = 4.33$. The observed number of 1's is 1.85 SEs below expected. Eliminate the "number of 1's."

    The observed value for the number of 2's is 54. The expected value is $100 \times 1/2 = 50$. The SE is $\sqrt{100 \times (1/2) \times (1/2)} = 5$. The observed number of 2's is 0.8 SEs above expected: the "number of 2's" goes into the blank.

    (b) The observed value for the number of 5's is 29. The expected value is 25. The SE is 4.33. The observed number of 5's is 0.92 SEs above expected. Eliminate the "number of 5's."

The observed value for the sum of the draws is $17+108+145 = 270$. The average of the box is 2.5; the SD is 1.5. The expected value for the sum is $100\times2.5 = 250$. The SE for the sum is $\sqrt{100}\times1.5 = 15$. The observed value is 1.33 SEs above the expected value: the "sum of the draws" goes into the blank.

If this is unfamiliar ground, you might want to read chapter 17 in Freedman-Pisani-Purves (1998).

9. Model.

10. a and b are unobservable parameters; ϵ_i is an unobservable random variable; Y_i is an observable random variable.

11. The observed value of a random variable.

12. (a) $\sum_1^n (x_i - \bar{x}) = \left(\sum_1^n x_i\right) - n\bar{x} = n\bar{x} - n\bar{x} = 0$.

   (b) Just square it out:

$$\sum_1^n (x_i - c)^2 = \sum_1^n \left[(x_i - \bar{x}) + (\bar{x} - c)\right]^2$$

$$= \sum_1^n \left[(x_i - \bar{x})^2 + (\bar{x} - c)^2 + 2(x_i - \bar{x})(\bar{x} - c)\right].$$

   But $\sum_1^n \left[2(x_i - \bar{x})(\bar{x} - c)\right] = 2(\bar{x} - c)\sum_1^n (x_i - \bar{x}) = 0$ by (a).
   And $\sum_1^n (\bar{x} - c)^2 = n(\bar{x} - c)^2$.

   (c) Use (b): $(\bar{x} - c)^2 \geq 0$, with a minimum in c at $c = \bar{x}$.

   (d) Put $c = 0$ in (b).

13. Sample mean: see 12(c).

14. Part (a) follows from equation (4); part (b), from (5). Part (c) follows from equation (1). For part (d),

$$(x_i - \bar{x})(y_i - \bar{y}) = x_i y_i - \bar{x} y_i - x_i \bar{y} + \bar{x}\,\bar{y}.$$

So

$$\text{cov}(x, y) = \frac{1}{n}\sum_{i=1}^n (x_i y_i - \bar{x} y_i - x_i \bar{y} + \bar{x}\,\bar{y})$$

$$= \frac{1}{n}\sum_{i=1}^n x_i y_i - \bar{x}\frac{1}{n}\sum_{i=1}^n y_i - \bar{y}\frac{1}{n}\sum_{i=1}^n x_i + \bar{x}\,\bar{y}$$

$$= \frac{1}{n}\sum_{i=1}^n x_i y_i - \bar{x}\,\bar{y} - \bar{x}\,\bar{y} + \bar{x}\,\bar{y}$$

$$= \frac{1}{n} \sum_{i=1}^{n} x_i y_i - \bar{x}\,\bar{y}.$$

Part (e). Put $y = x$ in (d) and use (c).

No answers supplied for exercises 15–16.

17. (a) $P(X_1 = 3 \mid X_1 + X_2 = 8)$ equals

$$\frac{P(X_1 = 3, X_2 = 5)}{P(X_1 + X_2 = 8)} = \frac{1/36}{5/36} = 1/5.$$

(b) $P(X_1 + X_2 = 7 \mid X_1 = 3) = P(X_2 = 4 \mid X_1 = 3) = 1/6$.

(c) Conditionally, X_1 is 1, 2, 3, 4, or 5 with equal probability, so the conditional expectation is 3.

Generally, $P(A|B) = P(A \text{ and } B)/P(B)$. If X_1 and X_2 are independent, conditioning on X_1 doesn't change the distribution of X_2. Exercise 17 prepares for chapter 4. If the material is unfamiliar, you might wish to read chapters 13–15 in Freedman-Pisani-Purves.

18. Each term $|x_i - c|$ is continuous in c. The sum is too. Suppose $x_1 < x_2 < \cdots < x_n$ and $n = 2m + 1$ with $m > 0$. (The case $m = 0$ is pretty easy; do it separately.) The median is x_{m+1}. Fix j with $m + 1 \le j < n$. Let $x_j < c < x_{j+1}$. Now

$$f(c) = \sum_{i=1}^{j} (c - x_i) + \sum_{i=j+1}^{n} (x_i - c).$$

So f is linear on the open interval (x_j, x_{j+1}), with slope $j - (n - j) = 2j - n > 0$. (The case $c > x_n$ is similar and is omitted.) That is why f increases to the right of the median. The slope increases with j, so f is convex. The argument for the left side of c is omitted.

Chapter 3 Matrix Algebra

Exercise Set A

1. r_i is $1 \times n$, c_j is $n \times 1$, and $r_i \times c_j$ is the ij th element of $A \times B$.

No answers supplied for exercises 2–4. Exercise 2 is one explanation for non-commutativity: if f, g are mappings, seldom will $f(g(x)) = g(f(x))$.

5. $M'M = \begin{pmatrix} 14 & -9 \\ -9 & 18 \end{pmatrix}$ and $MM' = \begin{pmatrix} 10 & 7 & 7 \\ 7 & 5 & 6 \\ 7 & 6 & 17 \end{pmatrix}$.

Both matrices have the same trace. In fact, $\text{trace}(AB) = \text{trace}(BA)$ when both products are defined, as will be discussed later (exercise B11).

6. $\|u\| = \sqrt{6} = 2.45$ and $\|v\| = \sqrt{21} = 4.58$. The vectors are not orthogonal: $u'v = v'u = 1$. The outer product is

$$uv' = \begin{pmatrix} 1 & 2 & 4 \\ 2 & 4 & 8 \\ -1 & -2 & -4 \end{pmatrix}.$$

The trace is 1. Again, $\text{trace}(uv') = \text{trace}(v'u)$.

Exercise Set B, Chapter 3

No answers supplied for exercises 2–8 or 11–13.

1. The adjoint is $\begin{pmatrix} 2 & 1 & -7 \\ -2 & 1 & 5 \\ 2 & -1 & -1 \end{pmatrix}$.

9. (a) Choose an i in the range $1, \ldots, m$ and a k in the range $1, \ldots, p$. The ikth element of MN is $q = \sum_j M_{ij} N_{jk}$. So q is the kith element of $(MN)'$. Also, q is the kith element of $N'M'$.

   (b) For the first claim, $MNN^{-1}M^{-1} = MI_{p \times p}M^{-1} = MM^{-1} = I_{p \times p}$, as required. For the second claim, $(M^{-1})'M' = (MM^{-1})' = I'_{p \times p} = I_{p \times p}$, as required; use part (a) for the first equality.

10. Let c be $p \times 1$. Suppose X has rank p. Why does $X'X$ have rank p? If $X'Xc = 0_{p \times 1}$, then—following the hints—$c'X'Xc = 0 \Rightarrow \|Xc\|^2 = 0 \Rightarrow Xc = 0_{n \times 1} \Rightarrow c = 0_{p \times 1}$ because X has rank p. Conversely, suppose $X'X$ has rank p. Why does X have rank p? If $Xc = 0_{n \times 1}$, then $X'Xc = 0_{p \times 1}$, so $c = 0_{p \times 1}$ because $X'X$ has rank p.

Terminology. Suppose X is $n \times p$ with $n \geq p$. As noted in the text, X has "full rank" if its rank is p.

14. Answers are given only for parts (l) and (m).
    (l) No: X has row rank p, so there is a non-trivial $n \times 1$ vector c with $c'X = 0_{1 \times p} \Rightarrow X'c = 0_{p \times 1} \Rightarrow XX'c = 0_{n \times 1}$, so XX' isn't invertible.
    (m) No: X isn't square because $p < n$, and only square matrices are invertible.

15. Because X is a column vector, $X'Y = X \cdot Y$ and $X'X = \|X\|^2$; substitute into the formula for $\hat{\beta}$.

16. Substitute into 15. We've derived 2B12(c) from a more general result.

17. f is the residual vector when we regress Y on M. So $f \perp M$ by 14(g). Likewise, g is the residual vector when we regress N on M, so $g \perp M$. Next, e is a linear combination of f and g, so $e \perp M$. And $e \perp g$: by 15, e is the residual vector when we regress f on g. So $e \perp g + M\hat{\gamma}_2 = N$. Consequently, $e \perp X = (M \ N)$. We're almost there: $Y = M\hat{\gamma}_1 + f = M\hat{\gamma}_1 + g\hat{\gamma}_3 + e = M\hat{\gamma}_1 + (N - M\hat{\gamma}_2)\hat{\gamma}_3 + e = M(\hat{\gamma}_1 - \hat{\gamma}_2\hat{\gamma}_3) + N\hat{\gamma}_3 + e$ with $e \perp X$. QED by 14(k). This result is sometimes called the "Frisch-Waugh" theorem by econometricians.

18. The rank is 1, because there is one free column (or row).

Exercise Set C, Chapter 3

No answers supplied for exercises 1–4.

5. The first assertion follows from the previous exercise, but here is a direct argument: $c'U - E(c'U) = c'[U - E(U)]$, so

$$\operatorname{var}(c'U) = E\{c'[U - E(U)][U - E(U)]'c\}$$
$$= c'E\{[U - E(U)][U - E(U)]'\}c.$$

For the second assertion, $U + c - E(U + c) = U - E(U)$, so

$$[U + c - E(U + c)][U + c - E(U + c)]' = [U - E(U)][U - E(U)]'.$$

Take expectations.

6. $\overline{U}$ is a scalar random variable, while $E(U)$ is a fixed 3×1 vector. The mean is one thing and expectation is another, although "mean" is often used to signify expectation.

7. Neither proposition is true: $P(\xi \perp \zeta) = 1$ does not imply $\xi \perp\!\!\!\perp \zeta$, and $\xi \perp\!\!\!\perp \zeta$ does not imply $P(\xi \perp \zeta) = 1$. (Notation: $\perp\!\!\!\perp$ means independence.)

Comment. Suppose ζ has a probability density, so $P(\zeta \in H) = 0$ for any 6-dimensional hyperplane H. If $P(\xi \perp \zeta) = 1$, then ξ and ζ cannot be independent, because $P(\zeta \in x^\perp \mid \xi = x) = 1$, where $x^\perp$ is the 6-dimensional hyperplane of vectors orthogonal to x. Conditioning on ξ changes the distribution of ζ.

8. $\operatorname{var}(\xi) = E\{[\xi - E(\xi)]^2\}$ and $\operatorname{cov}(\xi, \zeta) = E\{[\xi - E(\xi)][\zeta - E(\zeta)]\}$. But $E(\xi) = E(\zeta) = 0$.

9. $\operatorname{cov}(\xi) = E\{[\xi - E(\xi)][\xi' - E(\xi')]\}$. But $E(\xi) = E(\xi') = 0$.

10. (a) True. The pairs are identically distributed, and therefore have the same covariance.

(b) False. $\text{cov}(\xi_i, \zeta_i)$ is a theoretical quantity, computed from the joint distribution. By contrast, $\frac{1}{n}\sum_{i=1}^{n}(\xi_i - \bar{\xi})(\zeta_i - \bar{\zeta})$ is the sample covariance. Comment: when the sample is large, the sample covariance will be close to the theoretical $\text{cov}(\xi_i, \zeta_i)$.

11. (i) $\frac{1}{|\sigma|}f\left(\frac{x-\mu}{\sigma}\right)$ for $-\infty < x < \infty$. If $\sigma = 0$ then $\sigma X + \mu \equiv \mu$ so the "density" is point mass at μ.

(ii) $\frac{f(\sqrt{x}) + f(-\sqrt{x})}{2\sqrt{x}}$ for $0 < x < \infty$. If the function f is smooth, the density in (ii) is $f'(0)$ at $x = 0$.

The calculus may be confusing. We'll go through (i) when $\sigma < 0$. Let $Y = \sigma X + \mu$. Then $Y < y$ if $X > y^*$ where $y^* = -(y - \mu)/|\sigma|$. So $P(Y < y) = \int_{y^*}^{\infty} f(x)dx$. Differentiate with respect to y, using the chain rule. The density of Y at y is $|\sigma|^{-1}f(y^*)$, and $y^* = (y - \mu)/\sigma$.

Exercise Set D, Chapter 3

1. The first matrix is positive definite; the second, non-negative definite.

2. For (a), let c be a $p \times 1$ vector. Then $c'X'Xc = \|Xc\|^2 \geq 0$, so $X'X$ is non-negative definite. If $c'X'Xc = 0$ for $c \neq 0_{p\times1}$, then $Xc = 0_{n\times1}$ and X is rank deficient: a linear combination of its columns vanishes. Contradiction. So $X'X$ is positive definite. (Cf. exercise B10.) Part (b) is similar.

Comment. If $p < n$, then XX' cannot be positive definite: there is an $n \times 1$-vector $c \neq 0_{n\times1}$ with $c'X = 0_{1\times p}$; then $c'XX'c = 0$. See exercise B14(1).

3. $\|Rx\|^2 = (Rx)'Rx = x'R'Rx = x'x$.

4. Let x be $n \times 1$ and $x \neq 0_{n\times1}$. To show that $x'Gx > 0$, define $y = R'x$. Then $y \neq 0_{n\times1}$, and $x = Ry$. Now $x'Gx = y'R'GRy = y'R'RDR'Ry = y'Dy = \sum_{i=1}^{n} D_{ii}y_i^2 > 0$.

No answers supplied for exercises 5–6.

7. Theorem 3.1 shows that $G = RDR'$, where R is orthogonal and D is a diagonal matrix all of whose diagonal elements are positive. Then $G^{-1} = RD^{-1}R'$, $G^{1/2} = RD^{1/2}R'$, and $G^{-1/2} = RD^{-1/2}R'$ are positive definite: exercises 4–6.

8. Let $\mu = E(U)$, a 3×1 vector. Then $\text{cov}(U) = E[[U - \mu)(U - \mu)']$, a 3×3 matrix, call it M. So $0 \leq \text{var}(c'U) = c'Mc$ and M is non-negative definite. See exercise 3C5. If there is a 3×1 vector $c \neq 0$

with $c'Mc = 0$, then $\text{var}(c'U) = 0$, so $c'U = E(c'U) = c'\mu$ with probability 1.

Exercise Set E, Chapter 3

1. (a) Define U as in the hint. Then $\text{cov}(U) = G^{1/2}\text{cov}(V)G^{1/2} = G^{1/2}G^{1/2} = G$. See exercise 3C4: G is symmetric.
   (b) Try $\alpha + G^{1/2}V$.

2. Check that $E(RU) = 0$ and $\text{cov}(RU) = R\text{cov}(U)R' = R\sigma^2 I_{n\times n}R' = \sigma^2 RR' = \sigma^2 I_{n\times n}$ by exercises 3C3–4. Then use theorem 3.2. (A more direct proof shows that the density of RU equals the density of U, because R preserves lengths in Euclidean n-space; the change-of-variables formula is needed for integrals, in order to push this through.)

3. If ξ and ζ are jointly normal, the first proposition is good; if not, not. The second proposition is true: if ξ and ζ are independent, their covariance is 0. (In this book, all expectations, variances, covariances . . . exist unless otherwise stated.)

4. Answer omitted.

5. $E(\xi + \zeta) = \alpha + \beta$. $\text{var}(\xi + \zeta) = \text{var}(\xi) + \text{var}(\zeta) + 2\text{cov}(\xi, \zeta) = \sigma^2 + \tau^2 + 2\rho\sigma\tau$; here, ρ is the correlation between the random variables ξ, ζ. Any linear combination of jointly normal variables is normal.

6. The expected number of heads is 500. The variance is $1000 \times \frac{1}{2} \times \frac{1}{2} = 250$. The SE is $\sqrt{250} = 15.81$. The range 475–525 is -1.58 to 1.58 in standard units, so the chance is almost the area under the normal curve between -1.58 and 1.58, which is 0.886.

Comment. The exact chance is 0.893, to three decimals. The normal curve is an excellent approximation. With the coin and other variables taking integer values, if the range is specified as "inclusive" you could add 0.5 at the right and subtract 0.5 at the left to get even better accuracy. This is the "continuity correction." See, for instance, chapter 18 in Freedman-Pisani-Purves. (If e.g. the variable takes only even values, or it takes fractional values, life gets more complicated.)

7. $\hat{p} = 102/250 = 0.408$, $\widehat{\text{SE}} = \sqrt{0.408 \times 0.592/250} = 0.031$.
   NB. Variance has the wrong size and the wrong units. Take the square root of the variance to get the SE.

8. (a) 0.031. (b) $0.408 \pm 2 \times 0.031$. That's what the SE does for a living.

9. $\sigma^2 = 1/2$. If e.g. $x > 0$, then $P(Z < x) = 0.5 + 0.5\Psi(x/\sqrt{2})$.

10. This is a special case of exercise 2.

Chapter 4   Multiple Regression

Exercise Set A

1. (ii) is true by assumption (5); $\epsilon \perp X$ is possible, but unlikely.

2. (i) is true by exercise 3B14(g). Since e is computed from X and Y, (ii) will be false in general.

3. No. Unless there's a bug in the program, $e \perp X$. This has nothing to do with $\epsilon \perp\!\!\!\perp X$.

Comments on exercises 1–3. In this book, orthogonality ($\perp$) is about a pair of vectors, typically deterministic: $u \perp v$ if their inner product is 0, meaning the angle between them is $90°$. Independence ($\perp\!\!\!\perp$) is about random variables or random vectors: if $U \perp\!\!\!\perp V$, the conditional distribution V given U doesn't depend on U. If U, V are random vectors, then $P(U \perp V) = 1$ often precludes $U \perp\!\!\!\perp V$, because the behavior of V depends on U. In some probability texts, if W_1 and W_2 are random variables, $W_1 \perp W_2$ means $E(W_1 W_2) = 0$.

4. (a) $e \perp X$, so e is orthogonal to the first column in X, which says that
$$\sum_i e_i = 0.$$
   (b) No. If the computer does the arithmetic right, the sum of the residuals *has* to be 0, which says nothing about assumptions behind the model.
   (c) $\sum_i \epsilon_i$ is around $\sigma \sqrt{n}$ by the central limit theorem (section 3.5).

5. First, $\epsilon' \epsilon = \sum_i \epsilon_i^2$, so $E(\epsilon' \epsilon | X) = \sum_i E(\epsilon_i^2 | X)$. But $E(\epsilon_i | X) = 0$. So $E(\epsilon_i^2 | X) = \text{var}(\epsilon_i | X) = \sigma^2$ and $\sum_i E(\epsilon_i^2 | X) = n\sigma^2$. See (4), and exercise 3C8. Next, $\epsilon \epsilon'$ is an $n \times n$ matrix, whose ij th element is $\epsilon_i \epsilon_j$. If $i \neq j$, then $E(\epsilon_i \epsilon_j | X) = E(\epsilon_i \epsilon_j) = E(\epsilon_i) E(\epsilon_j) = 0 \times 0 = 0$ by independence. If $i = j$, then $E(\epsilon_i^2 | X) = \text{var}(\epsilon_i | X) = \sigma^2$ as before.

6. The second column in the table (lengths) should be the observed values of Y_i in equation (2.7), for $i = 1, 2, \ldots, 6$. Cross-references: equation (2.7) is equation (7) in chapter 2.

7. Look at equation (1.1) to see that $\beta = \begin{pmatrix} a \\ b \\ c \\ d \end{pmatrix}$, so $p = 4$. Next, look at table 1.3. There are 32 lines in the table, so $n = 32$. There is an intercept in the equation, so put a column of 1's as the first column of the design matrix X. Subtract 100 from each entry in table 1.3. After the subtraction, columns 2, 3, 4 of the table give you columns 2, 3, 4

in the design matrix X; column 1 of the table gives you the observed values of Y.

The first column in the design matrix is all 1's, so $X_{41} = 1$. The fourth union is Chelsea. The second column in X is ΔOut, which also happens to be the second column in the table. So $X_{42} = 21 - 100 = -79$ and $Y_4 = 64 - 100 = -36$. The estimated coefficient $\hat{b}$ of ΔOut will be the second entry in $\hat{\beta} = (X'X)^{-1}X'Y$, because b is the second entry in β; that in turn is because ΔOut is the second thing in equation (1.1)—right after the intercept.

Exercise Set B, Chapter 4

1. True.

2. (i) True. (ii) Before data collection, $\overline{Y}$ is a random variable; afterwards, it's the observed value of a random variable.

3. True.

4. (i) True. (ii) Before data collection, the sample variance is a random variable; afterwards, it's the observed value of a random variable. (In the exercise, we divided by n; for some purposes, it might be better to divide by $n - 1$: most often, it doesn't matter which divisor you use.)

5. $\hat{\beta} - \beta = (X'X)^{-1}X'\epsilon$ and $e = (I - H)\epsilon$. See equations (8) and (13). Condition on X. The joint distribution of $(X'X)^{-1}X'\epsilon$ and $(I - H)\epsilon$ doesn't depend on β: there's no β in the formula.

6. Use formulas (10) and (11). Cf. lab 3 below.

7. Use formulas (10) and (11). Cf. exercise 14 below.

8. Formula (i) is the regression model, with parameters β and random errors ϵ.

9. (i) is silly: at least in frequentist statistics, parameters don't have covariances. (ii) is true if X is fixed, otherwise, trouble. (iii) is true. (iv) is false. On the left, given X, we have a fixed quantity. On the right, $\hat{\sigma}^2$ is still random given X, because $\hat{\sigma}^2$ depends on ϵ through Y, and $\epsilon \perp\!\!\!\perp X$. (v) is true.

10. (b) is true; (a) is false (or incomplete at best).

11. (a) is silly, because—given X—the right hand side is random and the left hand side isn't. (b) is true: $\hat{Y} = X\hat{\beta}$, so $E(\hat{Y}|X) = XE(\hat{\beta}|X) = X\beta$.

12. Let H be the hat matrix. Exercise 3B9 shows that $H = I$. Then $\hat{Y} = HY = Y$. Attention: this will not work if $p < n$.

13. Let $e = Y - X\hat{\beta}$ be the residual vector. Now $Y = X\hat{\beta} + e$. But $\bar{e} = 0$, because e is orthogonal to the first column of X. So $\bar{Y} = \overline{X\hat{\beta}} + \bar{e} = \bar{X}\hat{\beta}$. When we're taking averages over rows, $\hat{\beta}$ can be viewed as constant—it's the same for every row. (If we were talking about expectations, $\hat{\beta}$ would *not* be constant.)

14. The design matrix has a column of 1's and then a column of X_i's. Call this matrix M. It will be convenient to use "bracket notation." For instance, $\langle X \rangle = n^{-1} \sum_1^n X_i$, $\langle XY \rangle = n^{-1} \sum_1^n X_i Y_i$, and so forth. By exercise 2B14(d)-(e), with var and cov applied to data variables,

$$\langle X^2 \rangle = \text{var}(X) + \langle X \rangle^2 \text{ and } \langle XY \rangle = \text{cov}(X, Y) + \langle X \rangle \langle Y \rangle. \qquad (*)$$

Then

$$M'M = \begin{pmatrix} n & \sum_i X_i \\ \sum_i X_i & \sum_i X_i^2 \end{pmatrix} = n \begin{pmatrix} 1 & \langle X \rangle \\ \langle X \rangle & \langle X^2 \rangle \end{pmatrix}$$

and

$$M'Y = n \begin{pmatrix} \langle Y \rangle \\ \langle XY \rangle \end{pmatrix}.$$

With the help of equation $(*)$, it is easy to check that

$$\det(M'M) = n^2(\langle X^2 \rangle - \langle X \rangle^2) = n^2 \text{var}(X).$$

So

$$(M'M)^{-1} = \frac{1}{n\text{var}(X)} \begin{pmatrix} \langle X^2 \rangle & -\langle X \rangle \\ -\langle X \rangle & 1 \end{pmatrix}$$

and

$$(M'M)^{-1} M'Y = \frac{1}{\text{var}(X)} \begin{pmatrix} \langle X^2 \rangle \langle Y \rangle - \langle X \rangle \langle XY \rangle \\ \langle XY \rangle - \langle X \rangle \langle Y \rangle \end{pmatrix}.$$

Now to clean up. The slope is the 2,1 element of $(M'M)^{-1}M'Y$, which is

$$[\langle XY \rangle - \langle X \rangle \langle Y \rangle]/\text{var}(X).$$

By $(*)$,

$$\langle XY \rangle - \langle Y \rangle \langle Y \rangle = \text{cov}(X, Y).$$

So

$$\text{slope} = \text{cov}(X, Y)/\text{var}(X) = r s_Y / s_X$$

as required. The intercept is the 1,1 element of $(M'M)^{-1}M'Y$, which is

$$[\langle X^2 \rangle \langle Y \rangle - \langle X \rangle \langle XY \rangle]/\text{var}(X). \qquad (**)$$

Use (∗) again to see that

$$\langle X^2\rangle\langle Y\rangle - \langle X\rangle\langle XY\rangle = [\text{var}(X) + \langle X\rangle^2]\langle Y\rangle - \langle X\rangle[\text{cov}(X, Y) + \langle X\rangle\langle Y\rangle]$$
$$= \text{var}(X)\langle Y\rangle - \langle X\rangle\text{cov}(X, Y)$$

because the terms with $\langle X\rangle^2\langle Y\rangle$ cancel. Substitute into (∗∗):

$$\text{intercept} = \langle Y\rangle - \left[\langle X\rangle\text{cov}(X, Y)/\text{var}(X)\right] = \langle Y\rangle - \text{slope} \cdot \langle X\rangle$$

as required. The variance of the estimated slope is σ^2 times the 2,2 element of $(M'M)^{-1}$, namely,

$$\sigma^2/[n\text{var}(X)]$$

as required. Since $\langle X^2\rangle = \text{var}(X) + \langle X\rangle^2$ by (∗), the 1,1 element of $(M'M)^{-1}$ is

$$\frac{\langle X^2\rangle}{n\text{var}(X)} = \frac{1}{n}\left[1 + \frac{\langle X\rangle^2}{\text{var}(X)}\right].$$

The variance of the estimated intercept is obtained on multiplication by σ^2, which completes the argument. (The variance of an estimate ... applies variance to a random variable, not data.)

Exercise Set C, Chapter 4

1. Let $c \neq 0_{p\times1}$ be $p\times1$. Then $d = Xc \neq 0_{n\times1}$ because X has full rank. So $c'X'G^{-1}Xc = d'G^{-1}d > 0$, because G^{-1} is positive definite (exercise 3D7). Thus, $X'G^{-1}X$ is positive definite and hence invertible—use exercise 3D7 again.

2. For part (a), $E(c'\hat{\beta}|X) = c'E(\hat{\beta}|X) = c'\beta$: indeed, by (9) and (18), $E(\hat{\beta}|X) = \beta + (X'X)^{-1}X'E(\epsilon|X) = \beta$. The proof of theorem 3 can be pushed through, starting from (18):

$$\text{cov}(\hat{\beta}|X) = (X'X)^{-1}X'E(\epsilon\epsilon'|X)X(X'X)^{-1}$$
$$= (X'X)^{-1}X'(\sigma^2 I_{n\times n})X(X'X)^{-1} = \sigma^2(X'X)^{-1}.$$

Then

$$\text{var}(c'\hat{\beta}|X) = E[c'(\hat{\beta} - \beta)(\hat{\beta} - \beta)'c|X]$$
$$= c'E[(\hat{\beta} - \beta)(\hat{\beta} - \beta)'|X]c$$
$$= c'\text{cov}(\hat{\beta}|X)c = \sigma^2 c'(X'X)^{-1}c.$$

Next, $\mathrm{var}(\hat{\beta}_1 - \hat{\beta}_2|X) = \mathrm{var}(\hat{\beta}_1) + \mathrm{var}(\hat{\beta}_2) - 2\mathrm{cov}(\hat{\beta}_1, \hat{\beta}_2)$, i.e., the 1,1 element of $\sigma^2(X'X)^{-1}$, plus the 2,2 element, minus two times the 1,2 element. This is a very useful fact: see, e.g., section 5.3.

The answer to part (b) is omitted, being very similar to (a).

3. To set this up in the GLS framework, stack the U's on top of the V's:

$$
Y = \begin{pmatrix} U_1 \\ U_2 \\ \vdots \\ U_m \\ V_1 \\ V_2 \\ \vdots \\ V_n \end{pmatrix}.
$$

The design matrix X is an $(m+n)\times1$ column vector of 1's. The random error vector ϵ is $(m+n) \times 1$, as in the hint. The parameter α is scalar. The matrix equation is $Y = X\alpha + \epsilon$. Condition (18) does not hold, because $\sigma^2 \neq \tau^2$. Condition (19) holds. The $(m+n)\times(m+n)$ matrix G vanishes off the diagonal. Along the diagonal, the first m terms are all σ^2. The last n terms are all τ^2. So G^{-1} is also a diagonal matrix. The first m terms on the diagonal are $1/\sigma^2$ while the last n terms are $1/\tau^2$. Check that

$$
X'G^{-1}Y = \frac{1}{\sigma^2} \sum_{i=1}^m U_i + \frac{1}{\tau^2} \sum_{j=1}^n V_i,
$$

a scalar; and

$$
X'G^{-1}X = \frac{m}{\sigma^2} + \frac{n}{\tau^2},
$$

another scalar. Use (22):

$$
\hat{\alpha}_{\mathrm{GLS}} = \frac{\dfrac{1}{\sigma^2} \displaystyle\sum_{i=1}^m U_i + \dfrac{1}{\tau^2} \displaystyle\sum_{j=1}^n V_i}{\dfrac{m}{\sigma^2} + \dfrac{n}{\tau^2}}.
$$

This is real GLS not feasible GLS: the covariance matrix G is given, rather than estimated from data. Notice that $\hat{\alpha}_{\mathrm{GLS}}$ is a weighted average: observations with bigger variances get less weight.

If σ^2 and τ^2 are unknown, they can be estimated by the sample variances; the estimates are plugged into the formula above. Now we have feasible GLS not real GLS. (This is actually one-step GLS, and iteration is possible.)

4. Part (a) follows from (21): $G_{ii} = \lambda c_i$, and the off-diagonal elements vanish. Multiply across by $G^{-1/2}$ and do least squares on the transformed model. Part (b) follows from example 2, with $\Gamma_{ii} = c_i$, and $\Gamma_{ij} = 0$ when $i \neq j$. (Remember, $\hat{\beta}_{OLS}$ minimizes the function $\gamma \to \|Y - X\gamma\|^2$.)

5. To set this up in the GLS framework, let Y stack the Y_{ij}'s. Put the 3 observations for subject #1 on top; then the 3 observations for #2; ...; at the bottom, the 3 observations for #800. It will be a little easier to follow the math if we write $Y_{i,j}$ instead of Y_{ij}:

$$Y = \begin{pmatrix} Y_{1,1} \\ Y_{1,2} \\ Y_{1,3} \\ Y_{2,1} \\ Y_{2,2} \\ Y_{2,3} \\ \vdots \\ Y_{800,1} \\ Y_{800,2} \\ Y_{800,3} \end{pmatrix}.$$

This Y is 2400×1. Next, the parameter vector β stacks up the 800 fixed effects a_i, followed by the parameter b:

$$\beta = \begin{pmatrix} a_1 \\ a_2 \\ \vdots \\ a_{800} \\ b \end{pmatrix}.$$

This β is 801×1. The design matrix M has to be 2400×801. The first 800 columns have *dummy variables* for each subject. A dummy variable is 0 or 1. Column 1, for instance, is a dummy variable for subject #1. Column 1 equals 1 for subject #1, and is 0 for all other subjects. This column stacks 3 ones on top of 2397 zeros:

$$\begin{pmatrix} 1 \\ 1 \\ 1 \\ 0 \\ 0 \\ 0 \\ \vdots \\ 0 \\ 0 \\ 0 \end{pmatrix}.$$

Column 2 equals 1 for subject #2, and is 0 for all other subjects. This column stacks 3 zeros, then 3 ones, then 2394 zeros:

$$\begin{pmatrix} 0 \\ 0 \\ 0 \\ 1 \\ 1 \\ 1 \\ \vdots \\ 0 \\ 0 \\ 0 \end{pmatrix}.$$

And so forth. Column 800 equals 1 for subject #800, and is 0 for all other subjects. Column 800 stacks 2397 zeros on top of 3 ones:

$$\begin{pmatrix} 0 \\ 0 \\ 0 \\ 0 \\ 0 \\ 0 \\ \vdots \\ 1 \\ 1 \\ 1 \end{pmatrix}.$$

The 801st—and last—column in the design matrix stacks up the Z_{ij}:

$$\begin{pmatrix} Z_{1,1} \\ Z_{1,2} \\ Z_{1,3} \\ Z_{2,1} \\ Z_{2,2} \\ Z_{2,3} \\ \vdots \\ Z_{800,1} \\ Z_{800,2} \\ Z_{800,3} \end{pmatrix}.$$

Let's call this design matrix X (surprise). Here's what X looks like when you put the pieces together:

$$X = \begin{pmatrix} 1 & 0 & \cdots & 0 & Z_{1,1} \\ 1 & 0 & \cdots & 0 & Z_{1,2} \\ 1 & 0 & \cdots & 0 & Z_{1,3} \\ 0 & 1 & \cdots & 0 & Z_{2,1} \\ 0 & 1 & \cdots & 0 & Z_{2,2} \\ 0 & 1 & \cdots & 0 & Z_{2,3} \\ \vdots & \vdots & \ddots & \vdots & \vdots \\ 0 & 0 & \cdots & 1 & Z_{800,1} \\ 0 & 0 & \cdots & 1 & Z_{800,2} \\ 0 & 0 & \cdots & 1 & Z_{800,3} \end{pmatrix}.$$

The matrix equation is $Y = X\beta + \epsilon$. The dummy variables work with the fixed effects a_i and get them into the equation.

Assumption (18) isn't satisfied, because different subjects have different variances. But (19) is OK. The 2400×2400 covariance matrix G is diagonal. The first 3 elements on the diagonal are all σ_1^2, corresponding to subject #1. The next 3 are all σ_2^2, corresponding to subject #2. And so forth. If we knew the σ's, we could use GLS. But we don't. Instead, we use feasible GLS.

(i) Fit the model by OLS and get residuals e.
(ii) Estimate the σ_i^2. For instance, $\hat{\sigma}_1^2 = (e_1^2 + e_2^2 + e_3^2)/2$, $\hat{\sigma}_2^2 = (e_4^2 + e_5^2 + e_6^2)/2, \ldots, \hat{\sigma}_{800}^2 = (e_{2398}^2 + e_{2399}^2 + e_{2400}^2)/2$. It's better to use 2 as the divisor, rather than 3, if you plan to get SEs from (25b); for estimation, the divisor doesn't matter.

(iii) String the $\hat{\sigma}_i^2$ down the diagonal to get $\hat{G}$.
(iv) Use (25a) to get the one-step GLS estimator.
(v) Iterate if desired.

Exercise Set D, Chapter 4

1. $\hat{\beta}$ is the sample mean and $\hat{\sigma}^2$ is the sample variance, where you divide by $n-1$ rather than n. By theorem 6, the sample mean and sample variance are independent, $\hat{\beta} - \beta$ is $N(0, \sigma^2/n)$, and $\hat{\sigma}^2$ is distributed as $\sigma^2 \chi_{n-1}^2/(n-1)$. Finally, $\sqrt{n}(\hat{\beta} - \beta)/\hat{\sigma}$ is t with $n-1$ degrees of freedom.

Comments. (i) $\widehat{\text{SE}}$ of $\hat{\beta}$ is $\hat{\sigma}/\sqrt{n}$. (ii) The joint distribution of $\hat{\beta} - \beta$ and $\hat{\sigma}^2$ doesn't depend on β, by exercise 4B5. (iii) When $p = 1$ and the design matrix is just a column of 1's, theorem 4.6 gives the joint distribution for the sample mean and variance of $X_1, \ldots, X_n$, the X_i being IID normal variables—a result of R. A. Fisher's. (iv) The result doesn't hold without normality. However, for $\hat{\beta}$ and t, the central limit comes to the rescue when the sample is reasonably large. The distribution of $\hat{\sigma}^2$ generally depends on fourth moments of the parent distribution.

2. (a) True: $3.79/1.88 \doteq 2.02$.
   (b) True: $P < 0.05$. If you want to compute P, see page 282.
   (c) False: $P > 0.01$.
   (d) This is silly. In frequentist statistics, probability applies to random variables not parameters. Either $b = 0$ or $b \neq 0$.
   (e) Like (d).
   (f) True. This is what the P-value means. Contrast with (e).
   (g) True. This is what the P-value means. Contrast with (d).
   (h) This is silly. Like (d). Confidence intervals are for a different game.
   (i) False: see (j)-(k).
   (j) True. You need the model to justify the probability calculations.
   (k) True. The test assumes the model $Y_i = a + bX_i + Z_i\gamma + \epsilon_i$, with all the conditions on the ϵ_i's. The test only asks whether $b = 0$ or $b \neq 0$. The hypothesis that $b = 0$ doesn't fit the data as well as the other hypothesis, $b \neq 0$.

If exercise 2 covers unfamiliar ground, read chapters 26–29 in Freedman-Pisani-Purves (1998); confidence intervals are discussed in chapters 21 and 23.

3. The philosopher is a little mixed up. The null hypothesis must involve a statement about a model. Commonly, the null restricts a parameter in the model. For example, here's a model for the philosopher's coin. The tosses of the coin are independent. In the first 5000 tosses, the coin lands

heads on each toss with probability p_1. In the last 5000 tosses, the coin lands heads on each toss with probability p_2. Null: $p_1 = p_2$. (That's the restriction.) Alternative: $p_1 \neq p_2$. Here, p_1 and p_2 are parameters, not relative frequencies in the data. The data are used to test the null, not to formulate the null.

If $|\hat{p}_1 - \hat{p}_2|$ is larger than what can reasonably be explained by "random fluctuations," we reject the null. The estimates $\hat{p}_1$, $\hat{p}_2$ are relative frequencies in the data, not parameters. The philosopher didn't pick up the distinction between parameters and estimates.

Comment. The null is about a model, or the connection between data and a model. See chapters 26 and 29 in Freedman-Pisani-Purves (1998).

4. Both statements are false. It's pretty safe to conclude that $\beta_2 \neq 0$, but if you want to know how big it is, or how big $\hat{\beta}_2$ is, look at $\hat{\beta}_2$. The significance level P is small because $t = \hat{\beta}_2/\widehat{SE}$ is big. That could be because $\hat{\beta}_2$ is big, or because $\widehat{SE}$ is small (or both). For more discussion, see chapter 29 in Freedman-Pisani-Purves (1998)

Exercise Set E, Chapter 4

1. Use the t-test. This is a regression problem with $p = 1$. The design matrix is a column of 1's. See exercise 4D1. (The F-test is OK too: $F = t^2$, with 1 and $n - 1$ degrees of freedom.)

2. This is exercise 1, in disguise: $\delta_i = U_i - \alpha$.

3. For small n, don't use the methods of this chapter. With reasonably large n, the central limit theorem will take care of things.

Comment. Without normality, if n is small, you might consider using "nonparametric methods." See Lehmann (1998).

4. Use the F-test with $n = 32$, $p = 4$, $p_0 = 2$. (See example 4, where $p = 5$.) The errors would need to be IID with mean 0 and finite variance. Normality would help but is not essential.

5. $\|X\hat{\beta}\|^2 + \|e\|^2 = \|Y\|^2 = \|X\hat{\beta}^{(s)}\|^2 + \|e^{(s)}\|^2.$

6. Georgia's null hypothesis has $p_0 = p - 1$: all coefficients vanish but the intercept. Her $\hat{\beta}^{(s)}$ consists of $\overline{Y}$ stacked on top of $p - 1$ zeros. In the numerator of the F-statistic,

$$\|X\hat{\beta}\|^2 - \|X\hat{\beta}^{(s)}\|^2 = \|X\hat{\beta}\|^2 - n\overline{Y}^2 = n\,\mathrm{var}(X\hat{\beta}) :$$

exercise 4B13. But $\mathrm{var}(X\hat{\beta}) = R^2\mathrm{var}(Y)$ and $\mathrm{var}(e) = (1 - R^2)\mathrm{var}(Y)$ by (15)-(16). The numerator of the F-statistic therefore equals

$$\frac{\|X\hat{\beta}\|^2 - \|X\hat{\beta}^{(s)}\|^2}{p - 1} = \frac{n}{p - 1}R^2\mathrm{var}(Y).$$

Because $\bar{e} = 0$, the denominator of F is

$$\frac{\|e\|^2}{n-p} = \frac{n}{n-p}\text{var}(e) = \frac{n}{n-p}(1 - R^2)\text{var}(Y).$$

So

$$F = \frac{n-p}{p-1}\frac{R^2}{1-R^2}.$$

7. We know $Y = \hat{Y} + e$ with $e \perp X$. So $e \perp \hat{Y}$ and therefore $\frac{1}{n}\sum_i \hat{Y}_i Y_i = \frac{1}{n}\sum_i \hat{Y}_i^2$. Since $Y = \hat{Y} + e$ and $\sum_i e_i = 0$, we also know that $\frac{1}{n}\sum_i \hat{Y}_i = \frac{1}{n}\sum_i Y_i$. Now use exercise 2B14, where cov and var are applied to data variables:

$$\text{cov}(\hat{Y}, Y) = \left(\frac{1}{n}\sum_i \hat{Y}_i Y_i\right) - \left(\frac{1}{n}\sum_i \hat{Y}_i\right)\left(\frac{1}{n}\sum_i Y_i\right)$$

$$= \left(\frac{1}{n}\sum_i \hat{Y}_i^2\right) - \left(\frac{1}{n}\sum_i \hat{Y}_i\right)^2$$

$$= \text{var}(\hat{Y}).$$

The squared correlation coefficient between $\hat{Y}$ and Y is

$$\frac{\text{cov}(\hat{Y}, Y)^2}{\text{var}(\hat{Y})\text{var}(Y)} = \frac{\text{var}(\hat{Y})^2}{\text{var}(\hat{Y})\text{var}(Y)} = \frac{\text{var}(\hat{Y})}{\text{var}(Y)} = R^2.$$

Exercise Set F, Chapter 4

1. The $\hat{\beta}$'s are dependent.
2. $e \perp \hat{Y}$ so $\|Y\|^2 = \|\hat{Y}\|^2 + \|e\|^2$. From the definition, $1 - R^2 = (\|Y\|^2 - \|\hat{Y}\|^2)/\|Y\|^2 = \|e\|^2/\|Y\|^2$.

Discussion questions, Chapter 4

1. Let $\epsilon_i = X_i - E(X_i)$. Use OLS to fit the model

$$\begin{pmatrix} X_1 \\ X_2 \\ X_3 \end{pmatrix} = \begin{pmatrix} 1 & 1 \\ 1 & 2 \\ 2 & 1 \end{pmatrix}\begin{pmatrix} \alpha \\ \beta \end{pmatrix} + \begin{pmatrix} \epsilon_1 \\ \epsilon_2 \\ \epsilon_3 \end{pmatrix}.$$

There is no intercept.

2. The random errors are independent from one subject to another; residuals are dependent. Random errors are independent of the design matrix X: residuals are dependent on X. Residuals are orthogonal to X, random errors are going to project into X, at least by a little.

  For instance, suppose there is an intercept in the equation, i.e., a column of 1's in X. The sum of the residuals is 0: that creates dependence across subjects. The sum of the random errors will not be 0 exactly—that's non-orthogonality. Since the residuals have to be orthogonal to X, they can't generally be independent of X.

3. If there is an intercept in the equation, the sum of the residuals has to be 0; or if the column space of the design matrix includes the constant vectors. Otherwise, the sum of the residuals will usually differ from 0.

4. (a) is false and (b) is true (section 4.4). Teminology: in the regression model $Y = X\beta + \epsilon$, the disturbance term for the ith subject is ϵ_i.

5. Conditionally on X, the Y_i are independent but not identically distributed. For instance, $E(Y_i|X) = X_i\beta$ differs from one i to another. Unconditionally, if the rows of X are IID, so are the Y_i; if the rows of X are dependent and differently distributed, so are the Y_i.

6. All the assertions are true—assuming the original equation is OK. Take part (c), for instance. The OLS assumptions would still hold, the true coefficient of the extra variable being 0. (We're tacitly assuming that the new design matrix would still have full rank.)

7. The computer can find $\hat{\beta}$ all right, but what is $\hat{\beta}$ estimating? And what do the standard errors mean? (The answers might well be, nothing.)

8. R^2 measures goodness of fit. It does not measure validity. See text for discussion and examples.

9. The F-test assumes the truth of the model, and tests whether some group of coefficients are all 0. The F-test can be used to see if a smaller model is OK, given that a larger model is OK. (But then how do you test the larger model? Not with the F-test?!?)

10. If $r = \pm 1$, then column 2 $= c \times$ column 1 $+ d$. Since the columns have mean 0 and variance 1, $c = \pm 1$ and $d = 0$, so the rank is 1. Suppose $|r| < 1$. Let $M = [u, v]$, i.e., column 1 is u and column 2 is v. Then

$$M'M = n \begin{pmatrix} 1 & r \\ r & 1 \end{pmatrix}, \quad \det(M'M) = n^2(1 - r^2),$$

$$(M'M)^{-1} = \frac{1}{n}\frac{1}{1 - r^2} \begin{pmatrix} 1 & -r \\ -r & 1 \end{pmatrix}.$$

Here, we know $\sigma^2 = 1$. So

$$\text{var}(\hat{a}) = \text{var}(\hat{b}) = \frac{1}{n}\frac{1}{1-r^2},$$

$$\text{var}(\hat{a} - \hat{b}) = \frac{1}{n}\frac{2(1+r)}{1-r^2} = \frac{1}{n}\frac{2}{1-r},$$

$$\text{var}(\hat{a} + \hat{b}) = \frac{1}{n}\frac{2(1-r)}{1-r^2} = \frac{1}{n}\frac{2}{1+r}.$$

See exercise 4C2(a). If r is close to 1, then $\text{var}(\hat{a} + \hat{b})$ is reasonable, but the others are ridiculously large—especially $\text{var}(\hat{a} - \hat{b})$. When collinearity is high, you cannot separate the effects of the variables.

Comment. If r is close to -1, then $a - b$ is the parameter that can be reasonably well estimated. When there are several explanatory variables, the issue is the multiple R^2 between each variable and all the others. If one of these R^2's is high, we have collinearity problems.

11. (a) and (b) are false, (c) is true.

12. Both assertions are generally false. By (9), $E(\hat{\beta}|X) - \beta = (X'X)^{-1}X'\gamma$. This won't be 0, unless $\gamma \perp X$. According to (20), $\text{cov}(\hat{\beta}|X) = (X'X)^{-1}X'GX(X'X)^{-1}$. There is no σ^2 in this formula: the ϵ_i have different variances, which appear on the main diagonal of G.

If you want (a) and (b) to be true, you need $\gamma \perp X$ and $G = \sigma^2 I_{n\times n}$.

13. $\hat{\beta}_1$ is biased, $\hat{\beta}_2$ unbiased. This follows from exercise 3B17, but here is a better argument. Let $c \neq 0$. Suppose that $\gamma_i = c$ for all i. The bias in $\hat{\beta}$ is $c(X'X)^{-1}X'1_{n\times 1}$ by (9). Let u be the $p \times 1$ vector which is all 0's, except that $u_1 = 1$. Then $c(X'X)^{-1}X'1_{n\times 1} = cu$, because $X'1_{n\times 1} = X'Xu$. This in turn follows from the fact that $Xu = 1_{n\times 1}$: the first column of X is all 1's.

Comment. There is some opinion that $E(\epsilon_i) \neq 0$ is harmless, only biasing $\hat{\beta}_1$. True enough—if $E(\epsilon_i)$ is the same for all i. Otherwise, there are problems. That is the message of questions 12–13.

14. (a), (b), (c) are true: the central limit theorem helps with (c), because you have 96 degrees of freedom. (d) is false: with 4 degrees of freedom, you need normal errors to use t.

15. If W_i is independent of X_i, dropping it from the equation creates no bias, but will probably increase the sampling error: the new disturbance term is $W_i b + \delta_i$, with larger variance than the old one. If W_i and X_i are

dependent, Tom's estimate is subject to *omitted-variable bias*, because the disturbance term $W_i b + \delta_i$ is correlated with X_i,

Here are details on bias. Write X for the vector whose ith coordinate is X_i; likewise for Y and W. From exercise 3B15, his estimator will be $\tilde{a} = X \cdot Y / \|X\|^2$. Now $X \cdot Y = a\|X\|^2 + bX \cdot W + X \cdot \delta$. So

$$\tilde{a} - a = bX \cdot W / \|X\|^2 + X \cdot \delta / \|X\|^2.$$

By the law of large numbers, $X \cdot W \doteq nE(X_i W_i)$, and $\|X\|^2 \doteq nE(X_i^2)$. By the central limit theorem, $X \cdot \delta$ will be something like $\sqrt{nE(X_i^2)E(\delta_i^2)}$ in size. With a large sample, $X \cdot \delta / \|X\|^2 \doteq 0$. Tom is left with omitted-variables bias that amounts to $bE(X_i W_i)/E(X_i^2)$: his regression of Y on X picks up the effect of the omitted variable W.

Other issues. If you set up the design matrix with one column for X and another for W, no column of 1's—i.e., no intercept—and a row for each observation, the OLS assumptions are satisfied: no intercept is needed. If X and W are perfectly correlated, the computer will complain: the design matrix only has rank 1. See question 10. There is a similar problem if X_i or W_i has variance 0.

Terminology. "Fitting a regression equation," "fitting a model," and "running a regression" are (slightly) colorful synonyms for computing OLS estimates.

16. The assertions about $Q'Q$ and $Q'Y$ follow from the law of large large numbers. For example, the 2,1 element in $Q'Y/n$ is $\frac{1}{n}\sum_{i=1}^{n} W_i Y_i \rightarrow E(W_i Y_i)$. Write L and M for the limiting matrices, so $Q'Q/n \rightarrow L$ and $Q'Y/n \rightarrow M$. Check that

$$L = \begin{pmatrix} 1 & c \\ c & c^2 + d^2 \end{pmatrix} \quad \text{and} \quad M = \begin{pmatrix} a \\ ac + d \end{pmatrix}.$$

For example, $M_{21} = ac + d$ because

$$E(W_i Y_i) = E\big[(cX_i + d\delta_i)(aX_i + \delta_i)\big] = ac + d.$$

Since $\det L = d^2$,

$$L^{-1} = \frac{1}{d^2}\begin{pmatrix} c^2 + d^2 & -c \\ -c & 1 \end{pmatrix} \quad \text{and} \quad L^{-1}M = \begin{pmatrix} a - (c/d) \\ 1/d \end{pmatrix}.$$

If Dick includes a variable that is correlated with the error term, his estimator will have *endogeneity bias*, which in this example is $-c/d$.

Putting endogenous variables into the equation is bad. It's quite hard to tell whether a variable is endogenous or exogenous, so putting extra variables into the equation is risky. ("Exogenous" variables are independent of error terms; "endogenous" variables are dependent on error terms.)

17. X_i and Y_i have covariance 0. If Harry regresses Y_i on X_i, the slope will be 0, up to sampling error. He will conclude there is no relationship. This is because he fitted a straight line to curved data. Of course, if he regressed Y_i on X_i and X_i^2, he'd be a hero.

18. Putting Z into the equation likely reduces the sampling error in the estimates, and guards against omitted-variables bias. On the other hand, if you do put in Z, endogeneity bias is a possibility.
     (: Omitted-variables bias + Endogeneity bias = Scylla + Charybdis :)

19. The X_i all have the same distribution—normal with mean μ and variance 2. The X_i are not independent: they have U in common. $\overline{X}$ is $N(\mu, 1 + \frac{1}{n})$. Thus, $\overline{X} - \mu$ is around 1. Next, $X_i - \overline{X} = V_i - \overline{V}$, so s^2 is the sample variance of $V_1, \ldots, V_n$ and $s^2 \sim \chi_{n-1}^2/(n-1) \doteq 1$. Sampling error in $\overline{X}$ is much larger than $s/\sqrt{n}$.
     Moral. Without independence, $s/\sqrt{n}$ isn't good for much.

20. (a) is true and (b) is false, as shown in 19. You need to assume independence to make the usual statistical calculations.

21. (a) is true. With reasonably large samples, normality doesn't matter so much. The central limit theorem takes care of things. However, as shown in 19, claim (b) is false—even for normal random variables. You need to assume independence to make the usual statistical calculations.

22. Bias is likely and the standard errors are not trustworthy. Let $\{X_i, Y_i : i = 1, \ldots, n\}$ be the sample. Nothing says that $E(Y_i|X_i) = a + bX_i$. For instance, suppose that in the population, $y_i = x_i^3$. Condition (18) won't hold.

Comment. The bias will be small when n is large. Even then, don't trust the standard errors: see (20) for the reason.

23. Lung cancer rates were going up for one reason, the population was increasing for another. This is association not causation.

Comment. Lung cancer death rates for men increased rapidly from 1950 to 1990 and have been coming down since then; cigarette smoking peaked in the 1960s. Women started smoking later and stopped later, so their death rates peaked about 10 years after the men. The population was increasing steadily.

(: Maybe crowding affects women more than men :)

24. Answer omitted.

25. Something is wrong. The SE for the sample mean is $\sqrt{110/25} \doteq 2.10$, so $t \doteq 5.8/2.1 \doteq 2.76$ and $P \doteq 0.01$.

26. The social scientist is a little mixed up. The whole point of GLS is to downweight observations with high variance—and on the whole, those are the observations that are far from their expected values. Feasible GLS tries to imitate real GLS: that means downweighting discrepant observations. If $\hat{G}$ in (25) is a good estimate for G in (19), then FGLS works like a champ. If not, not.

27. (i) is the model and (ii) is the fitted equation; b is the parameter and 0.755 is the estimate; ϵ_i is an unobservable error term and e_i is an observable residual.

28. The sample mean (iv) is an unbiased estimator for $E(X_1)$.

29. The statements are true, except for (c) and (e).

30. Analysis by treatment received can be severely biased, if the men who accept screening are different from the ones who decline. Analysis by intention to treat is the way to go (section 1.2).

Comment. Data in the paper can be used to do the intention-to-treat analysis. Screening has no effect on the death rate. Apparently, the kind of men who accept screening are at lower risk from the disease than those who refuse. See the table below; also see Ruffin (1999). US Preventive Services Task Force (2002) recommends against routine PSA screening.

| | Invitation Group | | | Control Group | | |
| | Number of | | Death | Number of | | Death |
	men	deaths	rate	men	deaths	rate
Screened	7348	10	14	1122	1	9
Not screened	23785	143	60	14231	74	52
Total	31133	153	49	15353	75	49

Data from figure 4 in Labrie et al (2004); deaths due to prostate cancer.

Chapter 5 Path Models

Exercise Set A

1. Answer omitted.

2. a is a parameter; the numbers are all estimates. The 0.753 is an estimate for the standard deviation of the error term η in equation (3).

3. True. Variables are standardized, so the residuals automatically have mean 0. The variance is the mean square. With these diagrams, it is conventional to divide by n not $n - p$: this is fine if n is large and p is small, which is the case here.

4. In matrix notation, after fitting, we get $Y = X\hat{\beta} + e$ where $e \perp X$. So $\|Y\|^2 = \|X\hat{\beta}\|^2 + \|e\|^2$. In particular, $\|e\|^2 \le \|Y\|^2$ and $\|e\|^2/n \le \|Y\|^2/n$. Since Y is standardized, $\|Y\|^2/n = 1$.

Comments. (i) Here, it is irrelevant that X is standardized. (ii) If we divide by $n - p$ rather than n, then var(e) may exceed 1.

5. The SD. Variance is the wrong size (section 2.4).

6. These arrows were eliminated by assumption. (On the other hand, if you put them and compute the path coefficients from table 1, they're pretty small.) There could in some sense be an arrow from Y to U, because people train themselves in order to get certain kinds of jobs. (A dedicated path analyst might respond by putting in "plans" as a latent variable driving education and occupation.)

7. Intelligence and motivation are mentioned in the text—as are mothers. Other possibilities include race, religion, area of residence, ...

8. When a variable is lumpy, $Y = X\beta + \epsilon$ isn't a good representation, because $X_i\beta + \epsilon_i$ can usually take a lot of different values—β varies and ϵ_i is random additive noise—whereas Y_i takes only a few values.

Exercise Set B, Chapter 5

1. v is for data, σ^2 is for random variables.

2. Write the model as $y_i = a + bx_i + \epsilon_i$; the ϵ_i are IID with mean 0 and variance σ^2, for $i = 1, \ldots, n$. The fitted equation is

$$y_i = \hat{a} + \hat{b}x_i + e_i. \tag{$*$}$$

Now

$$v = \frac{1}{n}\sum_{i=1}^{n}(x_i - \bar{x})^2 \quad \text{and} \quad s^2 = \frac{1}{n}\sum_{i=1}^{n}e_i^2.$$

Next, $\bar{y} = \hat{a} + \hat{b}\bar{x}$ because $\bar{e} = 0$. Then $y_i - \bar{y} = \hat{b}(x_i - \bar{x}) + e_i$. The sample covariance between x and y is

$$\frac{1}{n}\sum_{i=1}^{n}(x_i - \bar{x})(y_i - \bar{y}) = \frac{1}{n}\sum_{i=1}^{n}(x_i - \bar{x})[\hat{b}(x_i - \bar{x}) + e_i] = \hat{b}v$$

because $\bar{e} = 0$ and $e \perp x$. Similarly, the sample variance of y is $\mathrm{var}(y) = \hat{b}^2 v + s^2$. The standardized slope is the correlation between x and y, namely,

$$\frac{\mathrm{cov}(x, y)}{\sqrt{\mathrm{var}(x)\mathrm{var}(y)}} = \frac{\hat{b}v}{\sqrt{v(\hat{b}^2 v + s^2)}} = \frac{\hat{b}\sqrt{v}}{\sqrt{\hat{b}^2 v + s^2}}.$$

3. Suppose b is positive (as it would be for a spring). If σ^2 is small, the right side of (9) will be nearly 1, which tells us that the data fall along a straight line. This is fine as far as it goes, but is not informative about the stretchiness of the spring.

Exercise Set C, Chapter 5

1. The -0.35 is an estimate. It estimates the parameter β_2 in (10).

No answers supplied for exercises 2–4.

5. Lumpiness makes linearity a harder sell; see exercise A8.

6. $Y_i = \hat{a} + \hat{b}U_i + \hat{c}V_i + e_i$, so $\bar{Y} = \hat{a} + \hat{b}\bar{U} + \hat{c}\bar{V}$ and $Y_i - \bar{Y} = \hat{b}(U_i - \bar{U}) + \hat{c}(V_i - \bar{V}) + e_i$. Then

$$\frac{Y_i - \bar{Y}}{s_Y} = \hat{b}\frac{s_U}{s_Y}\frac{U_i - \bar{U}}{s_U} + \hat{c}\frac{s_V}{s_Y}\frac{V_i - \bar{V}}{s_V} + \frac{e_i}{s_Y}.$$

The standardized coefficients are $\hat{b}s_U/s_Y$ and $\hat{c}s_V/s_Y$.

Comment. If you normalize $\hat{\sigma}^2$ by $n - p$, the t-statistics for $\hat{b}$ are the same whether you standardize or not. Ditto for $\hat{c}$. If you want standardized coefficients to estimate parameters, the setup is explained in

http://www.stat.berkeley.edu/users/census/standard.pdf

Exercise Set D, Chapter 5

1. (a) $450 + 30 = 480$.   (b) $450 + 60 = 510$.
    (c) $450 + 30 = 480$.   (d) $450 + 60 = 510$.

Comment. You get the same answer for both subjects. That assumption is built into the response schedule.

2. All she needs is observational data on hours of coaching and Math SAT scores for a sample of coachees—*if* she's willing to assume the response schedule and exogeneity of coaching hours. Exogeneity is the additional

assumption. She would estimate the parameters by running a regression
of Math SAT scores on coaching hours.

Comments. (i) Response schedules and exogeneity are very strong assumptions. People do experiments because these assumptions seem unrealistic.

(ii) The constant intercept is particularly unattractive here. Some researchers might try a *fixed-effects* model, $Y_{i,x} = a_i + bx + \delta_i$. The intercept a_i varies from one coachee to another, and takes individual ability into account. Absent repeated measures, it might be assumed that there was no relationship between a_i and the amount of coaching taken by i—although this assumption, like the constancy of b, is not completely plausible. The "no-relationship" assumption could also be implemented in a *random-effects* model, where a_i is chosen at random from a population of possible intercepts. This is equivalent to the model we began with, with a the average of the possible intercepts; $a_i - a$ goes into δ_i. In many contexts, random-effects models are fanciful (Berk and Freedman 2003).

Exercise Set E, Chapter 5

1.   There would be two free arrows, one pointing into X and one into Y, representing the error terms in the equations for X and Y, respectively. The curved line represents association. There are two equations: $X_i = a + bU_i + cV_i + \delta_i$ and $Y_i = d + eU_i + fX_i + \epsilon_i$. We assume that the δ's are IID with mean 0 and variance σ^2; the ϵ's are IID with mean 0 and variance τ^2; the δ's are independent of the ϵ's. The parameters are a, b, c, d, e, f, also σ^2 and τ^2. You need U_i, V_i, X_i, Y_i for many subjects i, with the U's and V's independent of the δ's and ϵ's (exogeneity). You regress X on U, V, with an intercept; then Y on U, X, again with an intercept. There is no reason to standardize.

For causal inference, you would need to assume response schedules:

$$X_{i,u,v} = a + bu + cv + \delta_i, \tag{$*$}$$
$$Y_{i,u,v,x} = d + eu + fx + \epsilon_i. \tag{$**$}$$

There is no v on the right hand side of $(**)$ because there is no arrow leading directly from V to Y. You would need the usual assumptions on the error terms, and exogeneity.

You could conclude qualitatively that X affects (or doesn't affect) Y, depending on the significance of $\hat{f}$. You could conclude quantitatively that if X is increased by one unit, other things being held equal (namely, U and V), then Y would go up $\hat{f}$ units.

2. Answer omitted.

3. You just regress Z on X and Y. Do not standardize: for instance, you want to estimate e. The coefficients have a causal interpretation, in view of the response schedule. See section 5.4.

4. (a) False: no arrow from V to Y. (b) True. (c) True. (d) False.

5. (a) Use (17). The answer is b.
   (b) Use (18). The answer is $(13 - 12)d + (5 - 2)e = d + 3e$.

Comments. (i) By assumption, intervening doesn't change the parameters. (ii) The effects would be estimated from the data as $\hat{b}$ and $\hat{a} + 3\hat{e}$, respectively.

6. Disagree. The test tries to tell you whether an effect is zero or non-zero. It does not try to tell you about the size of the effect. See exercise 4D4.

Discussion questions, Chapter 5

1. You don't expect much change in Mrs. Wang. What the 0.57 means is this. If you draw the graph of averages for the data (figure 2.2), the dots slope up, more or less following a line—the regression line. That line has slope 0.57. So, let's fix some number of years of education, call it x, and compare two groups of women:

   (i) all the women whose husband's educational level was x years, and
   (ii) all the women whose husband's educational level was $x + 1$ years.

   The second group should have higher educational level—higher by around 0.57 years, on average.

2. (a) True.   (b) True.   (c) True.   (d) True.   (e) False.
   The computer is a can-do machine. It runs the regressions whether assumptions are true or false. (Although even the computer has trouble if the design matrix is rank-deficient.) The trouble is this. If errors are dependent, the SEs that the computer spits out can be quite biased (section 4.4). If the errors don't have mean 0, bias in $\hat{\beta}$ is another big issue.

3. (a) country, 72.
   (b) IID, mean 0, variance σ^2, independent of the explanatory variables.
   (c) Can't get $\hat{a}$ or the other coefficients without the data. You can estimate the standardized equation from the correlations.
   (d) Controlling for the other variables reversed the sign.
   (e) The t-statistics (and signs) will be the same in the standardized equation and the raw equation—you're just changing the scale. See exercise 5C6.

(f) Not clear why the assumptions make sense, or where a response schedule would come from. What intervention are we talking about?? Even if we set all such objections to one side, it is very odd to have CV on the right hand side of the equation. Presumably, as modelers would see things, CV is caused by PO; if so, it's endogenous. If you regress PO on FI and EN only, then FI has a tiny beneficial effect. If you regress CV on PO, FI, and EN (or just on FI and EN), then FI has a strong beneficial effect. The data show that foreign investment is harmful only if you insist on a set of rather arbitrary assumptions.

4. Take two people i and j in the same ethnic group, living in the same town: $\delta_i = Y_i - X_i\beta$, $\delta_j = Y_j - X_j\beta$, and $\delta_i - \delta_j = (X_j - X_i)\beta$ because $Y_i = Y_j$. In this model, independence cannot be. The standard errors and significance levels aren't reliable. The analysis is off the rails.

5. The diagram unpacks into five regression equations. The first equation is GPQ $= a$ABILITY $+ \delta$. The second is PREPROD $= b$ABILITY $+ \epsilon$. And so forth. The usual assumptions are made about the error terms. The numbers on the arrows are estimated coefficients in the equations. For instance, $\hat{a} = 0.62$, $\hat{b} = 0.25$, etc.

The good news—from the perspective of Rodgers and Maranto—must be the absence of an arrow that goes directly from GPQ to CITES, and the small size of the path coefficients from GPQ to QFJ and QFJ to PUBS or CITES. People will cite your papers even if you didn't get your PhD from a "prestigious graduate program." The bad news seems to be that GPQ has a positive indirect effect on CITES through QFJ. If two researchers are equal on SEX and ABILITY, the one with the PhD from Podunk University will have fewer CITES.

The news is less than completely believable. First of all, this is a very peculiar sample. Who are the $86 + 76 = 162$ people with data? Second, what do the measurements mean? (For instance, ABILITY is all based on circumstantial evidence—where the subject did the undergraduate degree, what others thought of the subject as an undergraduate, etc.) Third, why should we believe any of the statistical assumptions? Just to take one example, PREPROD is going to be a small whole number $(0, 1, 2, \ldots)$, and mainly 0. How can this be the left hand side variable in a regression equation? Next, well, maybe that's enough.

6. The data are inconsistent—measurement error. Let a be the exact weight of A, b the exact weight of B, etc. It will be easier to use offsets from a kilogram, so $a = 53 \, \mu g$; b is the difference between the exact weight

of B and 1 kg, etc. The parameters are b, c, d. The first line in the table says $a + b - c - d + \delta_1 = 42$, where δ_1 is measurement error. So $b - c - d + \delta_1 = 42 - a = -11$. The second line in the table says $a - b + c - d + \delta_2 = -12$, so $-b + c - d + \delta_2 = -12 - a = -65$. And so forth. Weights on the left hand balance pan come in with a plus sign; on the right, with a minus sign. We set up the regression model in matrix form as follows:

$$\text{observed value of } Y = \begin{pmatrix} -11 \\ -65 \\ -43 \\ -12 \\ +36 \\ +64 \end{pmatrix}, \quad Y = \begin{pmatrix} +1 & -1 & -1 \\ -1 & +1 & -1 \\ -1 & -1 & +1 \\ +1 & +1 & -1 \\ +1 & -1 & +1 \\ -1 & +1 & +1 \end{pmatrix} \begin{pmatrix} b \\ c \\ d \end{pmatrix} + \begin{pmatrix} \delta_1 \\ \delta_2 \\ \delta_3 \\ \delta_4 \\ \delta_5 \\ \delta_6 \end{pmatrix}.$$

In the last three rows, A is on the right, so you have to add 53 to the difference, not subtract. Assume the δ's are IID with mean 0 and variance σ^2—which in this application seems pretty reasonable. OLS gives $\hat{b} = 33$, $\hat{c} = 26$, $\hat{c} = 44$. The SEs are all estimated as 17. (There is a lot of symmetry in the design matrix.) Units are μg.

Comment. You can do the regression with a pocket calculator, but it's easier on the computer.

7. Answer omitted.

8. The average response of the subjects assigned to treatment at level 0 is an unbiased estimate of α_0. Likewise for α_{10} and α_{50}. (This follows from question 7: the subjects assigned to treatment at level 10 are a simple random sample of the population, and so are the subjects assigned to treatment at level 50; the average of a simple random sample is an unbiased estimate of the population average.) You can't get α_{75} without assuming a functional form for the response schedule—another reason why people model things. On the other hand, if you get the functional form wrong. . . .

9. Randomization doesn't justify the model. Why would the response be linear? For example, suppose that in truth, $y_{i,0} = 0$, $y_{i,10} = 0$, $y_{i,50} = 3$, $y_{i,75} = 3$. There is some kind of threshold, then the effect saturates. If you fit a straight line to the data, you will look pretty silly. *If* the linear model is right, yes, you can extrapolate to 75.

10. Like 9.

11. If $E(X_i \epsilon_i) = 0$, OLS will be asymptotically unbiased. If $E(\epsilon_i | X_i) = 0$, OLS will be exactly unbiased. Neither of these conditions is given. For

instance, suppose $p = 1$, the Z_i are IID $N(0, 1)$. Let $X_i = Z_i$, $\epsilon_i = Z_i^3$, and $Y_i = X_i \beta + \epsilon_i = \beta Z_i + Z_i^3$, where β is scalar. By exercise 3B15 and the law of large numbers, the OLS estimator is

$$\hat{\beta} = \frac{\sum_1^n X_i Y_i}{\sum_1^n X_i^2} = \beta + \frac{\sum_1^n Z_i^4}{\sum_1^n Z_i^2} \rightarrow \beta + \frac{E(Z_i^4)}{E(Z_i^2)}.$$

The bias is about 3 because $E(Z_i^4) = 3$, $E(Z_i^2) = 1$. See end notes to chapter 4.

12. Experiments are the best, because they minimize confounding. However, they are expensive, and they may be unethical or impossible to do. Natural experiments are second-best. They're hard to find, and data collection is expensive. Modeling is relatively easy: you control (or look like you're controlling) for many confounders, and sometimes you get data on a good cross section of the population you're interested in. This point is worth thinking about, because in practice, investigators often have very funny samples to work with. On the other hand, models need a lot of assumptions that are hard to understand, never mind verify. See text for more examples and discussion.

13. False. The computer only cares whether the design matrix has full rank.

14. $\hat{\epsilon}$ must be 0, because $\hat{Y} = X\hat{\beta}$ by definition.

15. The OLS assumptions are wrong, so the formulas for SEs aren't trustworthy.

Discussion. The coefficient of X_i^2 in the definition of $\epsilon_i = X_i^4 - 3X_i^2$ makes $E(\epsilon_i) = 0$. Odd moments of X_i vanish by symmetry, so $E(X_i \epsilon_i) = 0$. The upshot is this. The ϵ_i are IID, and $E(X_i) = E(\epsilon_i) = E(X_i \epsilon_i) = 0$. So $E\{[Y_i - a - bX_i]^2\} = E\{[-a + (1-b)X_i + \epsilon_i]^2\} = a^2 + (b-1)^2 + \text{var}(\epsilon_i)$ is minimized when $a = 0$ and $b = 1$. In other words, the true regression line has intercept 0 and slope 1. The sample regression line is an estimate of the true regression line. But ϵ_i is totally dependent on X_i. So the OLS assumptions break down. In fact, the usual formula for the SE is off by a factor of 3 or 4 when applied to the slope. (This is easiest to see by simulation, but an analytic argument is possible.)

The scale factor 0.025 was chosen to get the high R^2, which can be computed using the normal moments (see end notes to chapter 4). Asymptotically, the sample R^2 is

$$\left[\frac{\text{cov}(X_i, Y_i)}{\text{SD}(X_i)\text{SD}(Y_i)} \right]^2,$$

which equals

$$\frac{\text{cov}(X_i, Y_i)^2}{\text{var}(X_i)\text{var}(Y_i)} = \frac{1}{1 + 0.025^2 E\left[\left(X_i^4 - 3X_i^2\right)^2\right]} = 0.9744.$$

Conclusion: R^2 measures goodness of fit, not validity of model assumptions. For other examples, see

http://www.stat.berkeley.edu/users/census/badols.pdf

16. The relationship is causal, but your estimates will be biased unless $\rho = 0$.

17. Choose (i) and (iv), dismiss the others. The null and alternative hypotheses constrain parameters in the model. See exercise 4D3.

18. 24.6, $\sqrt{29.4} \doteq 5.4$. Take the square root to get the SD.

19. The quote confuses bias with chance error. On average, across the various splits into treatment and control, the two groups are exactly balanced: no bias. With a randomized controlled experiment, there is no confounding. On the other hand, for any particular split, there is likely to be some imbalance. That will be part of the chance error in estimated treatment effects. Moreover, looking at a lot of baseline variables almost guarantees that some differences will be "significant" (section 4.9).

20. No. Use the Gauss-Markov theorem (section 4.4).

Chapter 6 Maximum Likelihood

Exercise Set A

1. No coincidence. When the random variables are independent, the likelihood function is a product, so the log likelihood function is a sum.

2. No answer supplied for (a). For (b), $P(U < y) = \int_{-\infty}^{y} \phi(z)dz$ and $P(-U < y) = P(U > -y) = \int_{-y}^{\infty} \phi(z)dz$, where ϕ is the standard normal density. The integrals are areas under ϕ, which is symmetric; the areas are therefore equal. More formally, change variables in the second integral, putting $w = -z$.

3. The MLE is S/n, where S is Binomial(n, p). This is only asymptotically normal. The mean is p and the variance is $p(1 - p)/n$.

4. The MLE is S/n, where S is Poisson$(n\lambda)$. This is only asymptotically normal. The mean is λ and the variance is λ/n. Watch it: S/n isn't Poisson.

Comment. The normal, Poisson, and binomial examples are exponential families in the "mean parameterization." In such cases, the MLE is unbiased and option (i) in the theorem gives the exact variance. Generally, the MLE is biased and the theorem only gives approximate variances.

5. $P\{\theta U/(1 - U) > x\} = P\{U > x/(\theta + x)\} = 1 - [x/(\theta + x)] = \theta/(\theta + x)$, so the density is $\theta/(\theta + x)^2$. This is one way to construct the random variables in example 4, section 6.1.

6. The likelihood is $\theta^n / \prod_1^n [(\theta + X_i)^2]$. So

$$L_n(\theta) = n \log \theta - 2 \sum_1^n \log(\theta + X_i).$$

Then

$$L'_n(\theta) = \frac{n}{\theta} - 2 \sum_1^n \frac{1}{\theta + X_i}.$$

$$\theta L'_n(\theta) = n - 2 \sum_1^n \frac{\theta}{\theta + X_i}$$

$$= n - 2 \sum_1^n \left(1 - \frac{X_i}{\theta + X_i}\right) = -n + 2 \sum_1^n \frac{X_i}{\theta + X_i}$$

as required. But $X_i/(\theta + X_i)$ is a decreasing function of θ. Finally, $\theta L'_n(\theta)$ tends to n as θ tends to 0, while $\theta L'_n(\theta)$ tends to $-n$ as θ tends to ∞. Hence $\theta L'_n(\theta) = 0$ has exactly one root.

7. The median is θ.

8. The Fisher information is $\theta^{-2} - 2\theta \int_0^\infty (\theta + x)^{-4} dx$.

9. Let $S = X_1 + \cdots + X_n$. The MLE for λ is S/n, so the MLE for θ is $(S/n)^2$. This is biased: $E[(S/n)^2] = [E(S/n)]^2 + \text{var}(S/n) = \lambda^2 + (\lambda/n) = \theta + (\sqrt{\theta}/n)$.

10. The MLE is $\sqrt{S/n}$. Biased.

Comment. Generally, if $\hat{\lambda}$ is the MLE for a parameter λ, and f is a smooth 1–1 function, $f(\hat{\lambda})$ is the MLE for $f(\lambda)$. Even if $\hat{\lambda}$ is unbiased, however, you should expect bias in $f(\hat{\lambda})$ unless f is linear. For math types, if X is a positive random variable with a finite mean, not a constant, then $E(\sqrt{X}) < \sqrt{E(X)}$. Generally, if f is strictly concave, $E(f(X)) < f(E(X))$: this is Jensen's inequality.

11. Use the MLE. The likelihood function is

$$\prod_{i=1}^{20} \exp(-\beta i) \frac{(\beta i)^{X_i}}{X_i!}.$$

You write down the log likelihood function, simplify, differentiate, set the derivative to 0, and solve: $\hat{\beta} = \sum_{i=1}^{20} X_i / \sum_{i=1}^{20} i = \sum_{i=1}^{20} X_i/210$.

Comment. In this exercise and the next one, the random variables are independent but not identically distributed. Theorem 1 can be extended to cover that case, although options (i) and (ii) for asymptotic variance get a little more complicated. For instance, (i) becomes $\left\{ -E_{\theta_0}\left[L_n''(\theta_0) \right] \right\}^{-1}$.

12. The log likelihood function $L(\alpha, \beta)$ is

$$-\frac{1}{2}\left[(X - \alpha - \beta)^2 + (Y - \alpha - 2\beta)^2 + (Z - 2\alpha - \beta)^2 + 3\log(2\pi) \right].$$

Maximizing L is the same as minimizing the sum of squared residuals. (Also see discussion question 4.1.)

Comment. In the OLS model with IID $N(0, \sigma^2)$ errors and a fixed design matrix of full rank, the MLE for β coincides with the OLS estimator and is therefore unbiased (theorem 4.2). The MLE for σ^2 is the mean square of the residuals, with division by n not $n - p$, and is therefore biased (theorem 4.4).

13. $c(\theta) = \theta$: that's what makes $\sum_{j=0}^{\infty} P_\theta\{X_i = j\} = 1$. Use the MLE to estimate θ. (You should write down the log likelihood function and differentiate it.)

14. $L_n(\theta) = -\sum_1^n |X_i - \theta| - 2\log n$. This is maximized (because the sum is minimized) when θ is the median. See exercise 2B18.

Exercise Set B, Chapter 6

1. All the statements are true, except for (c): the probability is 0.

2. (a) X_i is the 1×4 vector of covariates for subject i, namely, 1, ED_i, INC_i, MAN_i. And β is the 4×1 parameter vector in the probit model: see text.

     (b) random, latent.

     (c) The U_i should be IID $N(0, 1)$ and independent of the covariates.

     (d) sum, term, subject.

3. False. The difference in probabilities is
$$\Phi(0.29) - \Phi(0.19) = 0.61 - 0.57 = 0.04.$$

Exercise Set C, Chapter 6

1. $E(X) = \mu$ so μ is estimable—the estimator is X. Next, $\text{var}(X) = \sigma^2$. The distribution of X determines σ^2, so σ^2 is identifiable. Watch it: $\text{var}(X)$ is computed not from X but from the distribution of X.

2. Both parameters are estimable: $E(X_1) = \alpha$, $E[(X_2 - X_1)/9] = \beta$. (You would get smaller variances with OLS, but the exercise only asks for unbiased estimators.)

3. If the rank is p, then β is estimable—use OLS—hence identifiable. If the rank is $p - 1$, there will be a $\gamma \neq 0_{p \times 1}$ with $X\gamma = 0_{n \times 1}$. If β is any multiple of γ, we get the same distribution for $Y = X\beta + \epsilon = \epsilon$, so β is not identifiable. That is why the rank condition is important.

4. Let δ_i be IID with mean 0 and variance σ^2. Let $\epsilon_i = \mu_i + \delta_i$. Then $Y = (X\beta + \mu) + \delta$. So $X\beta + \mu$ is estimable. But the pieces $X\beta$, μ aren't separately identifiable. As long as $X\beta + \mu$ stays the same, so does the distribution of Y.

5. p^3 is identifiable. If $p^3 \neq q^3$, then $p \neq q$ and
    $$P_p(X_1 = 1) \neq P_q(X_1 = 1).$$
    However, p^3 is not estimable. For the proof, let g be a function on pairs of 0's and 1's. Then $E_p\{g(X_1, X_2)\}$ is
    $$p^2 g(1, 1) + p(1 - p)g(1, 0) + (1 - p)pg(0, 1) + (1 - p)^2 g(0, 0).$$
    This is a quadratic function of p, not a cubic.

6. The sum of two independent normal variables is normal, so $U + V$ is $N(0, \sigma^2 + \tau^2)$. Therefore, $\sigma^2 + \tau^2$ is identifiable, even estimable—try $(U + V)^2$ for the estimator—remember that $E(U) = E(V) = 0$. But σ^2 and τ^2 aren't separately identifiable. If you want to add something to σ^2, just subtract the same amount from τ^2; that won't change the distribution of $U + V$.

7. If W is $N(\mu, 1)$ and $X \sim |W|$, then $E(X^2) = E(|W|^2) = E(W^2) = \mu^2 + 1$.

8. This question is well beyond the scope of the book, but the argument is sketched, for whatever interest it may have.

    $|\mu|$ *is not estimable.* Let f be a Borel function on $(-\infty, \infty)$. Assume by way of contradiction that $E[f(\mu + Z)] = |\mu|$ for all μ, with Z being $N(0, 1)$: we can afford to set $\sigma^2 = 1$. So
    $$E[f(\mu + Z)] = \frac{1}{\sqrt{2\pi}} \int_{-\infty}^{\infty} f(\mu + z) \exp(-z^2/2) \, dz = |\mu|. \qquad (*)$$

Let $g(x) = f(x)e^{-x^2/2}$. Set $x = \mu + z$ in (*) to see that

$$\int_{-\infty}^{\infty} e^{\mu x} g(x)\, dx = \sqrt{2\pi}\, e^{\mu^2/2}|\mu|. \qquad (**)$$

The idea is to show that the left side of (**) is a smooth function of μ, which is impossible at $\mu = 0$: look at the right side of the equation! We plan to differentiate the left side of (**) with respect to μ, using difference quotients—the value at $\mu + h$ minus the value at μ—with $0 < h < 1$. Start with $0 < x < \infty$. Because $h \to e^{hx}$ is an increasing convex function of h for each x,

$$0 < \frac{e^{(\mu+h)x} - e^{\mu x}}{h} = e^{\mu x}\frac{e^{hx} - 1}{h} < e^{\mu x}(e^x - 1) < e^{(\mu+1)x}. \qquad (\dagger)$$

Similarly, for each $x < 0$, the function $h \to -e^{hx}$ is increasing and concave, so

$$0 < \frac{e^{\mu x} - e^{(\mu+h)x}}{h} = e^{\mu x}\frac{1 - e^{hx}}{h} < e^{\mu x}|x| < e^{(\mu-1)x}. \qquad (\ddagger)$$

Equation (**) says that $x \to e^{\mu x}g(x) \in L^1$ for all μ. So $x \to e^{\mu x}g^+(x) \in L^1$ and $x \to e^{\mu x}g^-(x) \in L^1$ for all μ, where g^+ is the positive part of g and g^- is the negative part. Then $x \to e^{(\mu\pm1)x}g^\pm(x) \in L^1$ for all choices of signs. Apply the dominated convergence theorem separately to four cases: (i) g^+ on the positive half-line, (ii) g^+ on the negative half-line, (iii) g^- on the positive half-line, (iv) g^- on the negative half-line. Equations ($\dagger$) and ($\ddagger$) make this work. The conclusion is, we can differentiate under the integral sign:

$$\frac{\partial}{\partial \mu}\int_{-\infty}^{\infty} e^{\mu x}g(x)\, dx = \int_{-\infty}^{\infty} e^{\mu x}xg(x)\, dx$$

where the integral on the right converges absolutely. If you look back at (**), there is a contradiction: $|\mu|$ is not differentiable at 0. The conclusion: $|\mu|$ is not estimable.

σ^2 *is not estimable from a sample of size* 1. Let f be a Borel function on $(-\infty, \infty)$. Assume by way of contradiction that $E[f(\mu + \sigma Z)] = \sigma^2$ for all μ, σ. Let $g(x) = f(-x)$; then $E[g(\mu + \sigma Z)] = \sigma^2$ for all μ, σ too. We may therefore assume without loss of generality that f is symmetric: if not, replace f by $[f(x) + g(x)]/2$. Take the case $\mu = 0$.

The uniqueness theorem for the Laplace transform says that $f(x) = x^2$ a.e. But now we have a contradiction, because $E[(\mu + \sigma Z)^2] = \mu^2 + \sigma^2$ not σ^2. For the uniqueness theorem, see p. 243 in Widder (1946).

Comment. Let Z be $N(0, 1)$. An easier version of the first part of exercise 8 might ask if there is a function f such that $E[f(\mu + \sigma Z)] = |\mu|$ for all real μ and all $\sigma \geq 0$. There is no such f. (We proved a stronger assertion, using only $\sigma = 1$.) To prove the weaker assertion—which will be easier—take $\sigma = 0$, concluding that $f(x) = |x|$ for all x. Then take $\mu = 0, \sigma = 1$ to get a contradiction. (This neat argument is due to Russ Lyons.)

Exercise Set D, Chapter 6

Most answers are omitted. For exercise 1, the last step is $P\{F(X) < y\} = P\{X < F^{-1}(y)\} = F(F^{-1}(y)) = y$. For exercise 7, the distribution is logistic. For exercise 11, one answer is sketched in the hints. Here is a more elegant solution, due to Russ Lyons. With φ as in exercise 9,

$$L_n(\beta) = \sum_i \varphi(X_i \beta) + \left(\sum_i X_i Y_i\right)\beta,$$

the last term being linear in β. If x, x^* are real numbers, then

$$\varphi\left(\frac{x + x^*}{2}\right) \geq \frac{\varphi(x) + \varphi(x^*)}{2} \tag{†}$$

by exercise 9, the inequality being strict if $x \neq x^*$. Let $\beta \neq \beta^*$. By (†),

$$\sum_i \varphi\left(X_i \frac{\beta + \beta^*}{2}\right) \geq \sum_i \frac{\varphi(X_i \beta) + \varphi(X_i \beta^*)}{2}. \tag{‡}$$

If X has full rank, there is an i with $X_i \beta \neq X_i \beta^*$, and the inequality in (‡) must be strict. Reminder: f is concave if $f[(x+x^*)/2] \geq [f(x)+f(x^*)]/2$, and strictly concave if the inequality is strict when $x \neq x^*$. If f is smooth, then f is strictly concave when $f''(x) < 0$.

Exercise Set E, Chapter 6

1. 0.777 is an estimate for the parameter α. This number is on the probit scale. Next, 0.041 is an estimate for another parameter in equation (1), namely, the coefficient of the dummy variable FEMALE (one of the covariates in X_i).

2. This number is on the probit scale. Other things being equal, students whose parents have some college education are less likely to graduate than students whose parents have a college degree. (Look at table 1 in

Evans and Schwab to spot the omitted category.) How much less likely?
The estimate is, 0.204 on the probit scale.

3. (a) α.

   (b) random, latent.

   (c) The U_i, V_i are IID as pairs across subjects i. They are bivariate
   normal. Each has mean 0 and variance 1. They are independent of
   all the covariates in both equations, namely, IsCat and X. But U_i, V_i
   are correlated within subject i. Let's call the correlation coefficient
   ρ, for future reference.

4. 0.859 estimates the parameter α in the two-equation model. This is
   supposed to tell you the effect of Catholic schools. The -0.053 estimates the parameter ρ: see 3(c) above. This correlation is small and
   insignificant, so—if the model is right—selection effects are trivial.

Comment. The 0.777 in exercise 1 and the 0.859 in exercise 4 both seem to be
estimating the same parameter α. Why are they different? Well, exercise 1 is
about the one-equation model and exercise 4 is about the two-equation model.
The models are different. The two estimates are similar because $\hat{\rho}$ is close
to 0.

5. sum, term, student.

6. The factor is

$$P\{U_{77} < -X_{77}b \text{ and } V_{77} > -X_{77}\beta\} = \int_{-X_{77}\beta}^{\infty} \int_{-\infty}^{-X_{77}b} \phi(u, v) \, du \, dv.$$

There's no a because this student isn't Catholic. There's no α because
this student didn't attend a Catholic high school.

7. The factor is

$$P\{U_{4039} < -a - X_{4039}b \text{ and } V_{4039} < -X_{4039}\beta\}$$
$$= \int_{-\infty}^{-X_{4039}\beta} \int_{-\infty}^{-a-X_{4039}b} \phi(u, v) \, du \, dv.$$

Notation. Integrals are read from the inside out. Take $\int_0^2 \int_0^1 \phi(u, v) \, du \, dv$.
First, you integrate with respect to u, over the range $[0, 1]$. Then you integrate
with respect to v, over $[0, 2]$. You might have to squint, to distinguish a from
α and b from β.

8. ρ is in ϕ: see equation (15).

9. Presumably, the two numbers got interchanged—a typo.

10. This is the standard deviation of the data—not the standard error:

$$\sqrt{0.97 \times 0.03} = 0.17.$$

The standard deviation is a useful summary statistic for quantitative data, not for 0's and 1's.

11. Unless the design matrix is a little weird, the MLE will be close to truth, so you'd nail α, β. But even if you know α, β, you don't know the latent variables. For example, suppose subject i went to Catholic school and graduated. According to the model, $V_i > -C_i\alpha - X_i\beta$. That's quite a range of possible values for V_i. In this respect, the probit model is less satisfying than the regression model.

Discussion Questions, Chapter 6

1. The MLE is generally biased, but not always. Compare exercises 6A2–4 with 6A9–10. Also see 6A12, and lab 11 below.

Comment. When the sample size is large, the bias is small, and so is the random error. (There are regularity conditions. . . .)

2. The response variables are independent conditionally on the covariates. The covariate vectors are allowed to be dependent across subjects. Covariates have to be linearly independent, i.e., perfect collinearity is forbidden: if one covariate was a linear combination of others, parameters would not be identifiable. Covariates do not have to be statistically independent, nor do they have to be orthogonal. From the modelers' perspective, that is a great advantage: you can disentangle effects even when the causes are all mixed up together in various ways.

3. (a)-(b)-(c) are true, but (d) is false. If the model is wrong, the parameter estimates may be meaningless. (What are we estimating?) Even if meaningful, the estimates are liable to be biased. Nor could we trust the standard errors printed out by the computer.

4. (a) False.   (b) True.   (c) False.   (d) True.   (e) False.   (f) True.

Comment. With respect to parts (a) and (b), the model does allow the effect of Catholic schools to be 0. The data are used to reject this hypothesis. If the model is right, the data show the effect to be large and positive.

5. Independence is violated. So is a more basic assumption—that a subject's response depends only on that subject's covariates and assignment.

6. (a) c. You could estimate the equations by maximum likelihood. (Here, coaching is binary—you either get it, or not; the response Y is continuous.)

(b) The response schedule is $Y_{i,x} = cx + V_i\beta + \sigma\epsilon_i$, where $x = 1$ means coaching, and $x = 0$ means no coaching. Nature generates the U, V, δ, ϵ according to the specifications given in the problem. If $U_i\alpha + \delta_i > 0$, she sets $X_i = 1$ and $Y_i = Y_{i,X_i} = c + V_i\beta + \sigma\epsilon_i$: subject i is coached, and scores Y_i on the SAT. If $U_i\alpha + \delta_i < 0$, Nature sets $X_i = 0$ and $Y_i = Y_{i,X_i} = V_i\beta + \sigma\epsilon_i$: subject i is not coached, and scores Y_i on the SAT. The two versions of Y_i differ by c, the effect of coaching. (You only get to see one version.)

(c) The concern is self-selection. If the smart kids choose coaching, and we just fit a response equation, we will over-estimate the effect of coaching. The assignment equation (if it's right) helps us adjust for self-selection. The parameter ρ captures the dependence between X_i and ϵ_i. This is just like Evans and Schwab, except that the outcome variable (SAT score) is continuous.

(d) In the selection equation, the scale of the latent variable is not identifiable, so Powers and Rock set it to 1. See text (section 6.2). In the response equation, there is a scale parameter σ.

Comment. Powers and Rock show, without any adjustment at all, that the effect of coaching is small. Their tables suggest that confounding makes the unadjusted effect an over-estimate. The models are decorative not essential. On the whole, the paper is persuasive as well as interesting.

7. (a) A dummy variable is 0 or 1 (section 5.6): D_{1992} is 1 for observations in 1992 and 0 for other observations; it's there in case 1992 was special in some way.

   (b) No. You have to take the interactions into account. If the Republicans buy 500 GRPs in year t and state i, then Rep.TV goes up by 5, and their share of the vote should go up by

$$5 \times \big[0.430 + 0.066 \times (\text{Rep. } AP - \text{Dem. } AP) + 0.032 \times UN + 0.006 \times RS\big]$$

where Rep. AP is evaluated in year t and state i, and likewise for the other variables.

Comment. All other factors are held constant, and we've suspended disbelief in the model. Shaw (1999, p. 352) interprets the coefficient 0.430 as meaning that a 500 GRP buy of TV time yields a 2.2 percentage point increase in votes.

8. Use logistic regression not OLS, because the response variable is binary. For the ith subject, let $Y_i = 1$ if that subject experienced a heart attack during the study period, else $Y_i = 0$. The sample size is

$$n = 6{,}224 + 27{,}034 = 33{,}258.$$

The number of variables is $p = 8$ because there is an intercept, a treatment variable, and six covariates. The design matrix X is $33{,}258 \times 8$.

Its ith row X_i is

[1 HRT$_i$ AGE$_i$ HEIGHT$_i$ WEIGHT$_i$ CIGS$_i$ HYPER$_i$ HICHOL$_i$]

where

HRT$_i$ = 1 if subject i was on HRT, else HRT$_i$ = 0,
AGE$_i$ is subject i's age,
HEIGHT$_i$ is subject i's height,
WEIGHT$_i$ is subject i's weight,
CIGS$_i$ = 1 if subject i was a smoker, else CIGS$_i$ = 0,
HYPER$_i$ = 1 if subject i had hypertension, else HYPER$_i$ = 0,
HICHOL$_i$ = 1 if subject i had high cholesterol levels,
else HICHOL$_i$ = 0.

The statistical model says that given the X's, the Y's are independent, and

$$\log \frac{P\{Y_i = 1|X\}}{1 - P\{Y_i = 1|X\}} = X_i \beta.$$

The crucial parameter is β_2, the HRT coefficient. The investigators want $\hat{\beta}_2$ to be negative (HRT reduces the risk) and statistically significant. The SE would be estimated from the observed information. Then a t-test would be made. For causal inference, we also want a response schedule and an exogeneity assumption.

The statistical model is needed to control for confounding. All the usual questions are left open. Why these variables and that functional form? Why are the coefficients constant across subjects? And so forth.

Comment. In this example, experimental evidence showed the observational data to be misleading (end notes for chapter 1 and chapter 6).

9. In an experiment, the investigator assigns the subjects to treatment or control. In an observational study, the subjects assign themselves (or are assigned by some third party). The big problem is confounding. Possible solutions include stratification and modeling. See text for discussion and examples.

10. The fraction of successes in the treatment group is an unbiased estimate of α^T. The fraction of successes in the control group is an unbiased estimate of α^C. The difference is an unbiased estimate of $\alpha^T - \alpha^C$.

11. Each model assumes linear additive effects on its own scale—look at the formulas. Randomization justifies neither model. Why would it justify one rather than the other, to say nothing of all the remaining possibilities? Just for example, treatment might help women not men. Neither model allows for this possibility.

12. Not a good idea. Here, one child's outcome may well depend on neigh-
    boring children's assignments. (Malaria is an infectious disease, trans-
    mitted by the Anopheles mosquito.)

13. Looks good so far.

14. Stratification is probably a better way to go—fewer assumptions. On the
    other hand, the groups might be heterogeneous. With more covariates
    used to define smaller groups, you may run out of data. Also, with
    stratification, there's no way to estimate what would happen with other
    values of covariates.

15. Maximum likelihood is a large-sample technique. With 400 observa-
    tions, she'd be fine. With 4, the advice is, think again.

16. The quote might be a little mixed up. White's correction is a way
    of taking heteroscedasticity into account when computing standard er-
    rors for OLS (end notes to chapter 4). The relevance to the MLE is
    not obvious. The Y_{it} will be heteroscedastic given the X's, because
    $\text{var}(Y_{it}|X) = P(Y_{it} = 0|X) \times [1 - P(Y_{it} = 0|X)]$ will depend on
    i and t. If the model is right—that's a whole other issue—the MLE
    automatically accounts for differences in $P(Y_{it} = 0|X)$ across i and t.

17. Sounds like non-identifiability.

18. Even if the model is right, and $c > 0$, the combined effect of left-wing
    power in country i and year t is

$$a \times \text{LPP}_{it} + b \times \text{TUP}_{it} + c \times \text{LPP}_{it} \times \text{TUP}_{it},   \quad (*)$$

which can be negative. It all depends on the size of a, b, c and LPP_{it},
TUP_{it}. Maybe the right wing has a point after all.

Comments. (i) With Garrett's model, the combined effect ($*$) of left-wing
power was to reduce growth rates for most years in most countries. (ii) The
ϵ_{it} are random errors, with mean 0; apparently, Garrett took these errors to
be IID in time, but allowed covariance across countries.

19. In this exercise, LPP, TUP, and the interaction don't matter—they are
    folded into Z. To create Garrett's design matrix M, which is 350×24,
    stack the data as in exercise 4C5, with 25 observations on country #1—
    ordered by year—at the top, then the observations for country #2, ...
    The first 14 columns of M are the country dummies; α_i is the coefficient
    of the dummy variable for country i. Take L to be a 24×24 matrix with
    1's along the main diagonal; the first 14 entries in the first column are
    also 1: all other entries are 0. You should check that ML gives Beck's

design matrix. Now

$$[(ML)'(ML)]^{-1}(ML)'Y = [L'(M'M)L]^{-1}L'M'Y$$
$$= L^{-1}(M'M)^{-1}(L')^{-1}L'M'Y$$
$$= L^{-1}(M'M)^{-1}M'Y.$$

If β is Garrett's parameter vector and β^* is Beck's, then $\hat{\beta}^* = L^{-1}\hat{\beta}$, so $\hat{\beta} = L\hat{\beta}^*$. (A more direct argument is possible too.)

20. In 1999, statisticians placed less reliance on the normal law of error than they did in 1899. (What will things look like in 2099?) Yule is playing a little trick on Sir Robert. If the OLS model holds, OLS estimates are unbiased. But why does the model hold? Choosing models is a rather subjective business that goes well beyond the data—especially when causation gets into the picture.

Chapter 7 The Bootstrap

Exercise Set A

1. True.

2. These statements are all true, illustrating the idea of the bootstrap. (Might be even better, e.g., to divide by 99 not 100, but we're not going to be fussy about details like that.)

3. (a) $\sigma/\sqrt{5000}$.   (b) $\sqrt{V}/\sqrt{100}$.   (c) $\sqrt{V}$.

    Reason for (b): $\overline{X}_{ave}$ is the average of 100 IID $\overline{X}_{(k)}$'s whose sample variance is V. In (c), there is no need for "around."

Exercise Set B, Chapter 7

1. Choose (ii). See text.

2. The parameters are the 10 regional intercepts a_j and the five coefficients b, c, d, e, f. These are unobservable. So are the random errors $\delta_{t,j}$. Otherwise, everything is observable.

3. The disturbance term for 1975 could have been different, and then energy consumption would have been different.

4. 0.281 is the one-step GLS estimate for the parameter f.

5. One-step GLS is biased, for estimating d, e, f: compare columns (A) and (C). The bias in $\hat{f}$, for instance, is highly significant, and comparable in size to the SE for $\hat{f}$. Not trivial.

6. Biased. Compare columns (D) and (E): see text.

7. Biased, although not as badly as the plug-in SEs. Compare columns (D) and (F): see text.

8. The bootstrap is not reliable with such a small sample. With 40 observations, maybe. But with 4?!? Maybe Paula needs another idea.

9. $\epsilon_i = Y_i - a - bY_{i-1}$ for all i. If $1 \le i < n$, then $Y_i = X_{i+1,2}$ and $Y_{i-1} = X_{i,2}$ can be computed from X. So, $\epsilon_n \perp\!\!\!\perp X$, but the earlier ϵ_i are completely dependent on X.

Chapter 8   Simultaneous Equations

Exercise Set A

1. a_1 should be positive. Supply increases with price. By contrast, a_2 and a_3 should be negative. When the price of labor and materials goes up, supply goes down.

2. b_1 should be negative. Demand goes down as price goes up. Next, b_2 should be negative. Demand goes down as the price of complements goes up. You're not going to spread butter on that ten-dollar piece of toast, because you're not going to eat that piece of toast in the first place. Finally, b_3 should be positive. Demand goes up as the price of substitutes goes up. When margarine costs $50 a pound, you throw caution to the winds and eat butter.

3. The law of supply and demand is built into the model as an assumption: Q_t and P_t are the market-clearing quantity and price. We got them by solving the supply and demand equations in year t, i.e., by finding the point where the supply curve crosses the demand curve. See figure 1, and equations (2)-(3).

   The supply curve is concave but not strictly concave. (It's linear.) The demand curve is convex but not strictly convex. (It's linear too.) For this reason among others, economists prefer log linear equations, like

   $$\log Q = a_0 + a_1 \log P + a_2 \log W + a_3 \log H + \delta_t,$$
   $$\log Q = b_0 + b_1 \log P + b_2 \log T + b_3 \log M + \epsilon_t.$$

4. Equation (2a) is the relevant one: (2b) says how consumers would respond. The reduced-form equations (3a) and (3b) say how quantity and price would respond if we manipulated the exogenous variables W_t, H_t, T_t, M_t. Notice that P_t does not appear on the right hand side of (3a); and Q_t does not appear on the right hand side of (3b).

Exercise Set B, Chapter 8

1. For part (a), let c be $p \times 1$. Then $c'Z'Zc = \|Zc\|^2 \geq 0$. If $c'Z'Zc = 0$, then $Zc = 0$ and $Z'Zc = 0$, so $c = 0$ because $Z'Z$ has full rank (this was given). Thus, $Z'Z$ is positive definite. The rest of part (a) follows from exercise 3D7. For (b), the matrix $(Z'Z)^{-1}$ is positive definite, so

$$c'X'Z(Z'Z)^{-1}Z'Xc \geq 0.$$

   Equality entails $Z'Xc = 0$, hence $c = 0$, because $Z'X$ has full rank (this was given). Thus, $X'Z(Z'Z)^{-1}Z'X$ is positive definite, hence, invertible (exercise 3D7).

2. (a) and (b) are true; (c), (d), and (e) are false.

Comment. With a large sample, the sample mean will be nearly the same as $E(U_i)$, and the sample variance will be nearly the same as $\mathrm{var}(U_i)$. But the concepts are different—and with small or medium-sized samples, so are the numbers (section 2.4).

Exercise Set C, Chapter 8

1. Don't do that without further thought. According to the model, price and quantity are endogenous. You might want to fit by OLS even so (section 8.8), but you have to consider endogeneity bias.

2. Statements (a)-(b)-(c) are all false, unless there is some miracle of cancellation. The OLS residuals are orthogonal to X, but IVLS isn't OLS. Statement (d) is true by definition (13).

3. OLS always gives a better fit: see exercise 3B14(j). You do IVLS only if there's a model you believe in, you want to estimate the parameters in that model, and are concerned about endogeneity bias.

   (: OLS may be ordinary, but it makes the least of the squares :)

4. Biased. IVLS isn't real GLS. We're pretending that $Z'X$ is constant. But that isn't right, at least, not exactly. As the sample size grows, the bias will (with any luck) get small.

Discussion. Here is an example to illustrate the setup in section 8.2, with $q = p = 1$. Let $(Z_i, \delta_i, \epsilon_i)$ be IID triplets for $i = 1, \ldots, n$. Each triplet consists of three independent random variables. Each random variable has mean 0 and positive variance. Let $X_i = aZ_i + b\delta_i + c\epsilon_i$ and $Y_i = \beta X_i + \delta_i$, where a, b, c, β are positive scalar parameters. We wish to estimate β, but X_i is endogenous because $a > 0$. The IVLS estimator for β is $\sum_{i=1}^{n} Z_i Y_i / \sum_{i=1}^{n} Z_i X_i = \beta + \eta$, where $\eta = \sum_{i=1}^{n} Z_i \delta_i / \sum_{i=1}^{n} Z_i X_i$.

You can work out the small-sample bias in IVLS, and the bias in $\hat{\sigma}^2$, at least to a good approximation. This takes the central limit theorem, the "delta method," and some effort. For details, see

http://www.stat.berkeley.edu/users/census/ivls.pdf

Exercise Set D, Chapter 8

1. $0.128 - 0.042 - 0.0003 \times 300 + 0.092 + 0.005 \times 11$
    $$+0.015 \times 12 - 0.046 + 0.277 + 0.041 + 0.336 = 0.931.$$

2. $- 0.042 - 0.0003 \times 300 + 0.092 + 0.005 \times 11$
    $$+0.015 \times 12 - 0.046 + 0.277 + 0.041 + 0.336 = 0.803.$$

Comment. In exercises 1 and 2, the parents live in district 1, so the universal-choice dummy is 0: its coefficient (-0.035) does not come into the calculation. Frequency of church attendance is measured on a scale from 1 to 7, with "never" coded as 1. The 0.931 is indeed too close to 1.00 for comfort. . . .

3. The difference is 0.128. This is the "effect" of school choice.

4. estimated expected probabilities. We're substituting estimates for parameters in (26), and replacing the latent variable V_i by its expected value, 0.

5. School size is a much bigger number than other numbers in the equation. For example, $-0.3 \times 300 = -90$. If the coefficient was -0.3, we'd be seeing a lot of negative probabilities.

6. No. The left hand side variable has to be a probability, not a 0–1 variable. Equation (2) in Schneider et al is about estimation, not modeling assumptions.

7. Some of the numbers line up between the sample and the population, but there are real discrepancies, e.g., on the educational level of parents in District 4. In the sample, 65% have a high school education or better, compared to 48% in the population. (The SE on the 65% is something like $100\% \times \sqrt{0.48 \times 0.52/333} \doteq 3\%$: this isn't a chance effect.) Schneider et al collected income data but elected not to use it. Why not? The intervention is left unclear in the paper, as is the model. The focus is on estimation technique.

Exercise Set E, Chapter 8

1. From (10),

$$\hat{\beta}_{\text{IVLS}} = \left[X'Z(Z'Z)^{-1}Z'X \right]^{-1} X'Z(Z'Z)^{-1}Z'Y = QY$$

where
$$Q = \left[X'Z(Z'Z)^{-1}Z'X\right]^{-1}X'Z(Z'Z)^{-1}Z'.$$

Now $QX = I_{p\times p}$ and $Y = X\beta + \delta$, so $QY = \beta + Q\delta$. Since Q is taken as constant (rather than random),

$$\text{cov}\{\hat{\beta}_{\text{IVLS}}|Z\} = \sigma^2 Q I_{n\times n} Q' = \sigma^2 Q Q' = \sigma^2\left[X'Z(Z'Z)^{-1}Z'X\right]^{-1}.$$

Evaluating QQ' is straightforward but tedious.

2. The $p \times p$ matrix $Z'X$ has full rank, by assumption (ii) in section 8.2. Hence, $Z'X$ is invertible. By (10),

$$\begin{aligned}
\hat{\beta}_{\text{IVLS}} &= \left[X'Z(Z'Z)^{-1}Z'X\right]^{-1}X'Z(Z'Z)^{-1}Z'Y\\
&= (Z'X)^{-1}(Z'Z)(X'Z)^{-1}X'Z(Z'Z)^{-1}Z'Y\\
&= (Z'X)^{-1}Z'Y.
\end{aligned}$$

Watch it. This only works when $q = p$. Otherwise, $Z'X$ isn't square.

3. Go with investigator #3, who is doing IVLS: exercise 2. Investigator #1 is doing OLS, which is biased. Investigator #2 is a little mixed up. To pursue that, we need some notation for the covariance matrix of $X_i, Z_i, \epsilon_i, Y_i$. This is a 4×4 matrix. The top left 3×3 corner in (∗) shows the notation and assumptions. For example, σ^2 is used to denote $\text{var}(\epsilon_i)$, ψ to denote $\text{cov}(X_i, Z_i)$, and θ to denote $\text{cov}(X_i, \epsilon_i)$. Since Z_i is exogenous, $\text{cov}(Z_i, \epsilon) = 0$. The last row (or column) is derived by math. For instance, $\text{var}(Y_i) = \beta^2\text{var}(X_i) + \text{var}(\epsilon_i) + 2\beta\text{cov}(X_i, \epsilon_i) = \beta^2 + \sigma^2 + 2\beta\theta$.

$$
\begin{array}{c}
\begin{array}{cccc}
X_i & Z_i & \epsilon_i & Y_i
\end{array}\\
\begin{array}{c}X_i\\Z_i\\\epsilon_i\\Y_i\end{array}
\left(
\begin{array}{cccc}
1 & \psi & \theta & \beta+\theta\\
\psi & 1 & 0 & \beta\psi\\
\theta & 0 & \sigma^2 & \sigma^2+\beta\theta\\
\beta+\theta & \beta\psi & \sigma^2+\beta\theta & \beta^2+\sigma^2+2\beta\theta
\end{array}
\right)
\end{array}
\qquad (*)
$$

For investigator #2, the design matrix M has a column of X's and a column of Z's, so

$$M'M/n \doteq \begin{pmatrix} 1 & \psi\\ \psi & 1 \end{pmatrix}, \qquad M'Y/n \doteq \begin{pmatrix} \beta+\theta\\ \beta\psi \end{pmatrix},$$

$$\begin{pmatrix} 1 & \psi\\ \psi & 1 \end{pmatrix}^{-1} = \frac{1}{1-\psi^2}\begin{pmatrix} 1 & -\psi\\ -\psi & 1 \end{pmatrix},$$

$$(M'M)^{-1}M'Y \doteq \begin{pmatrix} \beta + [\theta/(1 - \psi^2)] \\ -\theta\psi/(1 - \psi^2) \end{pmatrix}.$$

When n is large, the estimator for β suggested by investigator #2 is biased by $\theta/(1 - \psi^2)$. A much easier calculation shows the OLS estimator is biased by θ.

4. The correlation between Z and ϵ is not identifiable, so Z cannot be used as an instrument. Here are some details. The basic thing is the joint distribution of X_i, Z_i, ϵ_i. (These are jointly normal random variables, mean 0, and IID as triplets.) The joint distribution is specified by its 3×3 covariance matrix. In that matrix, $\text{var}(X_i)$, $\text{var}(Z_i)$ and $\text{cov}(X_i, Z_i)$ are almost determined by the data (n is large). Let's take them as known. For simplicity, let's take $\text{var}(X_i) = \text{var}(Z_i) = 1$ and $\text{cov}(X_i, Z_i) = \frac{1}{2}$. There are three remaining parameters in the joint distribution of X_i, Z_i, ϵ_i:

$$\text{cov}(X_i, \epsilon_i) = \theta, \quad \text{cov}(Z_i, \epsilon_i) = \phi, \quad \text{var}(\epsilon_i) = \sigma^2.$$

So the covariance matrix of X_i, Z_i, ϵ_i is

$$\begin{array}{c} \\ X_i \\ Z_i \\ \epsilon_i \end{array} \begin{array}{ccc} X_i & Z_i & \epsilon_i \\ \begin{pmatrix} 1 & \frac{1}{2} & \theta \\ \frac{1}{2} & 1 & \phi \\ \theta & \phi & \sigma^2 \end{pmatrix} \end{array}. \qquad (\dagger)$$

The other random variable in the system is Y_i, which is constructed from X_i, Z_i, ϵ_i and another parameter β: $Y_i = \beta X_i + \epsilon_i$. We can now make a complete list of the parameters in the system:

    (i) $\text{cov}(X_i, \epsilon_i) = \theta$,
    (ii) $\text{cov}(Z_i, \epsilon_i) = \phi$,
    (iii) $\text{var}(\epsilon_i) = \sigma^2$,
    (iv) β.

The random variable ϵ_i is not observable. The observables are X_i, Z_i, Y_i. The joint distribution of X_i, Z_i, Y_i determines—and is determined by—its 3×3 covariance matrix (theorem 3.2). This matrix can be computed from the four parameters:

$$\begin{array}{c} \\ X_i \\ Z_i \\ Y_i \end{array} \begin{array}{ccc} X_i & Z_i & Y_i \\ \begin{pmatrix} 1 & \frac{1}{2} & \beta + \theta \\ \frac{1}{2} & 1 & \frac{1}{2}\beta + \phi \\ \beta + \theta & \frac{1}{2}\beta + \phi & \beta^2 + \sigma^2 + 2\beta\theta \end{pmatrix} \end{array}. \qquad (\ddagger)$$

For example, the 2,3 element in the matrix (repeated as the 3,2 element) is supposed to be $\text{cov}(Y_i, Z_i)$. Let's check. We're given that $E(X_i) = E(Z_i) = E(Y_i) = E(\epsilon_i) = 0$. So

$$\text{cov}(Y_i, Z_i) = E(Y_i Z_i) = E[(\beta X_i + \epsilon_i) Z_i],$$

which is

$$\beta E(X_i Z_i) + E(Z_i \epsilon_i) = \beta \text{cov}(X_i, Z_i) + \text{cov}(Z_i, \epsilon_i) = \tfrac{1}{2}\beta + \phi.$$

The joint distribution of X_i, Z_i, Y_i determines—and is determined by—the following three things:

(a) $\beta + \theta$,
(b) $\tfrac{1}{2}\beta + \phi$,
(c) $\beta^2 + \sigma^2 + 2\beta\theta$.

That's all you need to fill out the matrix ($\ddagger$), and that's all you can get out of the data on X_i, Z_i, Y_i, no matter how large n is. There are three knowns: (a)-(b)-(c). There are four unknowns θ, ϕ, σ^2, β. Blatant non-identifiability.

To illustrate, let's start with the parameter values shown in column #2 of the following table.

	1	2	3
θ		$\frac{1}{2}$	$\frac{3}{2}$
ϕ		0	$\frac{1}{2}$
σ^2		1	3
β		2	1

Then (a) $\beta + \theta = 2.5$, (b) $\tfrac{1}{2}\beta + \phi = 1.0$, and (c) $\beta^2 + \sigma^2 + 2\beta\theta = 7.0$. Now, increase ϕ to $\frac{1}{2}$, as shown in column #3. Choose a new value for β so (b) doesn't change, a new θ so (a) doesn't change, and σ^2 so (c) doesn't change. The new values are shown in column #3 of the table. Both columns lead to the same numbers for (a), (b), (c), hence the same joint distribution for X_i, Z_i, Y_i. That already demonstrates non-identifiability, and there are many other possible choices of $\theta, \phi, \sigma^2, \beta$ leading to the same (a)-(b)-(c). With column #2, Z is exogenous: $\text{cov}(Z_i, \epsilon_i) = \phi = 0$. With column #3, Z is endogenous: $\text{cov}(Z_i, \epsilon_i) \neq 0$. Exogeneity cannot be determined from the joint distribution of the observables. That is the whole trouble with the exogeneity assumption.

Comments. (i) This exercise is similar to the previous one. In that exercise, $\text{cov}(Z_i, \epsilon_i) = 0$ because Z_i was given as exogenous; here, $\text{cov}(Z_i, \epsilon_i) = \phi$ is an important parameter because Z_i is likely to be endogenous. There, $\text{cov}(X_i, Z_i) = \psi$ was a free parameter; here, we chose $\psi = \frac{1}{2}$ (for no particular reason). There, we displayed the 4×4 covariance matrix of $X_i, Z_i, \epsilon_i, Y_i$. Here, we display two 3×3 covariance matrices. If you take $\phi = 0$ and $\psi = \frac{1}{2}$, the matrices (∗), (†), (‡) will all line up.

(ii) For a similar example in a discrete choice model, see

http://www.stat.berkeley.edu/users/census/socident.pdf

(iii) There is a lot of econometric theorizing about instrumental variables. What it boils down to is this. If you are willing to assume that some variables are exogenous, you can test the exogeneity of others.

5. This procedure is inconsistent: it gives the wrong answer no matter how much data you have. This is because you're estimating σ^2 with only $q - p$ degrees of freedom.

Discussion. In principle, you can work everything out for the following model, which has $q = 2$ and $p = 1$. Let $(U_i, V_i, \delta_i, \epsilon_i)$ be IID in i. The four-tuple $(U_i, V_i, \delta_i, \epsilon_i)$ is jointly normal. Each variable has mean 0 and variance 1. Although U_i, V_i, and (δ_i, ϵ_i) are independent, $E(\delta_i \epsilon_i) = \rho \neq 0$. Let $X_i = U_i + V_i + \epsilon_i$ and $Y_i = X_i \beta + \delta_i$. The unknown parameters are ρ and β. The observables are U_i, V_i, X_i, Y_i. The endogenous X_i can be instrumented by U_i, V_i. When n is large, $\hat{\beta}_{\text{IVLS}} \doteq \beta$; the residual vector from (4) is almost the same as δ. Now you have to work out the limitng behavior of the residual vector from (6), and show that it's pretty random, even with huge samples. For detail on a related example with $q = p = 1$, see

http://www.stat.berkeley.edu/users/census/ivls.pdf

The Computer Labs

Introduction

Labs are a key part of the course: the computations illustrate some of the main ideas. At Berkeley, labs are set up for MATLAB in a UNIX environment. The UNIX prompt is (usually) a percent sign. At the prompt, type `matlab`. After a bit, MATLAB will load. Its prompt is >>. If you type `edit` at the prompt, you get a program editor. Changes for WINDOWS are pretty straightforward: you can launch MATLAB from the start menu, and get the program editor by clicking on an icon in a toolbar. The directory names will look different.

Don't write MATLAB code or create data files in a word processing package like WORD, because formatting is done with a lot of funny characters that MATLAB finds indigestible. (You can work around this, but why bother?) In UNIX, `nedit` is a straight-ahead program editor. In WINDOWS, you can use `notepad` or `wordpad`, although `TextPad` is a better bet:

`http://www.textpad.com`

If you type `helpdesk` at the MATLAB prompt, you get a browser-based help facility, with demos and tutorials. If you only want help on a particular command, type `help` at the MATLAB prompt, followed by the name of the command. For instance, `help load`. This works if you know the name of the command. . . .

MATLAB runs interactively from the command prompt, and you can do a lot that way. After a while, you may want to store commands in a text file. This sort of file is called a "script file." Script files make it easier to edit and debug code. Script files end with the suffix `.m`, for instance, `demolab.m`. If you have that file on your system, type `demolab` at the MATLAB prompt. MATLAB will execute all the commands in the file. (There is an annoying technicality: the file has to be in your working directory, or on the search path: click on `File` and follow your nose, or type `help path` at the MATLAB prompt, or—if all else fails—look at the documentation.)

A lot of useful MATLAB features are illustrated in demolab.m, including "function files," which are special script files needed later in the course. Some people like to read computer documentation, and MATLAB has pretty good documentation. Or, you can just sit down at the keyboard and start fooling around. Some people like to look at code: demolab.m—listed in an appendix below—is for them.

When you are finished running MATLAB, type exit to end your session, or quit the window. (In UNIX, quitting a window is a much more final act than closing it.) Oh, by the way, what happens if your program goes berserk and you need to stop it? Just hit control-C: hold down the control-key, press C. That will return you to the command prompt. (Be patient, it may take a minute for MATLAB to notice the interrupt.)

Data sets used in the labs, and sample code, are available at

http://www.stat.berkeley.edu/users/census/data.zip

Numerics

Computers generally do "IEEE arithmetic," which isn't exactly arithmetic. There is roundoff error. MATLAB is usually accurate to 10^{-12}. It seldom does better than 10^{-16}, although it can. Here is some output:

```
>> (sqrt(2))^2-2

ans =

    4.4409e-016

>>(sqrt(4))^2-4

ans =

        0
```

4.4409e-016 is MATLAB's way of writing 4.4409×10^{-16}. This is roundoff error.

Lab 1

Summary Statistics and Simple Regression

In this lab, you will calculate some descriptive statistics for Yule's data and do a simple regression. The data are in table 1.3, and in the file

/StatMods/yule.dat

You need to subtract 100 from each entry to get the percent change. Refer to chapter 1 for more information, or to /StatMods/yuledoc.txt.

1. Compute the means and SDs of ΔPaup, ΔOut, ΔPop, and ΔOld.
2. Compute all 6 correlations between ΔPaup, ΔOut, ΔPop, and ΔOld.
3. Make a scatter plot of ΔPaup against ΔOut.
4. Run a regression of ΔPaup on ΔOut, i.e, find the slope and intercept of the regression line. You might also compute the SD of the residuals.

Useful MATLAB commands: `load`, `mean`, `std`, `corrcoef`, `plot(u,v,'x')`.

Lab 2

An Exercise with MATLAB

1. Create a 4×3 matrix X and a 4×1 vector Y:

$$X = \begin{pmatrix} 1 & -1 & 1 \\ 1 & 2 & 3 \\ 4 & 5 & 6 \\ 7 & 8 & 9 \end{pmatrix}, \qquad Y = \begin{pmatrix} 1 \\ 2 \\ 3 \\ 4 \end{pmatrix}.$$

2. Compute $X'X$, $X'Y$, det $X'X$, rank X, rank $X'X$.
3. Compute $(X'X)^{-1}$.
4. Write a single line of MATLAB code to compute

$$\hat{\beta} = (X'X)^{-1}X'Y.$$

   Report $\hat{\beta}$ as well as the code.

5. Let

$$A = \begin{pmatrix} 1 & 3 & 5 & 7 \\ -1 & 2 & 9 & -3 \\ 6 & 3 & 0 & 33 \end{pmatrix}.$$

   Compute trace AX and trace XA. Comment?

Useful MATLAB commands: `A'`, `A+B`, `A-B`, `A*B`, `det`, `inv`, `rank`, `trace`, `size`. To create a matrix, type `Q=[1 2 3; 4 5 6]`, or do it on two lines:

```
Q=[1 2 3
4 5 6]
```

Lab 3

Replicating Yule's Regression

In this lab, you will replicate Yule's regression equation for the metropolitan unions, 1871–81. See chapter 1. Fix the design matrix X at the values reported in table 1.3. (Subtract 100 from each entry to get the percent changes.) The data are in the file /StatMods/yule.dat. The file yuledoc.txt gives the variable names. Yule assumed

$$\Delta \text{Paup}_i = a + b \times \Delta \text{Out}_i + c \times \Delta \text{Old}_i + d \times \Delta \text{Pop}_i + \epsilon_i$$

for 32 metropolitan unions i. Suppose the errors ϵ_i are IID, with mean 0 and variance σ^2.

1. Estimate a, b, c, d, and σ^2.

2. Compute the SEs.

3. Are these SEs exact, or approximate?

4. Plot the residuals against the fitted values. (This is often a useful diagnostic: if you see a pattern, something is wrong with the model. You can also plot residuals against other variables, or time, or. . . .)

Useful MATLAB commands: ones(32,1), [A B].

For bonus points. If you get a different answer from Yule, why might that be?

Lab 4

Path Diagrams

In this lab, you will replicate part of Blau and Duncan's path model in figure 5.1. Equation (5.3) explains son's occupation in terms of father's occupation, son's education, and son's first job. Variables are standardized. Correlations are given in table 5.1.

1. Estimate the path coefficients in (5.3) and the standard deviation of the error term. How do your results compare with those in figure 5.1?

2. Compute SEs for the estimated path coefficients. (Assume there are 20,000 subjects.)

Lab 5

More Path Diagrams

In this lab, you will replicate Gibson's path diagram, which explains repression in terms of mass and elite tolerance (section 5.3). The correlation between mass and elite tolerance scores is 0.52; between mass tolerance scores and repression scores, -0.26; between elite tolerance scores and repression scores, -0.42. (Tolerance scores were averaged within state.)

1. Compute the path coefficients in figure 5.2.
2. Estimate σ^2. Gibson had repression scores for all the states. He had mass tolerance scores for 36 states and elite tolerance scores for 26 states. You may assume the correlations are based on 36 states—this will understate the SEs, by a bit—but you need to decide if p is 2 or 3.
3. Compute SEs for the estimates.
4. Compute the SE for the difference of the two path coefficients. You will need the off-diagonal element of the covariance matrix: see exercise 4C2(a). Comment on the result.

Note. Gibson used weighted correlations, but this makes almost no difference to the estimates. Also see exercises 5C2–4.

Lab 6

Simulation with MATLAB

1. Simulate observations on 32 IID normal variables X_i with mean $\mu = 15$ and variance $\sigma^2 = 100$.
2. Calculate the sample mean $\overline{X}$ and the sample SD $\hat{\sigma}$ of the data.
3. Repeat 1 and 2, 1000 times.
4. Plot a histogram of the 1000 $\overline{X}$'s.
5. Plot a histogram of the 1000 $\hat{\sigma}$'s.
6. Plot a scatter diagram of the 1000 pairs $(\overline{X}, \hat{\sigma})$.
7. Calculate the SD of the 1000 $\overline{X}$'s. How does this compare to $\sigma/\sqrt{32}$? Comment?

Useful MATLAB commands: `rand`, `randn`, `for...end`, `hist(x,25)`.

MATLAB loves matrices. It hates loops. Your code should include a couple of lines like

```
FakeData=randn(32,1000);
```

```
Aves=mean(FakeData);
```
Put in the semicolons, or you will spend a lot of time watching random numbers scroll by on the screen.

Random Numbers

Appearances notwithstanding, computers have no random elements. MATLAB generates "pseudo-random" numbers—numbers which look pretty random—by some clever numerical algorithm that is completely deterministic. One consequence may take you by surprise. With any particular release of the program, if you start a MATLAB session and type rand(1), you will always get the same number. (With Release 13, the answer is 0.9501.) In particular, you might get exactly the same results as all the other students in the class who are doing Lab 6. (Doesn't seem random, does it?) A work-around, if you care, is to burn some random numbers before doing a simulation—type x=rand(abcd,1);, where abcd is the last four digits of your telephone number.

Vectorizing Code

These days, computers are very, very fast. It may not pay to spend a lot of time writing tight code. On the other hand, if you are doing a big simulation, and it is running like molasses, getting rid of loops is good advice. If you have nested loops, make the innermost loop as efficient as you can.

Lab 7

The t-Test. Part 1.

Yule's model is described in chapter 1, and in Lab 3. Fix the design matrix X at the values reported in table 1.3. (Subtract 100 from each entry to get the percent changes.) Suppose the errors ϵ_i are IID $N(0, \sigma^2)$, where σ^2 is a parameter (unknown). Make a t-test of the null hypothesis that $b = 0$. What do you conclude? If you were arguing with Yule at a meeting of the Royal Statistical Society, would you want to take the position that $b = 0$ and he was fooled by chance variation?

The t-Test. Part 2.

In this part of the lab, you will do a simulation to investigate the distribution of

$$t = \hat{b}/\widehat{\text{SE}},$$

under the null hypothesis that $b = 0$.

1. Set the parameters in Yule's equation (Lab 3) as follows: $a = -40$, $b = 0$, $c = 0.2$, $d = -0.3$, $\sigma = 15$. Fix the design matrix X as in Part 1.

2. Generate 32 $N(0, \sigma^2)$ errors and plug them into the equation

$$\Delta\text{Paup}_i = -40 + 0 \times \Delta\text{Out}_i + 0.2 \times \Delta\text{Old}_i - 0.3 \times \Delta\text{Pop}_i + \epsilon_i,$$

to get simulated values for ΔPaup_i, with $i = 1, \ldots, 32$.

3. Regress the simulated ΔPaup on ΔOut, ΔPop, and ΔOld. Calculate $\hat{b}$, $\widehat{\text{SE}}$, and t.

4. Repeat 2 and 3, 1000 times.

5. Plot a histogram for the 1000 $\hat{b}$'s, a scatter diagram for the 1000 pairs $(\hat{b}, \hat{\sigma})$, and a histogram for the 1000 t's.

6. What is the theoretical distribution of $\hat{b}$? of $\hat{\sigma}^2$? of t? How close is the theoretical distribution of t to normal?

7. Calculate the mean and SD of the 1000 $\hat{b}$'s. How does the mean compare to the true b? ("True" in the simulation.) How does the SD compare to the true SE for $\hat{b}$?

You need to compute $(X'X)^{-1}$ only once, but $\hat{\sigma}^2$ many times. Your code will run faster with more matrices and fewer loops. (As they say, vectorize your code.) Try this:

```
beta=[-40 0 .2 -.3]'

sigma=15

betaSim=X\(X*beta*ones(1,1000)+sigma*randn(32,1000));
```

The backslash operator does the least squares fit.

For discussion. Would it matter if you set the parameters differently? For instance, you could try $a = 10$, $b = 0$, $c = 0.1$, $d = -0.5$ and $\sigma = 25$. What if you change b, for instance, to 0.5? What if the errors aren't normally distributed? The simulation in this lab is for the size of the test. How would you do a simulation to get the power of the test? (Size and power are defined below.)

A tangential issue. Plot a a scatter diagram for the 1000 pairs $(\hat{a}, \hat{b})$. What accounts for the pattern?

Hypothesis Testing

The discussion question in Part 2 of Lab 7 refers to *size* and *power*. To review these ideas, and put them in context, let θ be a parameter (or parameter vector). Write $\mathcal{P}_\theta$ for the probability distribution of the random variables in the model, when the parameter is θ. The null hypothesis is a set of θ's; the alternative is a disjoint set of θ's. Let T be a test statistic. We reject the null if $T > k$, where k is the *critical value*, chosen so that $\mathcal{P}_\theta(T > k) \leq \alpha$ for all θ in the null. Here α is the *size* or *level* of the test. *Power* is $\mathcal{P}_\theta(T > k)$ for θ in the alternative. This will depend, among other things, on k and θ.

From the Neyman-Pearson perspective, the ideal test maximizes power—the chance of rejecting the null when the null is false—while controlling the size, which is the chance of rejecting the null when the null is true.

With the t-test in Lab 7, the parameter vector θ is a, b, c, d, σ^2. The null is the set of θ's with $b = 0$. The alternative is the set of θ's with $b \neq 0$. The test statistic T is, surprise, $|t|$. If you want $\alpha = 0.05$, choose $k \doteq 2$. More precisely—with normal errors—you want the k such that the area beyond $\pm k$ under Student's t-density with 28 degrees of freedom is equal to 0.05. The answer to that riddle is 2.0484.... (See "Statistical Packages" below.) For our purposes, the extra precision isn't worth the bother: $k \doteq 2$ is just fine.

The *observed significance level* P or P_{obs} is $\mathcal{P}_\theta(T > T_{\text{obs}})$, where T_{obs} is the observed value of the test statistic. If you think of T_{obs} as random (i.e., before data collection), then P_{obs} is random. In Lab 7 and many similar problems, P_{obs} is uniform on $[0,1]$, provided θ satisfies the null hypothesis: $\mathcal{P}_\theta(P_{\text{obs}} < p) = p$ for $0 < p < 1$. If the null is $b \leq 0$ vs the alternative $b > 0$, then $T = t$ rather than $|t|$, and P_{obs} is uniform when $b = 0$; however, if $b < 0$ then $\mathcal{P}_\theta(P_{\text{obs}} < p) > p$ for $0 < p < 1$.

Lab 8

The F-Test. Part 1.

Yule's model is explained in chapter 1, and in Lab 3. Fix the design matrix X at the values reported in table 1.3. (Subtract 100 from each entry to get the percent changes.) Assume that the errors ϵ_i are IID $N(0, \sigma^2)$. Test the null hypothesis that $c = d = 0$. Use the F-test, as explained in section 4.8.

1. Fit the big model and the small model to Yule's data by OLS and compute the sums of squares that are needed for the test: $\|e\|^2$, $\|X\hat{\beta}\|^2$, and $\|X\hat{\beta}^{(s)}\|^2$.

2. Calculate the F-statistic. What do you conclude?

3.   Is $\|Y\|^2 = \|X\hat{\beta}^{(s)}\|^2 + \left(\|X\hat{\beta}\|^2 - \|X\hat{\beta}^{(s)}\|^2\right) + \|e\|^2$? Coincidence or math fact?

MATLAB tip. $X(:,1:2)$ picks off the first two columns in X. Colons are powerful.

The F-test. Part 2.

In this part of the lab, you will use simulation to investigate the distribution of the F-statistic for testing the null hypothesis that $c = d = 0$. You should consider two ways to set the parameters:

(i) $a = 8$, $b = 0.8$, $c = 0$, $d = 0$, $\sigma = 15$

(ii) $a = 13$, $b = 0.8$, $c = 0.1$, $d = -0.3$, $\sigma = 10$

Fix the design matrix X as in Part 1. Simulate data from each set of parameters to get the distribution of F.

For example, let's look at (i). Generate 32 ϵ's and use the equation

$$\Delta\text{Paup}_i = 8 + 0.8 \times \Delta\text{Out}_i + \epsilon_i$$

to get simulated data on ΔPaup. Calculate F. Repeat 1000 times and make a histogram for the values of F. You can take the ϵ's to be IID $N(0, 15^2)$.

Repeat for (ii). Which set of parameters satisfies the null hypothesis and which satisfies the alternative hypothesis? Which simulation tells you about size and which about power?

For discussion. Would it matter if you set the parameters in (i) differently? For instance, you could try $a = 13, b = 1.8, c = 0, d = 0$ and $\sigma = 25$. Would it matter if you set the parameters in (ii) differently? What if the errors aren't normally distributed?

Lab 9

Collinearity

In this lab, you will use simulation to examine the effect of collinearity. See discussion question 10 in chapter 4. To get started, you might think about $r = 0.3$ where collinearity is mild, and $r = 0.99$ where collinearity is severe. If you feel ambitious, also try $r = -0.3$ and $r = -0.99$.

1.   Simulate 100 IID picks (ξ_i, ζ_i) from a bivariate normal distribution, where $E(\xi_i) = E(\zeta_i) = 0$, $E(\xi_i^2) = E(\zeta_i^2) = 1$, and $E(\xi_i \zeta_i) = 0$. Use randn(100,2).

2. As data, these columns won't quite have mean 0, variance 1, or correlation 0. (Why not?) Cleaning up takes a bit of work.

    (a) Standardize ξ to have mean 0 and variance 1: call the result U.

    (b) Regress ζ on U. No intercept is needed. (Use the backslash operator \ to do the regression.) Let e be the vector of residuals. So $e \perp U$. Standardize e to have mean 0 and variance 1. Call the result W.

    (c) Let r be the correlation you want. Set $V = rU + \sqrt{1-r^2}W$.

    (d) Check that U and V have mean 0, variance 1, and correlation r—exactly. (Exactly? or up to roundoff error?)

2. Simulate $Y_i = U_i + V_i + \epsilon_i$ for $i = 1, \ldots, 100$, where the ϵ_i are IID $N(0, 1)$. Keep U and V fixed. Use `randn(100,1)` to get the ϵ's.

3. Fit the no-intercept regression equation

    $$Y_i = \hat{a}U_i + \hat{b}V_i + \text{residual}$$

    to your simulated data set.

4. Repeat 1000 times.

5. Plot histograms for $\hat{a}, \hat{b}, \hat{a} + \hat{b}$, and $\hat{a} - \hat{b}$.

6. There are four parameters of interest: $a, b, a + b, a - b$. What are their true values? Which parameter is easiest to estimate? Hardest? Discuss briefly.

MATLAB tip. `std(x)` divides by $n - 1$, but `std(x,1)` divides by n. You can work the lab either way: just be consistent.

Lab 10

Maximum Likelihood

In this lab, you will compute the MLE by numerical maximization of the log likelihood. Suppose that X_i are IID for $i = 1, 2, \ldots, 50$. Their common density function is $\theta/(\theta + x)^2$ for $0 < x < \infty$. The parameter θ is an unknown positive constant. See example 4 in section 6.1. Data on the X_i's are in the file `/StatMods/mle.dat`. Find the MLE for θ. It will be better to use the parameter $\phi = \log \theta$. If ϕ is real, $\theta = e^\phi > 0$, so the positivity constraint on θ is satisfied. This is a complicated lab, which might take two weeks to do.

1. Write down the log likelihood.

2. Find the MLE $\hat{\theta}$ by numerical maximization.

3. Put a standard error on $\hat{\theta}$. (See theorem 6.1, and exercise 6A8.)

Some useful MATLAB commands: `fminsearch`, `log`, `exp`.

`fminsearch` does minimization. (Minimizing $-f$ is the same as maximizing f, although it's a little more confusing.) Use the syntax

```
phiwant=fminsearch(@negloglike, start, [ ], x)
```

Here, `phiwant` is what you want—the parameter value that minimizes the negative log likelihood. The MLE for θ is `exp(phiwant)`. The at-sign @ is MATLAB's way of referring to functions. `fminsearch` looks for a local minimum of `negloglike` near the starting point, `start`. The log median of the data is a good choice for `start`. This particular negative likelihood function has a unique minimum (exercise 6A6). The rationale for the log median is exercise 6A7, plus the fact that $\phi = \log\theta$. Starting at the median or log median of the data is *not* a general recipe. Some versions of MATLAB may balk at @: if so, try

```
phiwant=fminsearch('negloglike',...)
```

You have to write `negloglike.m`. This is a function file that computes the negative log likelihood from `phi` and `x`, where `phi` is the parameter $\log\theta$ and `x` is the data—which you get from `mle.dat`. The call to `fminsearch` passes the data `x` to `negloglike.m`. It does not pass the parameter `phi` to `negloglike.m`: MATLAB will minimize over `phi`. The first line of `negloglike.m` should be

```
function negll=negloglike(phi,x)
```

The rest of the file is MATLAB code that computes `negll`—the negative log likelihood—from `phi` and `x`. At the end of `negloglike.m`, you need a line of code that sets `negll` to the value that you have computed from `phi` and `x`.

Just to illustrate syntax, here is a function file that computes $(u + \cos u)^2$ from u.

```
function youpluscosyoutoo=fun(u)
youpluscosyoutoo=(u+cos(u))^2;
```

You would save these two lines of code as `fun.m`. If at the MATLAB prompt—or in some other m-file—you type `fun(3)`, MATLAB will return $(3 + \cos 3)^2 = 4.0401$. If you type

```
fminsearch(@fun,1)
```

MATLAB will return -0.7391, the u that minimizes $(u + \cos u)^2$. The search started at 1.

Lab 11

Simulations for the MLE

In this lab, you will investigate the distribution of $\hat{\theta}$, the maximum likelihood estimate of θ, for the model in Lab 10. You should be able to reuse most of your code. This might be an occasion for loops.

1. Generate 50 IID variables U_i that are uniform on $[0, 1]$. Set $\theta = 25$ and $X_i = \theta U_i/(1 - U_i)$. According to exercise 6A5, you now have a sample of size 50 from the density $\theta/(\theta + x)^2$.

2. Find the MLE $\hat{\theta}$ by numerical maximization.

3. Repeat 1000 times.

4. Plot a histogram for the 1000 realizations of $\hat{\theta}$.

5. Calculate the mean and SD of the 1000 realizations of $\hat{\theta}$. How does the SD compare to $1/\sqrt{50 \cdot I_\theta}$? (The Fisher information I_θ is computed in exercise 6A8.) Comment?

6. *For bonus points.* Let $t = (\hat{\theta} - 25)/\widehat{SE}$, where $\widehat{SE}$ is computed either from the Fisher information as in point 5, or from observed information. Which version of t is more like a normal distribution?

7. *Double or quits on bonus points.* What happens to $\hat{\theta}$ if you double θ, from 25 to 50? What about Fisher information? observed information?

Lab 12

The Logit Model

In this lab, you will fit a logit model, using data from the 2001 Current Population Survey. The data are in `/StatMods/pac01.dat` The data cover 13,803 individuals 16 years of age or older, in the five Pacific states of the US. The variables and file layout are explained in `pac01doc.txt` in the same directory.

The dependent variable Y is 1 if the person is employed and at work (`LABSTAT` is 1). Otherwise, $Y = 0$. The explanatory variables are age, sex, race, and educational level. The following categories should be used:

Age: 16–19, 20–39, 40–64, 65 or above.

Sex: male, female. (Not much choice about this one.)

Race: white, non-white.

Educational level: not a high school graduate, a high school education but no more, more than a high school education.

For the baseline individual in the model, choose a person who is male, non-white, age 16–19, and did not graduate from high school.

1. What is the size of the design matrix?

2. Use `fminsearch` to fit the model; report the parameter estimates.

3. Estimate the SEs; use observed information.

4. What does the model say about employment?

5. Why use dummy variables for education, rather than EDLEVEL as a quantitative variable?

6. *For discussion.* Why might women be less likely to have LABSTAT = 1? Are LABSTAT codes over 4 relevant to this issue?

Where should `fminsearch` start looking? Read section 6.2! How to compute the log likelihood function and its derivatives? Work exercises 6D9–10.

Numerical Maximization

Numerical maximization is something of a black art. The more parameters there are, the blacker it gets. As a partial check on the algorithm, you can start the maximization from several different places. Another useful trick: if the computer tells you the max is at `[1.4517 0.5334 0.8515 ...]`, start the search again—from a nearby point, like `[1.5 0.5 0.8 ...]`.

Lab 13

Simultaneous Equations

In this lab, you will fit a model that has two simultaneous equations. The model is the one proposed by Rindfuss et al for determining a woman's educational level (ED) and age at first birth (AGE). The model is described in section 8.5; variables are defined in table 8.1. The correlation matrix is shown at the top of the next page. Also see

`/StatMods/rindcor.dat`

In `rindcor.dat`, the upper right triangle is filled with 0's: that way, MAT-LAB can read the file. You may need to do something about all those 0's.

	OCC	RACE	NOSIB	FARM	REGN	ADOLF	REL	YCIG	FEC	ED	AGE
OCC	1.000										
RACE	−.144	1.000									
NOSIB	−.244	.156	1.000								
FARM	−.323	.088	.274	1.000							
REGN	−.129	.315	.150	.218	1.000						
ADOLF	−.056	.150	−.039	−.030	.071	1.000					
REL	.053	−.152	.014	−.149	−.292	−.052	1.000				
YCIG	−.043	.030	.028	−.060	−.011	.067	−.010	1.000			
FEC	.037	.035	.002	−.032	−.027	.018	−.002	.009	1.000		
ED	.370	−.222	−.328	−.185	−.211	−.157	−.012	−.171	.038	1.000	
AGE	.186	−.189	−.115	−.118	−.177	.111	.098	−.122	.216	.380	1.000

Your mission, if you choose to accept it, is to estimate parameters in the standardized equations that explain ED and AGE. Variables are standardized to mean 0 and variance 1, so equations do not need intercepts. You do not have the original data, but can still use IVLS (section 8.2) or IISLS (section 8.4). IVLS might be easier. You have to translate equation (8.10) into usable form. For example, $Z'X/n$ becomes the $q \times p$ matrix of correlations between the instruments and the explanatory variables. See section 5.1.

Keeping track of indices is irritating. Here is a useful MATLAB trick. Number the variables from 1 through 11: OCC is #1, ..., AGE is #11. Let X consist, e.g., of variables 11, 2 through 8, and 1 (i.e., AGE, RACE , ..., YCIG, OCC). How do you get the correlation matrix M for X from the correlation matrix C for all the variables in the system? Nothing is easier. You get C by loading `rindcor.dat` and filling in the upper triangular part. Then you type

```
idx=[11 2:8 1];
M=C(idx',idx);
```

(Here, `idx` is just a name—ID numbers of variables in X.) Let Z consist, e.g., of variables 9, 2 through 8, and 1 (i.e., FEC, RACE , ..., YCIG, OCC). How do you get the matrix L of correlations between Z and X? You define `idz`—that's part of your job—then type `L=C(idz',idx)`. There is a row of L for each variable in Z, and a column for each variable in X.

Any comment on the coefficients of the control variables (OCC, ..., FEC)?

For bonus points

1. Rindfuss et al is reprinted at the back of the book. If your results differ from those in the paper (table 2), why might that be?

2. Find the asymptotic SEs. Hint: in equation (8.13),

$$\|Y - X\hat{\beta}_{IVLS}\|^2 = \|Y\|^2 + \beta'_{IVLS}(X'X)\beta_{IVLS} - 2(Y'X)\hat{\beta}_{IVLS}.$$

Lab 14 and Beyond

Additional Topics

Additional labs can be based on the data-snooping simulation (section 4.9, end notes to chapter 4); on discussion questions 15–17 in chapter 4, or 3, 6, 15 in chapter 5; on bootstrap example 4 in chapter 7; and on the IVLS simulations in section 8.8.

Statistical Packages

The labs are organized to help you learn what's going on underneath the hood when you fit a model. Statistical packages are organized to help you fit standard models with a minimum of fuss—although software designers have their own ideas about what "fuss" should mean to the rest of us. Recommended packages include the MATLAB statistics toolbox, R, and SAS. For instance, in release 13 of the MATLAB toolbox, you can fit a probit model by the command

```
glmfit(X,[Y ones(n,1)], 'binomial','probit')
```

Here, X is the design matrix, and Y is the response variable. MATLAB thinks of [Y ones(n,1)] as describing n binomial variables, each with 0 or 1 success out of 1 trial. The first column in [Y ones(n,1)] tells it the number of successes, and the second column tells it the number of trials. There is a quirk in the code: you don't put a column of 1's into X. MATLAB will do this for you, and two columns of 1's is one too many. In version 1.9.0 of R,

```
glm(Y~X1+X2,family=binomial(link=probit))
```

will fit a probit model. The response variable is Y, as above. There are two independent variables, X1 and X2. Again, an intercept is supplied for you. The formula with the tilde, Y~X1+X2, is just R's way of describing a model to itself: the dependent variable is Y; and there are two explanatory variables, X1 and X2. The family=binomial(link=probit) tells it you have a binomial response variable and want to fit a probit model. (Before you actually do this, please read *An Introduction to R*—click on Help in the R console, then on Manuals.)

What about statistical tables? The MATLAB statistics toolbox has "cdf" and "icdf" functions that replace printed tables for the normal, t, F, and a dozen other classical distributions. In R, check the section called "R as a set of statistical tables" in *An Introduction to R*. Looking things up in printed statistical tables is now like using a slide rule to multiply numbers.

Appendix: Sample MATLAB Code

This program has most of the features you will need during the semester. It loads a data file `small.dat` listed at the end. It calls a function file `phi.m` also listed at the end.

A script file—demolab.m

```
% demolab.m
% a line that starts with a percent sign
% is a comment
% at the UNIX prompt, type matlab...
% you will get the matlab prompt, >>
%
% you can type edit to get an editor
%
% help to get help
%
% helpdesk for a browser-based help facility
%
% emergency stop is  ....   control-c
%
% how to create matrices

x=[1 2
3 4
5 6]

y=[3 3
4 3
3 1]
```

```
disp('CR means carriage-return-- the "enter" key')
qq=input('hit cr to see some matrix arithmetic');

% this is a way for the program to get input,
% here it just waits until you press the enter key,
% so you can look at the screen....

% names can be pretty long and complicated

twice_x=2*x
x_plus_y=x+y
transpose_x=x'
transpose_x_times_y=x'*y

qq=input('hit cr to see determinants and inverses');

determinant_of_xTy=det(x'*y)
inverse_of_xTy=inv(x'*y)

disp('hit cr to see coordinatewise multiplication,')
qq=input('division, powers....  ');

x_dotstar_y=x.*y
x_over_y=x./y
x_squared=x.^2

qq=input('hit cr for utility matrices ');

ZZZ=zeros(2,5)
WON=ones(2,3)
ident=eye(3)

disp('hit cr to put matrices together--')
qq=input('concatenation-- use [ ] ');

concatenated=[ones(3,1) x y]

qq=input('hit cr to graph log(t) against t ...  ');
```

```
t=[.01:.05:10]';
% start at .01, go to 10 in steps of .05

plot(t,log(t),'x')
disp('look at the graph!!!')
disp(' ')
disp(' ')

disp('loops')
disp('if ...  then ...  ')
disp('MATLAB uses == to test for equality')
disp('MATLAB will print the perfect squares')
disp('from 1 to 50')
qq=input('hit cr to go ....  ');

for j=1:50 %sets up a loop

    if j==fix(sqrt(j))^2

        found_a_perfect_square=j
        % fix gets rid of decimals,
        % fix(2.4)=2, fix(-2.4)=-2

    end %gotta end the "if"

end %end the loop
% spaces and indenting make the code easier to read

qq=input('hit cr to load a file and get summaries');

load small.dat
ave_cols_12=mean(small(:,1:2))
SD_cols_12=std(small(:,1:2))

% small(:,1) is the first column of small...
% that is what the colon does
% small(:,1:2) is the first two columns
% matlab divides by n-1 when computing the SD

u=small(:,3);
```

```
v=small(:,4);

% the semicolon means, don't print the result

qq=input('hit cr for a scatterplot...   ');
plot(u,v,'x')

correlation_matrix_34=corrcoef(u,v)
% look at top right of the matrix
% for the correlation coefficient

disp('hit cr to get correlations')
qq=input('between all pairs of columns ');

all_corrs=corrcoef(small)

qq=input('hit cr for simulations ');

uniform_random_numbers=rand(3,2)
normal_random_numbers=randn(2,4)

disp('so, what is E(cos(Z)|Z>0) when Z is N(0,1)?')
qq=input('hit cr to find out ');
Z=randn(10000,1);
f=find(Z>0);
EcosZ_given_Z_is_positive=mean(cos(Z(f)))
trickier=mean(cos(Z(Z>0)))

disp('come let us replicate,')
qq=input('might be sampling error, hit cr ');
Z=randn(10000,1);
f=find(Z>0);
first_shot_was=EcosZ_given_Z_is_positive
replicate=mean(cos(Z(f)))

disp('guess there is sampling error....')

disp(' ')
disp(' ')
disp(' ')
```

```
disp('MATLAB has script files and function files ')
disp('mean and std are function files,')
disp('mean.m and std.m ')
disp('there is a function file phi.m')
disp('that computes the normal curve')

qq=input('hit cr to see the graph ');
u=[-4:.05:4];
plot(u,phi(u))
```

A function file—phi.m

```
% phi.m
% save this in a file called  phi.m
% first line of code has to look like this...
function y=phi(x)

y=(1/sqrt(2*pi))*exp(-.5*x.^2);
% at the end, you have to compute y--
% see first line of code
```

small.dat

```
1    2    2    4
4    1    3    8.5
2    2    5    1
8    9    7.5  0.5
3    3    4    2
7    7    0.5  3
```

Political Intolerance and Political Repression During the McCarthy Red Scare

James L. Gibson, University of Houston

Abstract

I test several hypotheses concerning the origins of political repression in the states of the United States. The hypotheses are drawn from the elitist theory of democracy, which asserts that repression of unpopular political minorities stems from the intolerance of the mass public, the generally more tolerant elites not supporting such repression. Focusing on the repressive legislation adopted by the states during the McCarthy era, I examine the relationships between elite and mass opinion and repressive public policy. Generally it seems that elites, not masses, were responsible for the repression of the era. These findings suggest that the elitist theory of democracy is in need of substantial theoretical reconsideration, as well as further empirical investigation.

James L. Gibson, 1988, Political Intolerance and Political Repression During the McCarthy Red Scare, *American Political Science Review* 82 (2): 511–29. © American Political Science Association, published by Cambridge University Press, reproduced with permission.

Over three decades of research on citizen willingness to "put up with" political differences has led to the conclusion that the U.S. public is remarkably intolerant. Though the particular political minority that is salient enough to attract the wrath of the public may oscillate over time between the Left and the Right (e.g., Sullivan, Piereson, and Marcus 1982), generally, to be much outside the centrist mainstream of U.S. politics is to incur a considerable risk of being the object of mass political intolerance.

At the same time, however, U.S. public policy is commonly regarded as being relatively tolerant of political minorities. Most citizens believe that all citizens are offered tremendous opportunities for the expression of their political preferences (e.g., McClosky and Brill 1983, 78). The First Amendment to the U.S. Constitution is commonly regarded as one of the most uncompromising assertions of the right to freedom of speech to be found in the world ("Congress shall make no law . . ."). Policy, if not public opinion, appears to protect and encourage political diversity and competition.

The seeming inconsistency between opinion and policy has not gone unnoticed by scholars. Some argue that the masses are not nearly so intolerant as they seem, in part due to biases in the questions used to measure intolerance (e.g., Femia 1975) and in part because the greater educational opportunity of the last few decades has created more widespread acceptance of political diversity (e.g., Davis 1975; Nunn, Crockett, and Williams 1978). Most, however, are willing to accept at face value the relative intolerance of the mass public and the relative tolerance of public policy but to seek reconciliation of the seeming contradiction by turning to the processes linking opinion to policy. Public policy is tolerant in the United States because the processes through which citizen preferences are linked to government action do not faithfully translate intolerant opinion inputs into repressive policy outputs. Just as in so many other substantive policy areas, public policy concerning the rights of political minorities fails to reflect the intolerant attitudes of the mass public.

Instead, the elitist theory of democracy asserts, policy is protective of political minorities because it reflects the preferences of elites, preferences that tend to be more tolerant than those of the mass public. For a variety of reasons, those who exert influence over the policymaking process in the United States are more willing to restrain the coercive power of the state in its dealings with political opposition groups. Thus there is a linkage between policy and opinion, but it is to *tolerant elite opinion*, not to *intolerant mass opinion*. Mass opinion is ordinarily not of great significance; public policy reflects elite opinion and is consequently tolerant of political diversity. The democratic character of the regime is enhanced through the political apathy and immobility of the masses, according to the elitist theory of democracy.[1]

The elitist theory nonetheless asserts that outbreaks of political repression—when they occur—are attributable to the mass public. While the preferences of ordinary citizens typically have little influence over public policy—in part, perhaps, because citizens have no real preferences on most civil liberties issues—there are instances in which the intolerance of the mass public becomes mobilized. Under conditions of perceived threat to the status quo, for example, members of the mass public may become politically active. In the context of the general propensity toward intolerance among the mass public, mobilization typically results in demands for political repression. Thus, the elitist theory of democracy hypothesizes that political repression flows from demands from an activated mass public.

The theory of "pluralistic intolerance"—recently proposed by Sullivan, Piereson, and Marcus (1979, 1982) and Krouse and Marcus (1984)—provides a nice explanation of the process through which mass intolerance is mobilized (see also Sullivan et al. 1985). The theory asserts that one of the primary causes of political repression is the *focusing* of mass intolerance on a specific unpopular political minority. To the extent that intolerance becomes focused, it is capable of being mobilized. Mobilization results in demands for political repression, demands to which policy makers accede. The authors claim support for their theory from recent U.S. history:

"During the 1950s, the United States was undoubtedly a society characterized by considerable consensus in target group selection. The Communist Party and its suspected sympathizers were subjected to significant repression, and there seemed to be a great deal of support for such actions among large segments of the political leadership as well as the mass public. . . . The political fragmentation and the proliferation of extremist groups in American politics since the 1950s has undoubtedly resulted in a greater degree of diversity in target group selection. If this is the case, such a situation is less likely to result in repressive action, even if the mass public is roughly as intolerant as *individuals* as they were in the 1950s (Sullivan, Piereson, and Marcus 1982, 85, emphasis in original)."

Thus both the elitist theory of democracy and the theory of pluralistic intolerance are founded upon assumptions about the linkage between opinion and policy.

Despite the wide acceptance of the elitist theory of democracy, there has been very little empirical investigation of this critical linkage between opinion and policy.[2] Consequently, this research is designed as an empirical test of the policy implications of the widespread intolerance that seems to characterize the political culture of the United States. Using data on elite and mass opinion and on public policy in the states, the linkage hypothesis is tested. My focus is on the era of the McCarthy Red Scare, due to its political and theoretical importance. Thus I assess whether there are any significant policy implications that flow from elite and mass intolerance.

Public Policy Repression
Conceptualization

A major impediment to drawing conclusions about the linkage between political intolerance and the degree of repression in U.S. public policy is that rigorous conceptualizations and reproducible operationalizations of policy repression do not exist. Conceptually, I define *repressive public policy* as statutory restriction on *oppositionist political activity* (by which I mean activities through which citizens, individually or in groups, compete for political power [cf. Dahl 1971]) upon some, but not all, competitors for political power.[3] For example, policy outlawing a political party would be considered repressive, just as would policy that requires the members of some political parties to register with the government while not placing similar requirements on members of other political parties. Though there are some significant limitations to this definition, there is utility to considering the absence of political repression (political freedom) as including unimpaired opportunities for all full citizens

1. to formulate their preferences
2. to signify their preferences to their fellow citizens and the government by individual and collective action
3. to have their preferences weighted equally in the conduct of the government, that is, weighted with no discrimination because of the content or source of the preference (Dahl 1971, 1–2).

That is the working definition to be used in this research.

Operationalizing Political Repression—the 1950s

There have been a few systematic attempts at measuring political repression as a policy output of government. Bilson (1982), for instance, examined the degree of freedom available in 184 polities, using as a measure of freedom the ratings of the repressiveness developed by Freedom House. Dahl provides system scores on one of his main dimensions of *polyarchy* (opportunities for political opposition) for 114 countries as they stood in about 1969 (Dahl 1971, 232). In their various research reports Page and Shapiro (e.g., 1983) measure civil rights and civil liberties opinions and policies in terms of the adoption of specific sorts of public policy. Typically, however, the endogenous concept in most studies of state policy outputs is some sort of expenditure variable. (See Thompson 1981 for a critique of this practice.) These earlier efforts can inform the construction of a measure of political repression in the policy outputs of the American states.

The measure of policy repression that serves as the dependent variable in this analysis is an index indicating the degree of political repression directed against the Communist party and its members during the late 1940s and 1950s. A host of actions against Communists was taken by the states, including disqualifying them from public employment (including from teaching positions in public schools); denying them access to the ballot as candidates, and prohibiting them from serving in public office even if legally elected; requiring Communists to register with the government; and outright bans on the Party. Forced registration was a means toward achieving these ends.

Of the fifty states, twenty-eight took none of these actions against Communists.[4] Two states—Arkansas and Texas—banned Communists from the ballot and from public employment, as well as banning the Party itself and requiring that Communists register with the government. Another five states adopted all three measures against the Communists, but did not require that they register with the government. Pennsylvania, Tennessee, and Washington did not formally bar Communists from public employment but did outlaw the party and forbade its members from participating in politics. The remaining twelve states took some, but not all, actions against the Communists. From these data, a simple index of political repression has been calculated. The index includes taking no action, banning Communists from public employment, banning Communists from running candidates and holding public office, and completely banning Communists and the Communist Party. A "bonus" score of .5 was given to those states requiring that Communists register with the government.[5] Table 1 shows the scores of the individual states on this measure.

This measure can rightly be considered to be a valid indicator of political repression by the states.[6] In asserting this I do not gainsay that the state has the right—indeed, the obligation—to provide for its internal security. Consequently, statutes that prohibit such actions as insurrection do not necessarily constitute political repression. For instance, Texas made it unlawful to "commit, attempt to commit, or aid in the commission of any act intended to overthrow" the Texas government (Art. 6689-3A, Sec. 5). This section proscribes action, not thought or speech, and is therefore not an appropriate measure of political repression. However, the next subsection of the statute made it illegal to "advocate, abet, advise, or teach by any means any person to commit" a revolutionary act. Indeed, even conspiracy to advocate is prohibited (Art. 6889-3A, Sec. 5 [3]). This is indeed a constraint on the speech of political minorities and therefore is treated as repressive. As the action prohibited moves beyond a specific, criminal behavior, the line between repressive and nonrepressive legislation becomes less clear. Gellhorn (1952) commented,

"Traditionally the criminal law has dealt with the malefactor, the one who himself committed an offense. Departing from this tradition is the recent tendency to ascribe criminal potentialities to a body of persons (usually, though not invariably, the Communists) and to lay restraints upon any individual who can be linked with the group. This, of course, greatly widens the concept of subversive activities, because it results, in truth, in forgetting about activities altogether. It substitutes associations as the objects of the law's impact. Any attempt to define subversion as used in modern statutes must therefore refer to the mere possibility of activity as well as to present lawlessness." (p. 360).

There can be little doubt as to the effectiveness of this anti-Communist legislation. Not only were the Communist Party U.S.A. and other Communist parties essentially eradicated, but so too were a wide variety of non-Communists. It has been estimated that of the work force of 65 million, 13 million were affected by loyalty and security programs during the McCarthy era (Brown 1958). Brown calculates that over 11 thousand individuals were fired as a result of government and private loyalty programs. More than 100 people were convicted under the federal Smith Act, and 135 people were cited for contempt by the House Un-American Activities Committee. Nearly one-half of the social science professors teaching in universities at the time expressed medium or high apprehension about possible adverse repercussions to them as a result of their political beliefs and activities (Lazarsfeld and Thielens 1958). Case studies of local and state politics vividly portray the effects of anti-Communist legislation on progressives of various sorts (e.g., Carleton 1985). The "silent generation" that emerged from McCarthyism is testimony enough to the widespread effects—direct and indirect—of the political repression of the era (see also Goldstein 1978, 369–96).

Nor was the repression of the era a function of the degree of objective threat to the security of the state. Political repression was just as likely to occur in states with virtually no Communists as it was to occur in states with large numbers of Communists.[7] The repression of Communists bore no relationship to the degree of threat posed by local Communists.

It might seem that the repression of Communists, though it is clearly repression within the context of the definition proffered above, is not

Table 1. Political Repression of Communists by American State Governments

State	Banned from Public Employment	Banned from Politics	Banned Outright	Scale Score
Arkansas	Yes	Yes	Yes	3.5
Texas	Yes	Yes	Yes	3.5
Arizona	Yes	Yes	Yes	3.0
Indiana	Yes	Yes	Yes	3.0
Massachusetts	Yes	Yes	Yes	3.0
Nebraska	Yes	Yes	Yes	3.0
Oklahoma	Yes	Yes	Yes	3.0
Pennsylvania	No	Yes	Yes	3.0
Tennessee	No	Yes	Yes	3.0
Washington	No	Yes	Yes	3.0
Alabama	Yes	Yes	No	2.5
Louisiana	Yes	Yes	No	2.5
Michigan	Yes	Yes	No	2.5
Wyoming	Yes	Yes	No	2.5
Florida	Yes	Yes	No	2.0
Georgia	Yes	Yes	No	2.0
Illinois	Yes	Yes	No	2.0
California	Yes	No	No	1.0
New York	Yes	No	No	1.0
Delaware	No	No	No	.5
Mississippi	No	No	No	.5
New Mexico	No	No	No	.5
Alaska	No	No	No	.0
Colorado	No	No	No	.0
Connecticut	No	No	No	.0
Hawaii	No	No	No	.0
Iowa	No	No	No	.0
Idaho	No	No	No	.0
Kentucky	No	No	No	.0
Kansas	No	No	No	.0
Maryland	No	No	No	.0
Maine	No	No	No	.0
Minnesota	No	No	No	.0
Missouri	No	No	No	.0
Montana	No	No	No	.0
North Carolina	No	No	No	.0
North Dakota	No	No	No	.0
New Hampshire	No	No	No	.0
New Jersey	No	No	No	.0
Nevada	No	No	No	.0
Ohio	No	No	No	.0
Oregon	No	No	No	.0
Rhode Island	No	No	No	.0
South Carolina	No	No	No	.0
South Dakota	No	No	No	.0
Utah	No	No	No	.0
Vermont	No	No	No	.0
Virginia	No	No	No	.0
West Virginia	No	No	No	.0
Wisconsin	No	No	No	.0

Note: The scale score is a Guttman score. A "bonus" of .5 was added to the scale added to the scale if the state also required that Communists register with the government. See note 4 for details of the assignments of scores to each state.

necessarily "antidemocratic" because the objects of the repression are themselves "antidemocrats." To repress Communists is to preserve democracy, it
might be argued. Several retorts to this position can be formulated. First, for
democracies to preserve democracy through nondemocratic means is illogical because democracy refers to a set of means, as well as ends (e.g., Dahl
1956, 1961, 1971; Key 1961; Schumpeter 1950). The means argument can
also be judged in terms of the necessity of the means. At least in retrospect
(but probably otherwise as well), it is difficult to make the argument that the
degree of threat to the polity from Communists in the 1940s and 1950s in any
way paralleled the degree of political repression (e.g., Goldstein 1978). Second, the assumption that Communists and other objects of political repression are "antidemocratic" must be considered as an empirical question itself
in need of systematic investigation. As a first consideration, it is necessary to
specify which Communists are being considered, inasmuch as the diversity
among those adopting—or being assigned—the label is tremendous. Merely
to postulate that Communists are antidemocratic is inadequate. Third, the repression of Communists no doubt has a chilling effect on those who, while
not Communists, oppose the political status quo. In recognizing the coercive
power of the state and its willingness to direct that power against those who
dissent, the effect of repressive public policy extends far beyond the target
group.

Public Opinion Intolerance

Conceptualization

"Political tolerance" refers to the willingness of citizens to support the extension of rights of citizenship to all members of the polity, that is, to allow
political freedoms to those who are politically different. Thus, "tolerance implies a willingness to 'put up with' those things that one rejects. Politically, it
implies a willingness to permit the expression of those ideas or interests that
one opposes. A tolerant regime, then, like a tolerant individual, is one that
allows a wide berth to those ideas that challenge its way of life" (Sullivan,
Piereson, and Marcus 1979, 784). Thus, political tolerance includes support
for institutional guarantees of the right to oppose the existing regime, including the rights to vote, to participate in political parties, to organize politically
and to attempt political persuasion. Though there may be some disagreement
about the operationalization of the concept, its conceptual definition is relatively noncontroversial (see Gibson and Bingham 1982).

Operationalization

The simple linkage hypothesis is that where the mass public is more intolerant, state public policy is more repressive. Though the hypothesis is simple,

deriving measures of mass intolerance is by no means uncomplicated. In-deed, the study of state politics continually confronts the difficulty of deriving measures of state public opinion. Though there are five general alternatives—ranging from simulations to individual state surveys—the only viable option for estimating state-level opinion intolerance during the McCarthy era is to aggregate national surveys by state.

The source of the opinion data is the Stouffer survey, conducted in 1954. This survey is widely regarded as the classic study that initiated inquiry into the political tolerance of elites and masses (even though earlier evidence ex-ists, e.g., Hyman and Sheatsley 1953). Two independent surveys were ac-tually conducted for Stouffer: one by the National Opinion Research Center (NORC) and the other by the American Institute for Public Opinion (AIPO-Gallup). This design was adopted for the explicit purpose of demonstrating the accuracy and reliability of public opinion surveys based on random sam-ples. Each agency surveyed a sample of the mass public and of the political elites.[8]

Stouffer created a six-point scale to indicate political intolerance (see Stouffer 1955, 262–69). The index is a Guttman scale based on the responses to fifteen items concerning support for the civil liberties of Communists, so-cialists, and atheists (see Appendix for details). The items meet conventional standards of scalability and are widely used today as indicators of political tol-erance (e.g., Davis 1975; Nunn, Crockett, and Williams 1978; McCutcheon 1985; and the General Social Survey, conducted annually by NORC).

The process of aggregating these tolerance scores by state is difficult be-cause the states of residence of the respondents in the Stouffer surveys were never entered in any known version of the data set. Through an indirect pro-cess, using the identity of the interviewer and the check-in sheets used to record the locations (city and state) of the interviews conducted by each in-terviewer, state of residence could be ascertained for the NORC half of the Stouffer data set. The respondents were aggregated by state of residence to create summary indicators of the level of intolerance in each of the states. The Appendix reports the means, standard deviations, and numbers of cases and primary sampling units for this tolerance scale for the states represented in the NORC portion of the Stouffer survey. Evidence that this aggregation process produces reasonably valid state-level estimates of political intolerance is also presented.

Aggregating the elite interviews to the state level is in one sense more perilous and in another sense less perilous. With a considerably small num-ber of subjects (758 in Stouffer's NORC sample), the means become more unstable. On the other hand, the aggregation is not done for the purpose of es-timating some sort of elite population parameter. The elites selected were in no sense a random sample of state elites, so it makes little sense to try to make inferences from the sample to some larger elite population. Instead, the elite samples represent only themselves. The Appendix reports the state means, standard deviations, and numbers of cases.

There is a moderate relationship between elite and mass opinion in the state ($r = .52$). To the extent that we would expect elite and mass opinion in the states to covary, this correlation serves to validate the aggregate measures of opinion. The substantive implications of this correlation are considered below.

The Simple Relationship between Opinion and Policy

Figure 1 reports the relationships between mass and elite political intolerance and the adoption of repressive public policies by the states. There is a modest bivariate relationship during the McCarthy era between mass opinion and repressive public policy. In states in which the mass public was more intolerant, there tended to be greater political repression, thus seeming to support the elitist theory. However, the relationship is somewhat stronger between elite opinion and repression. From a weighted least squares analysis incorporating both elite opinion and mass opinion, it is clear that it is elite preferences that most influence public policy. The *beta* for mass opinion is $-.06$; for elite opinion, it is $-.35$ (significant beyond .01).[9] Thus political repression occurred in states with relatively intolerant elites. Beyond the intolerance of elites, the preferences of the mass public seemed to matter little.

Figure 1. Relationships between Opinion and Policy

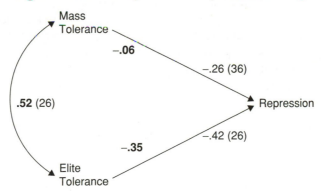

Note: Boldfaced entries are bivariate correlation coefficients, with pairwise missing data deletion. The nonboldfaced entries are standardized regression coefficients from a weighted least squares analysis using listwise missing data deletion. The numbers of cases are shown in parentheses.

Table 2. The Influence of Elite and Mass Opinion on the Repression of
Communists (Percentages)

Action	Elite Opinion Less Tolerant		Elite Opinion More Tolerant	
	Mass Opinion Less Tolerant	Mass Opinion More Tolerant	Mass Opinion less Tolerant	Mass Opinion More Tolerant
Adopted repressive legislation	71	100	33	39
Did not adopt repressive legislation	29	0	67	62
Total	100	100	100	101*
Number of cases	7	3	3	13

* Does not total 100 because of rounding error.

Table 2 reports a cross-tabulation of policy outputs with elite and mass opinion. The opinion variables have been dichotomized at their respective means. Though the number of cases shown in this table is small—demanding caution in interpreting the percentages—the data reveal striking support for the conclusion that elite opinion, not mass opinion, determines public policy. In eight of the ten states in which elites were relatively less tolerant, repressive legislation was adopted. In only six of the sixteen states in which elites were relatively more tolerant was repressive legislation passed. Variation in mass opinion makes little difference for public policy.[10]

It is a little surprising that elite opinion has such a significant impact on policy repression. After all, elites tend to be relatively more tolerant than the masses. Indeed, this finding is the empirical linchpin of the elitist theory of democracy.[11] This leads one to wonder just how much intolerance there was among the elites in the Stouffer data.

The survey data in fact reveal ample evidence of elite intolerance. For instance, fully *two-thirds* of the elites were willing to strip admitted Communists of their U.S. citizenship (Stouffer 1955, 43). Indeed, one reading of the Stouffer data is that elites and masses differed principally on the degree of proof of Communist party membership necessary before repression was thought legitimate. Much of the mass public was willing to accept a very low level of proof of party membership (e.g., innuendo), while many elites required a legal determination of Communist affiliation. Once convinced of the charge, however, elites were very nearly as intolerant of Communists as members of the mass public. Just as McClosky and Brill (1983) have more recently shown significant intolerance within their elite samples, there is enough intolerance among these state elites to make them the driving force in the repression of Communists. Thus it is plausible that elite intolerance was largely responsible for the repressive policies of the era.

At the same time, there is little evidence that the communism issue was of burning concern to the U.S. public. For instance, Stouffer reported that "the number of people who said [in response to an open-ended question] that they were worried either about the threat of Communists in the United States or about civil liberties was, even by the most generous interpretation of occasionally ambiguous responses, *less than 1%*" (Stouffer 1955, 59, emphasis in original). Only one-third of the subjects reported having talked about communism in the United States in the week prior to the interview, despite the fact that the Army-McCarthy hearings were in progress during a portion of the survey period. Stouffer asserted, "For most people neither the internal Communist threat nor the threat to civil liberties was a matter of universal burning concern. Such findings are important. They should be of interest to a future historian who might otherwise be tempted, from isolated and dramatic events in the news, to portray too vividly the emotional climate of America in 1954" (Stouffer 1955, 72).

The issue of communism in the United States was of much greater concern to the elites. Nearly two-thirds of them reported having talked about communism in the United States during the week prior to the interview. When asked how closely they followed news about Communists, fully 44% of the mass sample responded "hardly at all," while only 13% of the elite sample was as unconcerned (Stouffer 1955, 84). Just as elites typically exhibit greater knowledge and concern about public issues, they were far more attentive to the issue of domestic Communists.

Thus it is difficult to imagine that the repression of the 1950s was inspired by demands for repressive public policy from a mobilized mass public. Indeed, the most intense political intolerance was concentrated within that segment of the mass public *least* likely to have an impact on public policy (see also Gibson 1987). There can be no doubt that the mass public was highly intolerant in its attitudes during the 1950s. Absent issue salience, however, it is difficult to imagine that the U.S. people had mobilized sufficiently to have created the repression of the era.[12]

The actual effect of mass opinion may be masked a bit in these data, however. Perhaps it is useful to treat mass intolerance as essentially a constant across the states during the McCarthy era. Because the mass public was generally willing to support political repression of Communists, elites were basically free to shape public policy. In states in which the elites were relatively tolerant, tolerant policy prevailed. Where elites were relatively less tolerant, repression resulted. In neither case did mass opinion *cause* public policy. Instead, policy was framed by the elites. Nonetheless, the willingness of the mass public to accept repressive policies was no doubt important. Thus, the policy-making process need not be seen as a "demand–input" process with all its untenable assumptions but rather can be seen as one in which the preferences of the mass public—perhaps even the political culture of the state—set the broad parameters of public policy. In this sense, then, mass political intolerance "matters" for public policy.

We must also note that even if the broader mass public has little influence upon public policy, specialized segments of the public may still be important. For instance, there is some correlation ($r = .31$) between the number of American Legion members in the state and political repression.[13] Since the American Legion had long been in the forefront of the crusade against communism (see, e.g., American Legion 1937), it is likely that greater numbers of members in the state translated into more effective lobbying power. Thus particular segments of the mass public can indeed be mobilized for repressive purposes.

I should also reemphasize the strong correlation between elite opinion and mass opinion. This correlation may imply that elites are responsive to mass opinion or that they mold mass opinion or that elite opinion is shaped by the same sort of factors as shape mass opinion. Though it is not possible to disentangle the causal process statistically, there is some evidence that both elite and mass opinion reflect the more fundamental political culture of the state. The correlation between a measure of Elazar's state-level political culture and mass intolerance is $-.68$; for elite opinion the correlation is $-.66$. In states with more traditionalistic political cultures *both* mass and elites tend to be more intolerant. Moreover, there is some direct relationship between political culture and political repression ($r = .31$). Perhaps elite and mass preferences generally reflect basic cultural values concerning the breadth of legitimate political participation and contestation. In the moralistic political culture everyone should participate; only professionals should be active in the individualistic culture; and only the appropriate elite in traditionalistic political cultures (Elazar 1972, 101–2). Perhaps the political culture of the state legitimizes broad propensities toward intolerance, propensities that become mobilized during political crises.

One might also look at the data in Figure 1 from a very different perspective. Rather than mass opinion causing public policy, perhaps mass opinion *is caused by* policy (cf. Page, Shapiro, and Dempsey 1987). To turn the elitist theory on its head, it is quite possible that the U.S. mass public is intolerant precisely because they have been persuaded and reinforced by the intolerance of U.S. public policy. Through the intolerance of public policy, citizens learn that it is acceptable, if not desirable, to repress one's political enemies. Though I do not gainsay that there are significant norms in U.S. society supportive of political tolerance (see Sniderman 1975), in practice citizens have been taught by federal and state legislation that Communists should *not* be tolerated. It is not surprising that many citizens have learned the lesson well.[14]

This argument is somewhat at variance with those who argue that greater exposure to the dominant cultural norms in the United States contributes to greater political tolerance. If the norms are tolerant, then greater exposure should create tolerance. But greater awareness of *repressive* norms—as expressed in public policies—should be associated with greater *intolerance*. Thus the result of political activism, high self-esteem, and other qualities that make us assimilate social norms will vary according to the nature of the norms (see Sullivan et al. 1985).

The norms of U.S. politics are at once tolerant and intolerant. Certainly, no one can doubt that support for civil liberties is a widely shared value. The key question, however, is "civil liberties for whom?" The U.S. political culture has long distinguished between "true Americans" and others and has always been willing to deny civil liberties to those who are "un-American." Foreign "isms" have repeatedly become the bogeymen in ideological conflict in the United States. Thus, citizens learn that civil liberties are indeed important to protect, but only for those who have a "legitimate" right to the liberty.

Thus the initial evidence is that political repression during the McCarthy era was most likely initiated by elites even if the mass public in most states would have acquiesced. These findings are not compatible with the elitist views that mass intolerance threatens democracy and that elites are the carriers of the democratic creed.

The Political Culture of Intolerance and Repression

These findings may very well be limited to the specific historical era of McCarthyism. Due to the unavailability of historical data on elite and mass opinion it is difficult to judge whether earlier outbreaks of political repression can also be attributed to elite intolerance. Building on the discussion of political culture above, however, it is possible to give this issue further consideration.

Following World War I roughly one-half of the U.S. states adopted criminal syndicalism statutes.[15] For example, the statute adopted by California shortly after World War I defined the crime as "any doctrine or precept advocating, teaching or aiding and abetting the commission of crime, sabotage (which word is hereby defined as meaning willful and malicious physical

damage or injury to physical property), or unlawful acts of force and violence or unlawful methods of terrorism as a means of accomplishing a change
in industrial ownership or control, or effecting any political change" (Calif.
Statutes, 1919. Ch. 188, Sec. 1, p. 281). Though no opinion data exist for the
1920s, it is possible to examine the relationship between state-level political
culture and political repression during this earlier era.

The correlation between state political culture and the adoption of criminal syndicalism statutes is .40 ($N = 50$) indicating once again that more
traditionalistic states were more likely to engage in political repression. That
this correlation is slightly stronger than the coefficient observed for the 1950s
might speak to the breakdown of homogeneous state cultures as the population became more mobile in the twentieth century. In any event, we see in this
correlation evidence that the more detailed findings of the McCarthy era may
not be atypical.[16]

Discussion

What conclusions about the elitist theory of democracy and the theory of pluralistic intolerance does this analysis support? First, I have discovered no
evidence that political repression in the U.S. stems from demands from
ordinary citizens to curtail the rights and activities of unpopular political minorities. This finding differs from what is predicted by the elitist theory of
democracy. Second, I find some evidence of elite complicity in the repression
of the McCarthy era, a finding that is also incompatible with the elitist theory. Generally, then, this research casts doubt on the elitist theory of
democracy.

Nor are these findings necessarily compatible with the theory of pluralistic intolerance advocated by Sullivan, Piereson, and Marcus. Though political intolerance in the 1950s was widespread and highly focused, there seems
to have been little direct effect of mass opinion on public policy. Like the elitist theory of democracy, the theory of pluralistic intolerance places too much
emphasis on mass opinion as a determinant of public policy.

The "demand-input" linkage process implicitly posited by these theories
is probably their critical flaw. Early public opinion research that found high
levels of mass political intolerance too quickly assumed that mass intolerance
translated directly into public policy. The assumption was easy to make since
little was known of the processes linking opinions with policy. As linkage research has accumulated, however, the simple hypothesis relating opinion to
policy has become increasingly untenable. The justification for studying mass
political tolerance therefore cannot be found in the hypothesis that survey responses direct public policy.

At the same time, however, public opinion may not be completely irrelevant. Tolerance opinion strongly reflects the political cultures of the states, and, at least in the 1950s, political culture was significantly related to levels of political repression. Opinion is important in the policy process because it delimits the range of acceptable policy alternatives. It may well be that mass opinion is manipulated and shaped by elites; nonetheless, those who would propose repressive policies in California face a very different set of political constraints than those who propose repressive policies in Arkansas. This is not to say that repression is impossible—indeed, California has a long history of significant levels of political repression—but rather that the task of gaining acceptance for repression is different under differing cultural contexts.

For over three decades now, political scientists have systematically studied public policy and public opinion. Significant advances have been made in understanding many sorts of state policy outputs, and we have developed a wealth of information about political tolerance. To date, however, little attention has been given to repression as a policy output, and even less attention has been devoted to behavioral and policy implications of tolerance attitudes. The failure to investigate the linkage between opinion and policy is all the more significant because one of the most widely accepted theories in political science—the elitist theory of democracy—was developed on the basis of an assumed linkage between opinion and policy. I hope that this research, though only a crude beginning, will serve as an early step in continuing research into these most important problems of democracy.

Appendix: Measurement and Aggregation Error in the State-Level Estimates of Mass Political Intolerance

Measurement

The measure of political tolerance employed here is an index originally constructed by Stouffer. He used fifteen items to construct the scale. Eleven of the items dealt with communists; two with atheists (those who are against all churches and religion); and two with socialists (those favoring government ownership of all railroads and all big industries). Stouffer reported a coefficient of reproducibility of .96 for the scale, a very high level of reliability. He also reported that reproducibility was approximately the same at all educational levels.

I decided to use Stouffer's scale even though it includes items on atheists and socialists (1) in order to maintain comparability to Stouffer's research, (2) because an identical scale was created from a survey in 1973 that is very useful for assessment of aggregation error, and (3) because the scale is so reliable. Stouffer had a strong view of what his scale was measuring.

He asserted, "But again let it be pointed out, this scale does not measure ... tolerance *in general*. It deals only with attitudes toward certain types of nonconformists or deviants. It does not deal with attitudes toward extreme rightwing agitators, toward people who attack minority groups, toward faddists or cultists, in general, nor, of course, toward a wide variety of criminals. For purposes of this study, the tolerance of nonconformity or suspected nonconformity is *solely* within the broad context of the Communist threat" (Stouffer 1955, 54, emphasis in original).

The Stouffer measures of tolerance have recently been criticized (e.g., Sullivan, Piereson, and Marcus 1982). Perhaps the most fundamental aspect of this criticism is the assertion that the Stouffer items measure tolerance only for a specific group and thus are not generalizable. Because Stouffer was concerned only about intolerance of Communists, his findings may be time-bound; as the objects of mass displeasure evolve, the Communist-based approach to tolerance becomes less relevant and useful. This difficulty does not affect my analysis of policy and opinion from the 1950s, however, because Communists were probably a major disliked group for nearly all citizens in the survey. For instance, only 256 out of 4,933 of the mass respondents were willing to assert that someone believing in communism could still be a loyal U.S. citizen. Even if Communists were not the least-liked group for all U.S. citizens, they were certainly located in the "disliked-enough-not-to-tolerate" range for nearly everyone. Thus the Stouffer measure of tolerance is a valid and reliable indicator.

Aggregation Error

Table A-1 reports the state-level means, standard deviations, and numbers of cases for the aggregation of elite and mass opinion. Not all states are included in Table A-1 because survey respondents were not located in every state. Since the Stouffer survey was not designed to be aggregated by state, it is necessary to try to determine whether there is any obvious bias in the state-level estimates. A few empirical tests can be conducted that, while not assuaging all doubts about the aggregation process, may make us somewhat more comfortable about using the state means.

The Stouffer survey was replicated in 1973 by Nunn, Crockett, and Williams (1978). Their survey was very nearly an exact replication of the Stouffer survey. In terms of the indicators of tolerance, it was an exact replication. Nunn, Crockett, and Williams were even extremely careful to reproduce Stouffer's scaling methodology in creating a summary index of intolerance (pp. 179–91). Thus it is possible to aggregate the same scale variable by state and derive a measure of political tolerance for the early 1970s.

Table A-I. State Mean Tolerance Scores, Mass Public, and Elites, NORC
Stouffer Survey, 1954

| State | Mass Public | | | | Elites | | |
	Mean	Standard Deviation	Number of Cases	Number of PSUs	Mean	Standard Deviation	Number of Cases
California	4.47	1.50	174	4	5.09	1.43	65
Missouri	4.44	1.20	18	2	5.45	.69	11
New Jersey	4.41	1.43	61	1	4.90	1.28	60
Washington	4.33	1.44	52	2	5.14	.66	14
Iowa	4.26	1.42	23	1	–	–	–
Wisconsin	4.24	1.56	41	2	5.44	.87	25
Massachusetts	4.22	1.47	81	2	4.51	1.21	41
New York	4.21	1.40	273	6	5.06	1.08	81
Oregon	4.20	1.47	15	1	–	–	–
Colorado	4.13	1.46	23	1	5.29	1.33	14
Connecticut	4.12	1.17	17	1	5.17	.83	12
Nebraska	4.06	1.24	16	1	4.40	1.35	10
Minnesota	3.92	1.43	64	3	5.33	.96	27
Ohio	3.83	1.57	103	4	5.02	1.04	54
Illinois	3.81	1.55	86	2	4.97	1.39	39
Nevada	3.77	1.61	31	1	–	–	–
North Dakota	3.76	1.46	41	1	5.17	1.27	12
Pennsylvania	3.75	1.41	179	6	4.77	1.29	43
Michigan	3.75	1.34	163	4	4.92	1.26	38
Kansas	3.64	1.26	59	2	–	–	–
Florida	3.61	1.43	84	2	4.46	1.47	24
New Hampshire	3.58	1.71	19	1	5.36	1.03	11
Maryland	3.45	1.46	51	2	–	–	–
Idaho	3.45	1.65	22	1	5.15	1.07	13
Oklahoma	3.43	1.44	67	3	5.31	.85	13
Virginia	3.40	1.68	15	1	–	–	–
Indiana	3.36	1.32	129	5	4.61	1.40	36
Alabama	3.32	1.27	37	2	4.30	1.46	27
Texas	3.28	1.05	156	5	4.30	1.49	40
Louisiana	3.27	1.34	26	1	4.33	1.67	12
North Carolina	3.17	1.17	65	3	3.60	1.90	10
Tennessee	2.98	1.62	44	2	–	–	–
Georgia	2.86	1.39	50	3	–	–	–
Kentucky	2.86	1.25	22	1	4.77	1.39	26
West Virginia	2.34	.90	29	2	–	–	–
Arkansas	1.79	1.27	19	1	–	–	–
Average	3.65	1.40	65	2.3	4.88	1.22	29

With completely independent samples (including independent sampling
frames), one would not expect that there would be much of a correlation between the Stouffer and the Nunn, Crockett, and Williams state-level estimates.
Chance fluctuations in the distributions of primary sampling units (PSUs) per

state would tend to attenuate the correlation between the state-level estimates. (The average number of PSUs in Stouffer's NORC survey is 2.3; for the Nunn, Crockett, and Williams survey it is 7.8.) Yet the correlation between the estimates from the two surveys is a remarkable .63 ($N = 29$). If I were to exclude the 1973 estimate for Connecticut, an estimate that shows that state to be quite intolerant, then the correlation increases to .77 ($N = 28$). It is difficult to imagine an explanation for this correlation other than that it is due to a common correlation with the true score for the state.

I have also investigated the relationship between state sample size and number of primary sampling units and aggregation error. I first assumed that differences between the t_1 and t_2 estimates of state opinion were due to aggregation error. The residuals resulting from regressing t_2 opinion on t_1 opinion represent this error; if squared, the residuals represent the total amount of error. The correlations between the squared residuals and t_1 sample size and number of PSUs are $-.30$ and $-.27$. The correlations between the residuals and t_2 sample size and number of PSUs are $-.29$ and $-.29$. These correlations indicate that aggregation error is larger in states in which the number of subjects and number of PSUs is smaller—a not unexpected finding. However, since the relationships are modest, they do not undermine the basic aggregation procedure.

Another bit of evidence supporting the aggregation process comes from the correlations of tolerance and political culture. The correlation between Elazar's measure of political culture and average state tolerance in the 1950s is $-.68$. This correlation enhances my confidence in the utility of the state-level estimates.

Another, very different tack that can be taken is to estimate the error associated with the aggregation process. For each survey, I aggregated the proportion of the respondents having twelve or more years of formal education. These percentages can be compared to census estimates of the level of education in the state. The comparison is not perfect due to two considerations. First, the census data are themselves population estimates drawn from survey samples. Second, the census reports the percentage of residents over the age of twenty-five with twelve or more years of education. I assume that those with twelve or more years of education have a high school degree, although this might not be true for every single respondent. Moreover, it is not possible to isolate those respondents twenty-five years and older in the Stouffer survey. Nonetheless, the correlation for the 1950s data between the survey and census estimates of education is a substantial .72 ($N = 36$). While this correlation does not speak directly to the utility of the state-level estimates of tolerance, it does suggest that aggregation from the survey to the state is not completely inappropriate.

The correlation between elite opinion in the 1950s and elite opinion in the 1970s is .25 (.28 with a minimum-number-of-respondents requirement). That the correlation is not higher is a bit worrisome, although it is not difficult to imagine that there is greater flux in elite opinion over the two decades separating the two surveys than there is in mass opinion. Moreover, there were some slight differences in the composition of the elite samples drawn in 1954 and 1973.

As a means of assessing the validity of the aggregation of elite opinion, it is possible to compare elite tolerance with other elite attitudes. Erikson, Wright, and McIver (1987) have developed a separate measure of the degree of liberalism of state elites. The measure summarizes the ideological positions of the state's congressional candidates, state legislators, political party elites, and national convention delegates. As an overall index of the liberalism–conservatism of state elites, they take the average score of the Democrats and the Republicans. Thus each state receives a score indicating the degree of liberalism-conservatism of state elites. Though most of the indicators are drawn from the 1970s, the authors believe this to be a more stable attribute of state elites. According to their index, the most conservative elites are found in Mississippi; the most liberal elites are found in Massachusetts.

The correlation of state elite conservatism and political tolerance is $-.46$ ($N = 26$) for the Stouffer elites and $-.22$ ($N = 29$) for the Nunn, Crockett, and Williams elites. Though liberalism-conservatism is conceptually distinct from political tolerance, some solace can be taken in this correlation. The aggregation process seems not to have introduced unexpected or obviously biased estimates of state-level elite opinion.

References

The American Legion. 1937. *Isms: A Review of Alien Isms, Revolutionary Communism, and their Active Sympathizers in the United States*. 2d Edition. Indianapolis: author.

Bachrach, Peter. 1967. *The Theory of Democratic Elitism: A Critique*. Boston: Little, Brown.

Berelson, Bernard R., Paul F. Lazarsfeld, and William N. McPhee. 1954. *Voting*. Chicago: University of Chicago Press.

Bilson, John F. O. 1982. Civil Liberty: An Econometric Investigation. *Kyklos* 35:94–114.

Brown, Ralph S. 1958. *Loyalty and Security*. New Haven: Yale University Press.

Campbell, Donald T., and Donald W. Fiske. 1959. Convergent and Discriminant Validity by the Multitrait-Multimethod Matrix. *Psychological Bulletin* 56: 81–105.

Carleton, Donald E. 1985. *Red Scare!* Austin: Texas Monthly.

Chafee, Zechariah, Jr. 1967. *Free Speech in the United States*. Cambridge: Harvard University Press.

Dahl, Robert A. 1956. *A Preface to Democratic Theory*. Chicago: University of Chicago Press.

Dahl, Robert A. 1961. *Who Governs?* New Haven: Yale University Press.

Dahl, Robert A. 1971. *Polyarchy: Participation and Opposition*. New Haven: Yale University Press.

Davis, James A. 1975. Communism, Conformity, Cohorts, and Categories: American Tolerance in 1954 and 1972–73. *American Journal of Sociology* 81: 491–513.

Dowell, Eldridge. 1969. *A History of Criminal Syndicalism Legislation in the United States*. New York: Da Capo.

Dye, Thomas R. 1976. *Who's Running America: Institutional Leadership in the United States*. Englewood Cliffs, NJ: Prentice Hall.

Dye, Thomas R., and Harmon Zeigler. 1987. *The Irony of Democracy: An Uncommon Introduction to American Politics*. 7th ed. Monterey, CA: Brooks/Cole.

Elazar, Daniel. 1972. *American Federalism: A View from the States*. 2d ed. New York: Harper & Row.

Erikson, Robert S. 1976. The Relationship between Public Opinion and State Policy: A New Look Based on Some Forgotten Data. *American Journal of Political Science* 22:25–36.

Erikson, Robert S., Gerald C. Wright, Jr., and John P. McIver. 1987. Political Parties, Public Opinion, and State Policy. Presented at the annual meeting of the Midwest Political Science Association, Chicago.

Femia, Joseph V. 1975. Elites, Participation, and the Democratic Creed. *Political Studies* 27:1–20.

Gellhorn, Walter, ed. 1952. *The States and Subversion*. Ithaca, NY: Cornell University Press.

Gibson, James L. 1987. Homosexuals and the Ku Klux Klan: A Contextual Analysis of Political Intolerance. *Western Political Quarterly* 40: 427–48.

Gibson, James L., and Richard D. Bingham. 1982. On the Conceptualization and Measurement of Political Tolerance. *American Political Science Review* 76: 603–20.

Gibson, James L. and Richard D. Bingham. 1984. Skokie, Nazis, and the Elitist Theory of Democracy. *Western Political Quarterly* 37:32–47.

Glazer, Nathan. 1961. *The Social Basis of American Communism*. New York: Harcourt, Brace & World.

Goldstein, Robert Justin. 1978. *Political Repression in Modern America*. Cambridge, MA: Schenkman.

Hanushek, Eric A., and John E. Jackson. 1977. *Statistical Methods for Social Scientists*. New York: Academic.

Holbrook-Provow, Thomas M., and Steven C. Poe. 1987. Measuring State Political Ideology. *American Politics Quarterly* 15:399–416.

Hyman, Herbert H., and Paul B. Sheatsley. 1953. Trends in Public Opinion on Civil Liberties. *Journal of Social Issues* 9:6–16.

Jenson, Carol E. 1982. *The Network of Control: State Supreme Courts and State Security Statutes, 1920–1970*. Westport, CT: Greenwood.

Key, Valdimir O., Jr. 1961. *Public Opinion and American Democracy*. New York: Alfred A. Knopf.

Klehr, Harvey. 1984. *The Heyday of American Communism: The Depression Decade*. New York: Basic Books.

Klingman, David, and William W. Lammers. 1984. The "General Policy Liberalism" Factor in American State Politics. *American Journal of Political Science* 28:598–610.

Kornhauser, William. 1959. *The Politics of Mass Society*. Glencoe, IL: Free Press.

Krouse, Richard, and George Marcus. 1984. Electoral Studies and Democratic Theory Reconsidered. *Political Behavior* 6:23–39.

Lazarsfeld, Paul, and Wagner Thielens, Jr. 1958. *The Academic Mind*. Glencoe, IL: Free Press.

Library of Congress. Legislative Reference Service. 1965. *Internal Security and Subversion: Principal State Laws and Cases*. Prepared by Raymond J. Celanda. Washington: GPO.

Lipset, Seymour Martin. 1960. *Political Man*. New York: Doubleday.

McClosky, Herbert. 1964. Consensus and Ideology in American Politics. *American Political Science Review* 58:361–82.

McClosky, Herbert, and Alida Brill. 1983. *Dimensions of Tolerance: What Americans Believe about Civil Liberties*. New York: Russell Sage Foundation.

McCutcheon, Allan L. 1985. A Latent Class Analysis of Tolerance for Nonconformity in the American Public. *Public Opinion Quarterly* 49:474–88.

Nunn, Clyde Z., Harry J. Crockett, Jr., and J. Allen Williams, Jr. 1978. *Tolerance for Nonconformity*. San Francisco: Jossey-Bass.

Page, Benjamin I., and Robert Y. Shapiro. 1983. Effects of Public Opinion on Public Policy. *American Political Science Review* 77:175–90.

Page, Benjamin I., Robert Y. Shapiro, and Glenn R. Dempsey. 1987. What Moves Public Opinion? *American Political Science Review* 81:23–43.

Prendergast, William B. 1950. State Legislatures and Communism: The Current Scene. *American Political Science Review* 44:556–74.

Prothro, James W., and Charles M. Grigg. 1960. Fundamental Principles of Democracy: Bases of Agreement and Disagreement. *Journal of Politics* 22:276–94.

Rosenstone, Steven J. 1983. *Forecasting Presidential Elections*. New Haven: Yale University Press.

Schumpeter, Joseph. 1950. *Capitalism, Socialism, and Democracy*. New York: Harper & Row.

Shannon, David A. 1959. *The Decline of American Communism: A History of the Communist Party of the United States since 1945*. Chatham, NJ: Chatham Bookseller.

Sniderman, Paul M. 1975. *Personality and Democratic Politics*. Berkeley: University of California Press.

Stouffer, Samuel A. 1955. *Communism. Conformity, and Civil Liberties*. Garden City. NY: Doubleday.

Sullivan, John L., James Piereson, and George E. Marcus. 1979. A Reconceptualization of Political Tolerance: Illusory Increases, 1950s–1970s. *American Political Science Review* 73:781–94.

Sullivan, John L., James Piereson, and George E. Marcus. 1982. *Political Tolerance and American Democracy*. Chicago: University of Chicago Press.

Sullivan, John L., Michal Shamir, Patrick Walsh, and Nigel S. Roberts. 1985. *Political Tolerance in Context: Support for Unpopular Minorities in Israel, New Zealand, and the United States*. Boulder, CO: Westview.

Thompson, Joel A. 1981. Outputs and Outcomes of State Workmen's Compensation Laws. *Journal of Politics* 43:1129–52.

U.S. Congress. Senate. Committee on the Judiciary. 1956. *The Communist Party of the United States of America: What It Is. How It Works*. 84th Cong., 2d sess., S. Doc. 117.

Weissberg, Robert. 1978. Collective versus Dyadic Representation in Congress. *American Political Science Review* 72:535–47.

Wright, Gerald C., Jr., Robert S. Erikson, and John P. McIver. 1985. Measuring State Partisanship and Ideology with Survey Data. *Journal of Politics* 47:469–89.

Notes

This research has been conducted through the generous support of the National Science Foundation, SES84-21037. NSF is not responsible for any of the interpretations or conclusions reported herein. For research assistance, I am indebted to David Romero, James P. Wenzel, and Richard J. Zook. This is a revised version of a paper delivered at the 1986 annual meeting of the American Political Science Association, Washington, D.C., 1986. Several colleagues have been kind enough to comment on an earlier version of this article, including Paul R. Abramson, David G. Barnum, Lawrence Baum, James A. Davis, Thomas R. Dye, Heinz Eulau, George E. Marcus, John P. McIver, Paul M. Sniderman, Robert Y. Shapiro, and Martin P. Wattenberg. I am also indebted to Patrick Bova, librarian at NORC, for assistance with the Stouffer data.

    1. The elitist theory of democracy is actually an amalgam of the work of a variety of theorists, including Berelson, Lazarsfeld, and McPhee (1954); Kornhauser (1959); Lipset (1960); and Key (1961). The most useful analysis of the similarities and differences among the theories can be found in Bachrach 1967. Some elite theorists emphasize the dominance and control of public policy by elites, while other theorists emphasize the antidemocratic tendencies of the mass public. The single view most compatible with the hypotheses tested in this article is Kornhauser's (1959). The hypotheses are also to be found in Dye and Zeigler 1987 (see also Dye 1976). Earlier empirical work on the tolerance of elites and masses includes Berelson, Lazarsfeld, and McPhee 1954; Lipset 1960; Prothro and Grigg 1960; and McClosky 1964. A more recent analysis of some of the propositions of elitist theory can be found in Gibson and Bingham 1984.

    2. Linkage research is fairly common in other areas of substantive policy (e.g., Erikson 1976; Weissberg 1978), but the only rigorous investigation of civil liberties is that of Page and Shapiro (1983). They assessed the relationship between change in opinion and change in policy, and found that in eight of nine policy changes in the area of civil liberties there was opinion-policy congruence. They also found that state policies were more likely to be congruent with opinion than national policies, although the relationship did not hold in the multivariate analysis. Though their analysis was conducted at the national level, their findings seem to suggest that political repression results from demands from the mass public.

    3. This is similar to Goldstein's definition, "Political repression consists of government action which grossly discriminates against persons or organizations viewed as presenting a fundamental challenge to existing power relationships or key governmental policies, because of their perceived political beliefs" (1978, xvi).

    4. The source for these data is a 1965 study requested by a subcommittee of the Committee on the Judiciary in the U.S. Senate. See also Library of Congress, Legislative Reference Service, 1965; Gellhorn 1952; and Prendergast 1950. Care must be taken in using the Legislative Reference Service data because there are a variety of errors in the published report. Corrected data, based on an examination of all of the relevant state statutes, are available from the author.

    The scores shown in Table 1 reflect actions taken by the state governments between 1945 and 1965. The decision to limit the policy measures to this period is based on the desire to have some temporal proximity between the opinion and policy data. This decision has implications for the scores of three states. Kansas and Wisconsin both barred Communists from political participation in legislation adopted in 1941. This legislation is excluded from Table 1. Arkansas is shown as having banned Communists from public employment, from politics and outright. Only the outright ban was adopted in the 1945–65 period. Because a complete ban necessarily excludes Communists from public employment and from political participation, the score for Arkansas is shown as 3.5.

5. These three items scale in the Guttman sense. That is, nearly all of the states outlawing the Communist party also denied it access to the ballot and public employment. Nearly all of the states that denied Communists access to the ballot as candidates also made them ineligible for public employment. The registration variable does not, however, exhibit this pattern of cumulativeness. Registration seems to have been a means of enforcing a policy goal such as banning membership in the Party. Because registration can raise Fifth Amendment self-incrimination issues, some states chose not to require it. Statutes requiring registration are treated for measurement purposes as representing a greater degree of commitment to political repression, and for that reason the "bonus" points were added to the basic repression score.

6. Validity means not only that measures of similar concepts converge; measures of dissimilar concepts must also diverge (Campbell and Fiske 1959). Thus it is useful to examine the relationship between the repression measures and measures of other sorts of policy outputs. Klingman and Lammers (1984) have developed a measure of the "general policy liberalism" of the states. General policy liberalism is a predisposition in state public policies toward extensive use of the public sector and is thought to be a relatively stable attribute. I would expect that political repression is not simply another form of liberalism, and indeed it is not. The correlation between general policy liberalism and political repression during the 1950s is only −.18. Moreover, the relationship between repression and a measure of New Deal social welfare liberalism policy (see Holbrook-Provow and Poe 1987; Rosenstone 1983) is only −.22. Repression occurred in states with histories of liberalism just about as frequently as it did in states typically adopting conservative policies. Thus the measure of repression is not simply a form of political liberalism, a finding that contributes to the apparent validity of the measure.

7. This conclusion is based on figures compiled by Harvey Klehr on the size of the Communist Party U.S.A. during the 1930s (Klehr 1984, tbl. 19.1 and personal communication with the author, 21 May 1986). The data are from the Party's own internal record. Klehr believes the data to be reasonably reliable, and others seem to agree (see, e.g., Glazer 1961, 208, n. 3; and

Shannon 1959, 91). There is also a strong relationship between Party membership and votes for Communist candidates for public offices in the 1936 elections (as compiled by the American Legion 1937, 44), as well as a strong relationship with FBI estimates of Party membership in the states in 1951 (U.S. Senate, Committee on the Judiciary 1956, 34).

8. Stouffer defined elites as those who hold certain positions of influence and potential influence in local politics. The elite sample was drawn from those holding the following positions: community chest chairmen; school board presidents; library committee chairmen; Republican county chairmen; Democratic county chairmen; American Legion commanders; bar association presidents; chamber of commerce presidents; PTA presidents; women's club presidents; DAR regents; newspaper publishers; and labor union leaders.

9. Weighted least squares was used because I could not assume that the variances of the observations were equal. Following Hanushek and Jackson (1977, 151–52), I weighted the observations by the square root of the numbers of respondents within the state. The r-square from this analysis is .14. The regression equation with unstandardized coefficients is: $Y = 7.31 - .14(\text{mass opinion}) - 1.11(\text{elite opinion})$.

10. The data in Table 2 suggest that where the state elites are relatively less tolerant, increases in mass tolerance are associated with an increase in political repression. Caution must be exercised in interpreting the percentages, however, due to the small number of cases available. The data reveal that in five of the seven states with a relatively less tolerant mass public, repressive legislation was adopted, while in all three of the states with a relatively more tolerant mass public repressive legislation was adopted. In the context of the numbers of cases, I did not treat this difference as substantively significant.

11. It might be argued that elite opinion serves only to neutralize intolerant mass opinion. This suggests an interactive relationship between elite and mass opinion. Tests of this hypothesis reveal no such interaction. The impact of elite opinion on public policy is not contingent upon the level of tolerance of the mass public in the state.

12. Though it is a bit risky to do so, it is possible to break the policy variable into time periods according to the date on which the legislation was adopted. A total of sixteen states adopted repressive legislation prior to 1954; ten states adopted repressive legislation in 1954 or later. The correlations between pre-1954 repression and mass and elite tolerance, respectively, are −.05, and −.35. Where elites were more intolerant, policy was more repressive. Mass intolerance seems to have had little impact on policy.

The correlations change rather substantially for the post-1954 policy measure. There is a reasonably strong correlation between mass intolerance and repression ($r = -.32$) but little correlation with elite intolerance ($r = -.13$). If one were willing to draw conclusions based on what are surely relatively unstable correlations, based on limited numbers of observations, one might conclude that early efforts to restrict the political freedom of Communists were directed largely by elites, while later efforts were more likely to involve the mass public. The initiative for political repression therefore was with the elites, though the mass public sustained the repression once it was under way.

At the same time, however, the slight correlation between pre-1954 policy and mass intolerance suggests that mass opinion was not shaped by public policy. Where policy was more repressive, opinion was not more intolerant. The close temporal proximity here should give us pause in overinterpreting this correlation, however.

13. Note that Stouffer found that the leaders of the American Legion were the most intolerant of all leadership groups surveyed (Stouffer 1955, 52). Indeed, the commanders interviewed were only slightly less intolerant than the mass public.

14. At the same time, it should be noted that U.S. citizens became substantially more tolerant of Communists by the 1970s (e.g., Davis 1975; Nunn, Crockett, and Williams 1978). This too might reflect changes in public policy, as well as elite leadership of opinion. As the U.S. Supreme Court invalidated some of the most repressive state and federal legislation of the McCarthy era, and as U.S. political leaders (including Richard Nixon) sought improved foreign relations with Communist nations, it became less appropriate to support the repression of Communists. These comments illustrate, however, the difficulty of sorting out the interrelationships of opinion and policy and also reveal that many efforts to do so border on nonfalsifiability.

15. Between 1917 and 1920, twenty-four states adopted criminal syndicalism statutes. There is some ambiguity in published compilations about the number of states with such laws. Dowell (1969) lists twenty states with such legislation, not counting the three states that adopted but then repealed syndicalism laws. Dowell apparently overlooked Rhode Island, at least according to the compilations of Chafee (1967) and Gellhorn (1952). On the other hand, neither Chafee nor Gellhorn listed Colorado or Indiana as having such statutes (though Chafee did list the states that had repealed their legislation). This latter problem is in part a function of determining whether specific statutes should be classified as banning criminal syndicalism. By 1937, three states had repealed their statutes (although one of these—Arizona—apparently did so inadvertently during recodification). As of 1981, seven of these states still had the statutes on their books, and one additional state—Mississippi— had passed such legislation (Jenson 1982, 167–75). For purposes of this

analysis, Dowell's twenty-three states and Rhode Island are classified as having criminal syndicalism laws as of 1920.

16. It should also be noted that political culture is fairly stably related to mass political intolerance. Estimates of state opinion were derived from Roper data on an item about loyalty oaths asked in a 1937 survey. Opinion in more traditionalistic states was more supportive of mandatory loyalty oaths ($r = -.44$, $N = 47$). Similarly, the correlation between political culture and the state aggregates from the Stouffer replication in 1973 (see the Appendix) is $-.58$ ($N = 35$). These coefficients are nothing more than suggestive, but they do suggest that political intolerance is a relatively enduring attribute of state political culture.

Finishing High School and Starting College: do Catholic Schools Make a Difference?

William N. Evans, University of Maryland, Project Hope,
    and National Bureau of Economic Research
Robert M. Schwab, University of Maryland*

Abstract

In this paper, we consider two measures of the relative effectiveness of public and Catholic schools: finishing high school and starting college. These measures are potentially more important indicators of school quality than standardized test scores in light of the economic consequences of obtaining more education. Single-equation estimates suggest that for the typical student, attending a Catholic high school raises the probability of finishing high school or entering a four-year college by thirteen percentage points. In bivariate probit models we find almost no evidence that our single-equation estimates are subject to selection bias.

William N. Evans and Robert M. Schwab, 1995, Finishing High School and Starting College: Do Catholic Schools Make a Difference?, *The Quarterly Journal of Economics* 110 (4): 94–74. © 1995 by the President and Fellows of Harvard College and the Massachusetts Institute of Technology, reproduced with permission.

* We wish to thank Michael Cheng, Kamala Rajamani, Andrew Kochera, and Sheila Murray for excellent research assistance, and Lawrence Katz and two anonymous referees for helpful comments. We gratefully acknowledge the National Science Foundation which has supported this work under grant SBR9409499.

I. Introduction

More than ten years ago, James Coleman and his colleagues launched a national debate over the relative quality of public and Catholic schools [Coleman and Hoffer 1987; Coleman, Hoffer, and Kilgore 1982]. Based on their analysis of the High School and Beyond (*HS&B*) data, they concluded that Catholic school students scored significantly higher than public school students on standardized tests, even after controlling for differences in family characteristics. Catholic schools in their study appeared to be particularly effective with minority students.

Almost immediately, the Coleman results generated tremendous interest among both policy analysts and academics. Academic journals devoted special issues to their research on at least six different occasions (*Harvard Education Review* in 1991; *Phi Beta Kappa* in 1981; *Education Researcher* in 1981; and *Sociology of Education* in 1982, 1983, and 1985). Critics raised a number of issues about their work. Several papers showed that the estimated magnitude of the Catholic school effect was very sensitive to the choice of other independent variables (Lee and Bryk 1988; Noell 1982]. A number of papers questioned whether the results were driven by a selection bias. Since parents decide whether to send their children to public or Catholic schools, it is inappropriate to estimate the effect of Catholic schools on test scores with a single-equation model that treats school choice as an exogenous variable [Goldberger and Cain 1982]. Others argued that the increase in test scores between sophomore and senior years was so small that the Coleman results had little relevance in the debate over school choice [Murnane 1984; Alexander and Pallas 1985; Witte 1992].[1] Based on his review of the Coleman work and subsequent studies, Cookson [1993, p. 181] concluded that "... once the background characteristics of students are taken into account, student achievement is not directly related to private school attendance. The effects that were reported by Coleman and his associates are too small to be of any substantive significance in terms of incrementally improving student learning."

Most of Coleman's work and virtually all of the research that followed focused on the effects of Catholic schools on test scores.[2] In some ways it is surprising that test scores have received so much attention while other important education outcomes have not. Test scores have obvious limitations. It has often been argued that standardized tests in general may be culturally, racially, and sexually biased. Teachers may "teach to the test" and thus inflate scores [Henig 1994]. On the other hand, students often gain little by doing well on an exam and thus may not take the exam seriously. Standardized tests can only measure a student's ability to deal with a particular type of question and cannot measure a student's creativity or deeper problem-solving skills. The particular test included in the original Coleman work was a short and relatively simple exam, and the results may not be indicative of school performance. Perhaps most importantly, there is little evidence that raising test scores has important economic consequences. The impact of test scores on wages, for example, appears to be modest.[3]

This suggests that we consider alternative criteria to evaluate schools that have important economic consequences. Card and Krueger [1994] argue that measures of educational attainment such as completing high school and going on to college are particularly useful measures of schools' success. Unlike test scores, there is a great deal of evidence on the benefits of additional education. Only 65 percent of young male high school dropouts were employed in 1986 as compared with 85 percent of high school graduates [Markety 1988]. Between 1980 and 1985 the unemployment rate for males without a high school diploma was 35 percent higher than the rate for high school graduates and five times as large as the rate for college graduates [Murphy and Topel 1987]. The unemployment rate for young black males without high school degrees was over 40 percent for most of the 1980s. Wages and earnings are substantially lower for those high school dropouts who do find work. In 1987 the median yearly income for 25-to-34 year-old male full-time workers with a high school degree was 21.2 percent larger than the value for those who had not finished high school [Levy and Murnane 1992]. Hashimoto and Raisian [1985] and Weiss [1988] found that an extra year of education that leads to a high school degree has a much larger impact on wages than does an additional year of school that does not lead to a degree. Real wages for young male high school dropouts declined by 23 percent between 1979 and 1988, while young male college graduates experienced a 7 percent real wage increase over the same period [Bound and Johnson 1992]. High school dropouts are far more likely to commit crimes [Thornberry, Moore, and Christenson 1985] and to use illegal drugs [Mensch and Kandel 1988].

Thus, the debate over Catholic schools seems to have missed outcomes with important economic implications. In this paper we have gone back to the *HS&B* data and looked at the impact of a Catholic school education on the probability of, first, finishing high school, and, second, starting college. We have paid particular attention to the issue of selection bias. If students with more ability or students from families that place a higher value on education are more likely to attend Catholic schools, then single-equation models would overstate the effects of a Catholic school education. Therefore, the appropriate model must take this endogeneity into account. Because both of our outcome measures and the treatment variable (a Catholic school dummy) are dichotomous, we estimate a set of bivariate probit models.

Our major conclusions are as follows. We find a great deal of support for the argument that Catholic schools are more effective than public schools. Single-equation estimates suggest that for the typical student, attending a Catholic high school raises the probability of finishing high school or entering a four-year college by thirteen percentage points. Unlike single-equation estimates of the effect of Catholic schools on test scores, these results are qualitatively important and are robust. This Catholic school effect is very large. It is twice as large as the effect of moving from a one- to a two-parent family and two and one-half times as large as the effect of raising parents' education from a high school dropout to a college graduate. In models where we treat the decision to attend a Catholic school as an endogenous variable, we find almost no evidence of selection bias. Bivariate probit estimates of the average treatment effect of Catholic schools on high school graduation and entering college are very similar to single-equation probit estimates.

Our bivariate probit model is properly identified if there is at least one variable that is correlated with whether or not a student attends a Catholic school but is uncorrelated with a student's unobserved propensity to graduate from high school or start college. In most of our work we have used as our instrument a dummy variable that equals 1 if the student is from a Catholic family and 0 otherwise. The credibility of our bivariate probit results obviously hinges on our assumption that high school students who are Catholic are no more likely to graduate from high school or to begin college than students who are not Catholic. As we argue below, once we control for other observed factors, it appears that being Catholic is not an important determinant of most economic outcomes. We also present tests of overidentifying restrictions that indicate that our instruments are valid and additional results where we use the religious composition of the population in the county where a student attends school as an alternative instrument.

In the next section we describe the *HS&B* data set and the basic variables we have used in our analysis. In Section III we present single-equation probit estimates of high school completion and college entrance models. In that section we also present a number of sensitivity tests of our single-equation model. In Section IV we present bivariate probit models that treat the decision to attend a Catholic school as an endogenous variable. We present a brief summary and conclusions in the final section of the paper.

II. Data

Most of the data for our study were drawn from the *HS&B* survey, which began in the spring of 1980. The original sample was chosen in two stages. Over 1100 secondary schools were selected in the first stage. In the second up to 36 sophomores and 36 seniors were selected from each of the sample schools. Certain types of schools, including public schools with high percentages of Hispanic students and Catholic schools with high percentages of minority students, were oversampled. The original *HS&B* sample included more than 30,000 sophomores and 28,000 seniors. Follow-up surveys of a stratified random sample of the original sophomore cohort were conducted in 1982, 1984, and 1986. Our sample is drawn from the 13,683 students who were sophomores in 1980 and who were included in both the 1982 and 1984 follow-ups. We eliminated 389 students who attended private non-Catholic schools or whose education level in 1984 is unknown. Thus, our final sample includes 13,294 observations.

HS&B contains information on a wide range of topics including individual and family background, high school experiences, and plans for the future. Each student was also given a series of cognitive tests that measured verbal and quantitative ability. The sophomore cohort completed these tests in the initial 1980 survey and again in the first follow-up in 1982 (when most were seniors).[4] School questionnaires, which were completed by an official in each participating school, provided information about dropout rates, staff, educational programs, facilities, and services.

Table I presents definitions and summary statistics for some of the important variables we have used in our study.[5] We classify students as public or Catholic school students based on the school they attended as sophomores. Our study focuses on two measures of educational attainment: high school completion and the decision to begin college. We constructed both variables from the 1984 follow-up data when many of the 1980 *HS&B* sophomores would have been out of high school for two years. *HIGH SCHOOL GRADUATE* is a dummy variable that equals 1 if the student had completed high school by 1984. *COLLEGE ENTRANT* is a dummy variable that equals 1 if the student had enrolled in a four-year college by February of 1984 (and did not first enroll in a two-year college or a vocational training program). Since graduating high school is a precondition for starting college, all of our work defines the *COLLEGE ENTRANT* variable for only those students who have a high school degree.[6]

Most of the family characteristics require little explanation. As can be seen in Table I, data on family income and parents' education are missing in a significant number of cases. We suspect that these values are missing in a nonrandom sample of the population. For example, graduation rates among students where the parents' education is missing are ten percentage points lower than the rate for students where the education variable is available.[7] We looked at a number of strategies to deal with this missing data problem including the estimation of a model suggested by Griliches, Hall, and Hausman [1978] in which we treat nonreporting as an endogenous variable.

Table I. Summary Statistics: High School and Beyond Data Set

Variable name	Definition	Catholic school mean and (std. dev.)	Public school mean and (std. dev.)
High School Graduate	0–1 dummy variable, = 1 if student graduated from high school by February of 1984	0.97 (0.17)	0.79 (0.41)
College Entrant	0–1 dummy variable, = 1 if first postsecondary school attended was 4-year college	0.55[a] (0.50)	0.32[a] (0.47)
Catholic Religion	0–1 dummy variable, = 1 if the student is Catholic	0.79 (0.41)	0.29 (0.45)
% Catholic in County	Percent of the population in the county where the student attends school that is Catholic	31.65 (13.17)	22.37 (16.82)
Female	0–1 dummy variable, = 1 if student is female	0.56 (0.50)	0.50 (0.50)
Black	0–1 dummy variable, = 1 if student is black	0.15 (0.36)	0.13 (0.34)
Hispanic	0–1 dummy variable, = 1 if student is Hispanic	0.22 (0.41)	0.22 (0.41)
White	0–1 dummy variable, = 1 is student is white, non-Hispanic	0.61 (0.49)	0.58 (0.49)
Other Race	0–1 dummy variable, = 1 if student is other race	0.02 (0.15)	0.06 (0.24)
Family Income Missing	0–1 dummy variable, = 1 if family income is not reported	0.22 (0.41)	0.23 (0.42)
Family Income < $7000	0–1 dummy variable, = 1 if family income < $7000	0.03 (0.16)	0.07 (0.26)
Family Income $7000–$12,000	0–1 dummy variable, = 1 if family income ≥ $7000 and < $12,000	0.07 (0.26)	0.11 (0.31)
Family Income $12,000–$16,000	0–1 dummy variable, = 1 if family income ≥ $12,000 and < $16,000	0.12 (0.32)	0.15 (0.35)
Family Income $16,000–$20,000	0–1 dummy variable, = 1 if family income ≥ $16,000 and < $20,000	0.14 (0.35)	0.15 (0.35)
Family Income $20,000–$25,000	0–1 dummy variable, = 1 if family income ≥ $20,000 and < $25,000	0.16 (0.36)	0.13 (0.33)
Family Income $25,000–$38,000	0–1 dummy variable, = 1 if family income ≥ $25,000 and < $38,000	0.13 (0.33)	0.09 (0.29)
Family Income ≥ $38,000	0–1 dummy variable, = 1 if family income ≥ $38,000	0.14 (0.35)	0.07 (0.25)
Parent Education Missing	0–1 dummy variable, = 1 if parents' education not reported	0.09 (0.29)	0.19 (0.40)
Parent High School Dropout	0–1 dummy variable, = 1 if parents' highest education < high school graduate	0.23 (0.42)	0.30 (0.46)
Parent High School Graduate	0–1 dummy variable, = 1 if parents' highest education is high school graduate	0.19 (0.39)	0.20 (0.40)
Parent Some College	0–1 dummy variable, = 1 if parent's highest education is some college	0.28 (0.45)	0.19 (0.39)

Variable name	Definition	Catholic school mean and (std. dev.)	Public school mean and (std. dev.)
Parent College Graduate	0–1 dummy variable, = 1 if parents' highest education is college graduate	0.21 (0.41)	0.11 (0.31)
Single Mother	0–1 dummy variable, = 1 if student's household is headed by single mother	0.12 (0.32)	0.15 (0.35)
Single Father	0–1 dummy variable, = 1 if student's household is headed by single father	0.03 (0.17)	0.05 (0.21)
Natural Mother/ Stepfather	0–1 dummy variable, = 1 if student lives with natural mother and stepfather	0.04 (0.19)	0.06 (0.24)
Both Natural Parents	0–1 dummy variable, = 1 if student lives with both natural parents	0.76 (0.43)	0.62 (0.48)
Other Family Structure	0–1 dummy variable, = 1 if student's household has other structure	0.06 (0.24)	0.12 (0.32)
Age 16	0–1 dummy variable, = 1 if student is ≤ 16 years of age in February of 1982	0.03 (0.17)	0.03 (0.17)
Age 17	0–1 dummy variable, = 1 if student is 17 years of age in February of 1982	0.63 (0.48)	0.49 (0.50)
Age 18	0–1 dummy variable, = 1 if student is 18 years of age in February of 1982	0.32 (0.47)	0.40 (0.49)
Age 19+	0–1 dummy variable, = 1 if student is 19 years of age or older	0.02 (0.15)	0.08 (0.26)
Attends Religious Services Regularly	0–1 dummy variable, = 1 if student attends church at least twice a month	0.69 (0.46)	0.44 (0.50)
Attends Religious Services Occasionally	0–1 dummy variable, = 1 if student attends church occasionally	0.17 (0.38)	0.23 (0.42)
Never Attends Religious Services	0–1 dummy variable, = 1 if student never attends church	0.13 (0.34)	0.33 (0.47)
10th Grade Test Score	Student's sophomore score on standardized exam	30.06 (14.63)	24.53 (15.87)
Test Score Missing	0–1 dummy variable, = 1 if sophomore test score is missing	0.08 (0.28)	0.16 (0.37)
No. of obs.		10,767	2527

a. The COLLEGE ENTRANT means are conditional on having completed high school.

In the end we fell back on a straightforward approach of defining income and parents' education in terms of a set of dummy variables and including "missing data" as a category. We chose the highest income and highest education groups as reference categories in order to facilitate the interpretation of the results.

Table I shows that, compared with Catholic school students, public school students were more than seven times as likely to drop out of high school and were just over half as likely to start college. That table also indicates that the characteristics of Catholic school students suggest that they were more likely to succeed in school. Public school students scored lower on standardized tests and were far more likely to be eighteen years of age or older, to come from low-income families, to have parents who had not finished high school, and to live without their father. The basic question in this paper is whether Catholic schools still have an important impact on high school graduation and college entrance once we control for the effects of these measured differences across students as well as any unmeasured differences. Our sample includes significant numbers of Catholic students who attend public schools and non-Catholic students who attend Catholic schools, thus leaving open the possibility that we can separate the effects of religion from the effects of a religious education.

One simple yet informative test is to compare education outcomes across broad demographic and ability groups.[8] These results parallel the discussion in Coleman and Hoffer [1987, Chapter 4]. In Table II graduation and college entrance rates are computed by ability, family income, parents' education, sex, and race. The table shows that the probability that a public school student will graduate varied dramatically across groups. Among Catholic school students, however, these differences were small. For example, the graduation rate for public school students whose parents were high school dropouts was fourteen percentage points lower than the rate for public school students whose parents were college graduates. Among Catholic school students this difference was only four percentage points. As a consequence, the difference in graduation rates between Catholic and public school students is smallest among students with high test scores from high income, well-educated families. However, even for those groups, Catholic school students graduated at higher rates than their public school counterparts.

As one would expect, there is far more heterogeneity across across demographic groups in college entrance rates. Across all groups, however, Catholic school students were more likely to begin college. As with the high school graduation rates, the differences across sectors declines as ability, family income, and parents' education increase, but there are still large differences in college matriculation rates even for the top categories in all groups.[9]

III. Probit Models of Educational Attainment

The literature on the effect of Catholic schools on the probability of graduating from high school and going to college has rarely gone beyond the sort of

Table II. Educational Outcomes of High School Students by School Type

| Sample | HIGH SCHOOL GRADUATE | | COLLEGE ENTRANT[a] | |
	Public schools	Catholic schools	Public schools	Catholic schools
Full sample	0.79	0.97	0.32	0.55
Sophomore Test Score Missing	0.71	0.98	0.22	0.50
Sophomore Test First Quartile	0.63	0.91	0.11	0.25
Sophomore Test Second Quartile	0.80	0.96	0.19	0.40
Sophomore Test Third Quartile	0.89	0.98	0.37	0.56
Sophomore Test Fourth Quartile	0.95	0.99	0.62	0.78
Parent Education Missing	0.65	0.92	0.16	0.40
Parent H.S. Dropout	0.77	0.95	0.22	0.41
Parent H.S. Degree	0.82	0.97	0.30	0.54
Parent Some College	0.87	0.98	0.44	0.62
Parent College Graduate	0.91	0.99	0.61	0.67
Family Income Missing	0.74	0.97	0.25	0.48
Family Income < $7000	0.64	0.91	0.19	0.36
Family Income $7000–$12000	0.76	0.92	0.23	0.44
Family Income $12000–$16000	0.81	0.98	0.29	0.51
Family Income $16000–$20000	0.84	0.97	0.33	0.49
Family Income $20000–$25000	0.84	0.96	0.38	0.57
Family Income $25000–$38000	0.87	0.99	0.47	0.70
Family Income $38000	0.86	0.98	0.52	0.66
Female	0.80	0.97	0.33	0.53
Male	0.78	0.97	0.31	0.58
Black	0.76	0.95	0.33	0.62
Hispanic	0.76	0.93	0.21	0.45
White	0.81	0.99	0.35	0.56
Other Race	0.84	0.98	0.38	0.56

a. The COLLEGE ENTRANT means are conditional on having completed high school.

simple cross tabulations in Table II. In this section we extend this literature by examining the student's decision to complete high school or enter college by estimating a set of probit models.

A. Single-Equation Probit Models

In the high school graduation version of this model, let the indicator variable $Y_i = 1$ if student i completes high school, and let $Y_i = 0$ otherwise. The choice problem is described by the latent variable model.

$$Y_i^* = X_i \beta + C_i \delta + \epsilon_i, \qquad (1)$$

where Y_i^* is the net benefit a student receives from graduating high school, X_i is a vector of individual characteristics, C_i is a Catholic school dummy variable, and ϵ_i is a normally distributed random error with zero mean and unit variance. Students will only graduate from high school if the expected net benefits of completion are positive, and thus the probability that a student finishes high school is

$$\text{prob}[Y_i = 1] = \text{prob}[X_i\beta + C_i\delta + \epsilon_i > 0] = \Phi[X_i\beta + C_i\delta], \qquad (2)$$

where $\Phi[\ ]$ is the evaluation of the standard normal cdf.

In all of our high school graduation and college entrance probit models, we use the set of individual and family characteristics listed in Table I, dummy variables for urban and rural schools, and three indicators for census regions. Maximum likelihood estimates of the high school completion and college entrance models are reported in columns 1 and 3 of Table III. To measure the qualitative importance of all our right-hand-side variables, we report the marginal effect $\partial\text{prob}(Y_i = 1)/\partial X_i$ for a reference individual in columns 2 and 4.[10] For the *CATHOLIC SCHOOL* dummy variable, we also report at the bottom of Table III the "average treatment effect" which is the average difference between the probability that a student would graduate from high school if he or she attended a Catholic high school and the probability that student would graduate if he or she attended a public school. Thus, if n is the sample size and β and δ are the maximum likelihood estimates of the parameters in equation (2), then the average treatment effect equals $(1/n)\sum_i[\Phi(X_i\beta + \delta) - \Phi(X_i\beta)]$. We use the "delta" method to calculate the variance of the marginal effects and average treatment effects.

The results in Table III show that Catholic school students have a substantially higher probability of completing high school and entering a four-year college than do public school students. Our reference individual's probability of finishing high school would be twelve percentage points higher if she went to a Catholic school than if she went to a public school. The probability that she would enter college would be fourteen percentage points higher. To place these results in perspective, the impact of Catholic schools on high school completion is more than two and one-half times larger than the effect of moving from the lowest to the highest income group, 50 percent larger than the effect of moving from the lowest to the highest parents' education category, and three times as large as the impact of moving from a family headed by a single female to a two-parent family. The estimated marginal effects for *CATHOLIC SCHOOL* reported in Table III are roughly equal to the average treatment effects for the entire sample.[11]

Table III. Probit Estimates of *HIGH SCHOOL GRADUATE* and *COLLEGE ENTRANT* Models

Independent variable[a]	HIGH SCHOOL GRADUATE		COLLEGE ENTRANT	
	Probit coefficient	Marginal effect[b]	Probit coefficient	Marginal effect[b]
Catholic School	0.777	0.117	0.384	0.144
	(0.056)	(0.014)	(0.032)	(0.012)
Female	0.041	0.006	0.021	0.008
	(0.029)	(0.004)	(0.026)	(0.010)
Black	0.132	0.020	0.170	0.064
	(0.045)	(0.007)	(0.042)	(0.014)
Hispanic	0.080	0.012	−0.160	−0.060
	(0.037)	(0.006)	(0.036)	(0.014)
Other Race	0.346	0.052	0.316	0.118
	(0.067)	(0.011)	(0.060)	(0.022)
Family Income Missing	−0.111	−0.017	−0.382	−0.143
	(0.068)	(0.010)	(0.055)	(0.021)
Family Income < $7000	−0.300	−0.045	−0.484	−0.181
	(0.078)	(0.012)	(0.080)	(0.030)
Family Income $7000–$12,000	−0.121	−0.018	−0.408	−0.153
	(0.073)	(0.011)	(0.063)	(0.024)
Family Income $12,000–$16,000	−0.035	−0.005	−0.319	−0.119
	(0.072)	(0.011)	(0.056)	(0.021)
Family Income $16,000–$20,000	0.000	0.000	−0.283	−0.106
	(0.070)	(0.010)	(0.055)	(0.020)
Family Income $20,000–$25,000	−0.035	−0.005	−0.196	−0.073
	(0.072)	(0.011)	(0.055)	(0.021)
Family Income $25,000–$38,000	0.037	0.006	−0.025	−0.009
	(0.077)	(0.012)	(0.057)	(0.021)
Parent Education Missing	−0.730	−0.110	−0.916	−0.342
	(0.061)	(0.013)	(0.052)	(0.020)
Parent High School Dropout	−0.522	−0.078	−0.855	−0.320
	(0.058)	(0.011)	(0.043)	(0.017)
Parent High School Graduate	−0.375	−0.056	−0.602	−0.225
	(0.060)	(0.011)	(0.044)	(0.015)
Parent Some College	−0.204	−0.031	−0.290	−0.108
	(0.062)	(0.010)	(0.042)	(0.016)
Single Mother	−0.255	−0.038	−0.060	−0.023
	(0.041)	(0.007)	(0.042)	(0.016)
Single Father	−0.421	−0.063	−0.269	−0.101
	(0.063)	(0.010)	(0.069)	(0.026)
Natural Mother/Stepfather	−0.286	−0.043	−0.263	−0.098
	(0.056)	(0.009)	(0.060)	(0.023)
Other Family Structure	−0.155	−0.023	−0.060	−0.023
	(0.048)	(0.007)	(0.053)	(0.020)
Age 16	0.611	0.092	0.655	0.245
	(0.089)	(0.015)	(0.115)	(0.043)

Independent variable[a]	HIGH SCHOOL GRADUATE		COLLEGE ENTRANT	
	Probit coefficient	Marginal effect[b]	Probit coefficient	Marginal effect[b]
Age 17	1.025	0.154	0.718	0.268
	(0.050)	(0.014)	(0.087)	(0.033)
Age 18	0.699	0.105	0.603	0.225
	(0.050)	(0.012)	(0.088)	(0.033)
Attends Religious Services Regularly	0.321	0.048	0.299	0.112
	(0.035)	(0.006)	(0.035)	(0.014)
Attend Religious Services Occasionally	0.082	0.012	0.115	0.043
	(0.039)	(0.006)	(0.041)	(0.015)
Intercept	0.388		−0.683	
	(0.093)		(0.107)	
Average treatment effect of Catholic School	0.130		0.132	
	(0.007)		(0.011)	
Log Likelihood	−5155.26		−3297.87	

Asymptotic standard errors are in parentheses. The number of observations in the *HIGH SCHOOL GRADUATE* and *COLLEGE ENTRANT* models is 13,294 and 10,983, respectively.
a. Other exogenous variables include dummy variables for urban and rural schools, plus three regional dummy variables.
b. Marginal effects are calculated for a seventeen–year old white female, living with both natural parents where at least one parent has a high school degree and family income is between $16,000 and $20,000, attends church regularly, and lives in a suburban area in the south.

The other results in Table III are consistent with the literature in this field. Females, students from wealthier families, students with better educated parents, and students living with both natural parents are all more likely to graduate from high school and enter college. Students who are at least eighteen are far more likely to drop out of high school, largely because these students are more likely to have repeated a grade, a clear signal that they have struggled in school. The results on student age may also reflect, in part, the fact that compulsory education laws are not binding for older students [Angrist and Krueger 1991]. The effects of family income on high school graduation is large for students from families with incomes below $12,000 (conditional on parents' education), but increases in income beyond $12,000 seem to have little additional impact on the chances that a student will graduate. In contrast, the probability of college entrance increases monotonically as income rises. The results also show that although in the raw data blacks and Hispanics drop out at higher rates than do whites, once we control for observed characteristics these groups are actually more likely to finish high school.

B. Potential Omitted Variables Bias

In this section we ask whether our basic results are robust. Our primary concern here is that we have omitted important (measurable) characteristics of the student that are correlated with the Catholic school variable and that, as a consequence, we have overstated the benefits of a Catholic school education. The results of some of these sensitivity tests are shown in Table IV. We reproduce the basic results from Table III in the first line of Table IV.

We begin by asking whether including measures of student ability or achievement would change our basic finding. While we would certainly expect to find that better students are more likely to finish high school and start college, we are hesitant to include measures of ability or achievement in our basic model since they are potentially endogenous variables. Here we set these concerns aside for the moment and include in line (2) the student's sophomore score on the *HS&B* exams in the basic probit models. Not surprisingly, test score is an excellent predictor of both measures of educational attainment. The t-statistic on the test score variable is over 13 in both models. Including test score reduces the average treatment effect of Catholic schools from 13.0 percentage points in the dropout model to 10.0 and from 13.2 to 11.1 in the college model. While the effect of Catholic schools is still large in the second line of Table IV, we would argue that these models probably understate the true effect of Catholic schools. The sophomore test score is missing for over 1900 students. It is more likely to be missing for public school students and for students with the highest ex post probability of dropping out.[12] Excluding these observations from the data set would then drag the Catholic school coefficient downward. To illustrate this point more clearly, in line (3) we set the test score equal to zero if the score is missing and include a dummy that equals 1 if the score is missing but equals 0 otherwise. In this specification, including test scores has little impact on our basic conclusions. The average treatment effects in line (3) are very close to the average treatment effects in line (1).[13]

We noted above that Catholic school students are more likely to come from two-parent, high income, well-educated families; i.e., they have "better" observed characteristics. Moreover, they attend schools with peers who, on average, also have better observed characteristics. A number of authors have found that a range of social outcomes is correlated with the quality

Table IV. Probit Estimates of *HIGH SCHOOL GRADUATE* and *COLLEGE ENTRANT* Models

Model	Additional exogenous variables[a]	HIGH SCHOOL GRADUATE				COLLEGE ENTRANT			
		No. of obs.	Probit coefficient on *CATHOLIC SCHOOL*	Marginal effect[b]	Average treatment effect	No. of obs.	Probit coefficient on *CATHOLIC SCHOOL*	Marginal effect[b]	Average treatment effect
(1)		13,294	0.777 (0.056)	0.117 (0.014)	0.130 (0.007)	10,983	0.384 (0.032)	0.144 (0.012)	0.132 (0.011)
(2)	10th GRADE TEST SCORE	11,379	0.632 (0.061)	0.093 (0.013)	0.100 (0.008)	9,567	0.367 (0.036)	0.125 (0.012)	0.111 (0.011)
(3)	10th GRADE TEST SCORE and a dummy variable for missing test score	13,294	0.722 (0.059)	0.112 (0.014)	0.120 (0.007)	10,983	0.392 (0.034)	0.135 (0.012)	0.120 (0.011)
(4)	Measures of peer groups[c]	13,294	0.726 (0.058)	0.129 (0.016)	0.123 (0.007)	10,983	0.308 (0.034)	0.107 (0.012)	0.104 (0.012)
(5)	Dummy variables for whether family owns a calculator, encyclopedia, more than 50 books, or a typewriter	10,797	0.710 (0.062)	0.097 (0.014)	0.111 (0.007)	9,266	0.337 (0.035)	0.130 (0.013)	0.117 (0.012)
(6)	Dummy variables for whether family owns a calculator, encyclopedia, more than 50 books, or a typewriter, and dummy variables for whether these values are missing	13,294	0.764 (0.057)	0.106 (0.013)	0.128 (0.007)	10,983	0.382 (0.032)	0.146 (0.012)	0.130 (0.012)
(7)	State dummy variables	13,294	0.826 (0.058)	0.132 (0.016)	0.136 (0.007)	10,983	0.415 (0.034)	0.134 (0.011)	0.140 (0.012)
(8)	All of the variables in models (3), (4), (6), and (7)	13,294	0.735 (0.061)	0.137 (0.019)	0.120 (0.007)	10,983	0.369 (0.037)	0.111 (0.011)	0.108 (0.012)

Asymptotic standard errors are in parentheses.

a. Other right-hand-side variables include those listed in Table III.

b. Marginal effects are calculated for the individual defined in Table III. The medium public school test scores and average public school values for the peer group measures are used in the appropriate models. For models (5), (6), and (8), individuals are assumed to own all items listed. The reference state used in models (7) and (8) is the state with the most observations in the sample.

c. A set of seven variables that measure the percent of students in a high school whose parents fall into four education categories and whose family falls into three income categories.

of the peer group.[14] Therefore, it is possible that we have overstated the effect of Catholic schools by ignoring peer group effects. We have calculated a set of seven peer group measures for each school in our sample using data from all students in the first wave of *HS&B* (and thus in many cases these peer group measures are based on 72 students). Our peer group measures equal the proportion of students in a school whose parents fall into four education categories and whose family falls into three income categories.[15] In line (4) of Table IV we include these peer group measures in our basic probit. Although a number of the peer variables are statistically significant and ,indicate that better peer groups do increase the probability of completing high school and entering college, the coefficients on the *CATHOLIC SCHOOL* dummy variable and the average treatment effects change very little.[16]

A number of previous studies have found that measures of the family's inputs to education are important determinants of a student's score on standardized exams [Coleman, Hoffer, and Kilgore 1982; Coleman and Hoffer 1987; Noell 1982]. Coleman, for example, includes indicators for whether the student's family owns a calculator, an encyclopedia, more than 50 books, or a typewriter. As the results in line (5) indicate, including these variables does reduce the impact of a Catholic school education, but the Catholic school effect remains quite large. However, as with the test score data, there are many missing observations for these variables. Letting the indicator variables equal zero if the value is missing and including four dummy variables that equal one if the variable is missing, we see in line (6) that these four family measures have little impact on the average treatment effect.[17]

Given the variation in state labor market conditions, compulsory schooling laws and state support for higher education, it is possible that there are strong state effects in the models we have estimated. If these state effects are correlated with the probability of attending a Catholic high school, they may have led us to overstate the impact of a Catholic education on educational attainment. *HS&B* does not identify the state in which a student lives. We can, however, identify all of the students who live in the same state (although we do not know which state that is). The Local Labor Market Indicators for *HS&B* (1980–1982) supplemental file reports local labor market statistics at the county, MSA, and state level for the years 1980–1982. There are 51 unique values for the product of all state level unemployment rates for the three years. In line (7) of Table IV we include 50 state dummy variables in the basic probit models. The marginal and average treatment effects in this fixed-effects model are very similar to the estimates in line (1).[18]

Finally, we run one large model that includes the test scores and a dummy for missing test scores, the seven peer group measures, the four measures of home inputs into education and indicators for missing values, and 50 state dummy variables. Including all 67 of these variables decreases the average treatment effect of a Catholic education on high school completion and college entrance by 8 and 17 percent, respectively. For both dependent variables, however, the average treatment effect is still more than ten percentage points. Our results, therefore, appear to be robust to rather different model specification.

C. Catholic School Selectivity

Public schools must accept virtually all students who live within their attendance boundaries, and in general it is very difficult for most public schools to expel a student. Catholic schools, on the other hand, are free to select their students and to expel students because of poor behavior or poor academic performance. Thus, part of the Catholic school effect we have found could be due to the way Catholic schools choose their students. They are in a better position than public schools to avoid students who in the end are likely to drop out.[19]

The bivariate probit models we present in the next section of the paper can address this question. But we can also present some evidence on this point within our single-equation framework. *HS&B* asked school officials whether their schools used entrance exams as part of the admissions process and whether there was a waiting list for the school. If school selection does play an important role in explaining the success of Catholic schools, then we would expect Catholic schools that use entrance exams or that have waiting lists to have lower dropout rates than other Catholic schools. To test this hypothesis, we interacted the Catholic school dummy variable with these school characteristics. The results are presented in Table V. In both instances we do not find a pattern that is consistent with the school selection hypothesis. In all of the models in Table V, we are unable to reject the hypothesis that there is no difference in graduation or college entrance rates across types of Catholic schools.

D. Definition of the Dependent Variables

As a final sensitivity test in this section, we asked whether our results are robust to alternative definitions of the dependent variables. We have reestimated our models allowing for more inclusive measures of high school graduation and college completion. For example, we have estimated models where we count those with GED's and those who received diplomas after February of 1984 as high school graduates. Counting these students as high school graduates increases the sample average graduation rate to 90.4 percent

Table V. Probit Estimates of *HIGH SCHOOL GRADUATE* AND *COLLEGE ENTRANT* Models

Independent variables[a]	% of Catholic school students	HIGH SCHOOL GRADUATE				COLLEGE ENTRANT			
		Probit coefficient	Marginal effect[b]	Probit coefficient	Marginal effect[b]	Probit coefficient	Marginal effect[b]	Probit coefficient	Marginal effect[b]
Catholic Schools with Entrance Exams	0.842	0.778 (0.064)	0.117 (0.014)			0.388 (0.036)	0.145 (0.013)		
Catholic Schools without Entrance Exams	0.158	1.059 (0.178)	0.159 (0.030)			0.333 (0.072)	0.124 (0.027)		
Catholic Schools with Waiting Lists	0.512			0.850 (0.087)	0.128 (0.018)			0.378 (0.044)	0.141 (0.016)
Catholic Schools without Waiting Lists	0.488			0.782 (0.081)	0.118 (0.016)			0.379 (0.044)	0.141 (0.016)
−2 log likelihood test[c]		2.40		0.36		0.52		0.02	

Asymptotic standard errors are in parentheses. Answers to the entrance exam and waiting list questions were missing in some cases. The number of observations (mean of dependent variable) in the *HIGH SCHOOL GRADUATE* and *COLLEGE ENTRANT* models is 13,033 (0.821) and 10,735 (0.373), respectively.

a. Other right-hand-side variables include those listed in Table III.

b. The marginal effect is calculated for the individual described in Table III.

c. The test statistic is the statistic required to test the equality of the coefficients on the two types of Catholic schools. The test is asymptotically distributed as a χ^2 with one degree of freedom. The 95 percent critical value is 3.84.

and decreases the Catholic school average treatment effect to eight percentage points. Given the recent work of Cameron and Heckman [1993], who found that students earning a GED have poorer labor market outcomes than regular high school graduates, it is not clear that equating these two groups is appropriate. We also counted those who entered two-year colleges and those entering any college after February of 1984 as college entrants. This change in definition increases the mean of the dependent variable to 60 percent, but the Catholic school average treatment effect remains roughly twelve percentage points.[20]

IV. Testing for Selectivity Bias

All of the single-equation models we presented in the previous section treat the decision to attend Catholic schools as exogenous. As Goldberger and Cain [1982] and others argue (and Coleman acknowledges), selectivity bias is potentially the most serious problem in the literature on the effectiveness of private schools. The following example illustrates the nature of the error that could arise. Consider a child whose parents care a great deal about his welfare. We would expect this child to do well in school for two reasons. First, his parents will see that he attends a better than expected school and will be more willing to pay the cost of sending him to a private school. Second, he will succeed in part because of factors that cannot be observed but are under his parents' control. They will spend more time reading to him, they will stress the importance of good grades, and they will see that he does his homework. A single-equation model would mistakenly attribute all of this child's success to his private school. More formally, our results would be biased because the school choice variable in the high school completion and college entrance equations would be correlated with the error term. Similar problems will arise if Catholic schools are able to screen potential students on factors such as a personal interview or they expel students on the basis of poor behavior and academic performance.

A. A Bivariate Probit Model

In this section we outline a simple bivariate probit model that allows for these possibilities. Following the latent variable model in equation (1), suppose that the net benefits of attending Catholic school C_i^* can be written as

$$C_i^* = Z_i \gamma + \mu_i, \tag{3}$$

where Z_i is a vector of observables and μ_i is a random error. A family will enroll a child in a Catholic school if the net benefits are positive; i.e., if $C_i^* > 0$. To allow for the possibility that the unobserved determinants of a student's performance and the unobserved determinants of a family's decision to enroll their teenager in a Catholic school are correlated, we assume that ϵ_i and μ_i are distributed bivariate normal, with $E[\epsilon_i] = E[\mu_i] = 0$, var$[\epsilon_i]$ = var$[\mu_i]$ = 1 and cov$[\epsilon_i, \mu_i]$ = ρ. Because both decisions we model are dichotomous, there are four possible states of the world ($Y_i = 0$ or 1 and $C_i = 0$ or 1). The likelihood function corresponding to this set of events is therefore a bivariate probit.

This system is identified if at least one variable in Z_i is not contained in X_i, Initially, we use as our instrument a dummy variable CATHOLIC RELIGION that equals 1 if the student reports that she is Catholic and 0 otherwise. Subsequently, we consider alternative instruments such as whether a student attends school in a predominantly Catholic area and a set of instruments that we form by interacting CATHOLIC RELIGION with religious attendance variables. We look at the validity of these variables as instruments below.

The bivariate probit results are summarized in Table VI. We repeat the basic single-equation results from Table III in lines (1) and (6) of Table VI. In lines (2) and (7) we present the maximum likelihood (MLE) bivariate probit estimates using CATHOLIC RELIGION as an instrument and the same right-hand variables we use in the basic single-equation models. In both the high school graduate and college entrance models, the MLE estimates of the marginal effect of Catholic schools and the average treatment effect are quite close to the single-equation estimates. The MLE estimate of the correlation coefficient ρ is negative in the high school completion model and positive in the college model, but in both cases the estimate is small, imprecise, and thus statistically insignificant.

In the remainder of Table VI we look at the impact of adding state effects and tenth grade test scores (variables that appeared to be important when we looked at them in Table IV) to the bivariate probit model. These additional variables have little impact on our basic conclusions in the dropout model. The estimated average treatment effect in lines (3)–(5) is similar to the average treatment effect in (2). Our estimates of ρ are always statistically insignificant. Adding tenth grade test scores to the college models (regardless of whether we include state effects as well) reduces the average treatment effect and leads to an estimate of ρ which is positive and significantly different from zero. Even in these models, however, attending a Catholic high school increases the probability of entering college by more than seven percentage points.

The last column of Table VI presents estimates of a somewhat different econometric model. Although the bivariate probit model is straightforward to estimate, the model is substantially more complicated than a standard

Table VI. Maximum Likelihood Estimates of *HIGH SCHOOL GRADUATE* and *COLLEGE ENTRANT* Bivariate Probit Model Using *CATHOLIC RELIGION* as an Instrument

Model	Other variables[b] in X_i	MLE estimates of bivariate probit model				2SLS estimate of coefficient on *CATHOLIC SCHOOL*
		Coefficient on *CATHOLIC SCHOOL*	Marginal effect[c]	Average treatment effect	ρ	
High School Graduate[a]						
(1)		0.777	0.117	0.130		0.096[d]
		(0.056)	(0.014)	(0.007)		(0.008)
(2)		0.859	0.133	0.141	−0.053	0.127
		(0.115)	(0.022)	(0.014)	(0.067)	(0.024)
(3)	10th Grade Test Score and	0.678	0.078	0.114	0.028	0.103
	Test Missing	(0.126)	(0.018)	(0.017)	(0.072)	(0.024)
(4)	State Effects	0.911	0.142	0.144	−0.050	0.114
		(0.121)	(0.027)	(0.015)	(0.072)	(0.024)
(5)	10th Grade Test Score,	0.746	0.124	0.121	0.025	0.134
	Test Missing, and State	(0.132)	(0.028)	(0.016)	(0.077)	(0.030)
	Effects					
College Entrant[a]						
(6)		0.384	0.144	0.132		0.137[d]
		(0.032)	(0.012)	(0.011)		(0.011)
(7)		0.288	0.109	0.098	0.067	0.148
		(0.079)	(0.033)	(0.028)	(0.049)	(0.030)
(8)	10th Grade Test Score and	0.211	0.078	0.064	0.124	0.098
	Test Missing	(0.083)	(0.034)	(0.026)	(0.052)	(0.024)
(9)	State Effects	0.341	0.110	0.115	0.056	0.092
		(0.084)	(0.032)	(0.029)	(0.053)	(0.024)
(10)	10th Grade Test Score,	0.277	0.071	0.082	0.113	0.098
	Test Missing, and State	(0.090)	(0.026)	(0.027)	(0.046)	(0.028)
	Effects					

Asymptotic standard errors are in parentheses.
a. Models (1) and (6) are single-equation estimates from Table III. To estimate models (4), (5), (9), and (10), we deleted all states with no Catholic school students. The high school completion and college entrance models contain 10,120 and 8470 observations, respectively. Both models contain data from twenty states. Models (1), (2), and (3) contain 13,294 observations, and models (6), (7), and (8) contain 10,983 observations.
b. Other exogenous variables include those listed in Table III
c. Marginal effects are calculated for the individual defined in Table III
d. Estimated *CATHOLIC SCHOOL* coefficient from a linear probability model.

two-stage least squares (2SLS) model one could estimate if all potentially endogenous variables were continuous. Fortunately, Angrist [1991] has shown that instrumental variable estimation is a viable alternative to the bivariate probit model. In the notation of equation (1) Angrist showed in a

Monte Carlo study that if we ignore the fact that the dependent variable is dichotomous and estimate

$$Y_i = X_i\beta + C_i\delta + \epsilon_i \tag{4}$$

with instrumental variables (IV), the IV estimate of δ is very close to the estimated average treatment effects calculated in a bivariate probit model. A comparison of the third and fifth columns of Tables VI illustrate the Angrist result. The 2SLS estimates of the Catholic school effect and the average treatment effect are very similar in all of the models we have presented in that table. We will take advantage of this result below where we focus on the validity of our instruments.

B. The Validity of the Instruments

If *CATHOLIC RELIGION* is a valid instrument, then (i) it must be a determinant of the decision to attend a Catholic School, but (ii) it must not be a determinant of the decision to drop out of high school or to start college; i.e., it must not be correlated with the error term ϵ_i. Not surprisingly, it is easy to show that it meets the first test. In a probit model that explains the probability a student will attend a Catholic school, the t-statistic on the *CATHOLIC RELIGION* variable is 36.3. In a simple OLS model where *CATHOLIC SCHOOL* is regressed on *CATHOLIC RELIGION*, the R^2 is 0.16.

Thus, the credibility of our bivariate probit results turns on our assumption that high school students who are Catholic are no more likely to graduate from high school or to begin college than otherwise identical students who are not Catholic. There is little evidence from other studies that would suggest that there are important differences in the education levels of Catholics and non-Catholics. Taubman [1975, Table 3, p. 179], for example, found that the level of education of Jews and Protestants was not significantly different from the level of education of Catholics. Using the data appendix in Tomes [1984], we find that Catholics and non-Catholics have virtually the same average years of education (12.88 versus 12.64, respectively). However, in the raw *HS&B* data (that is, without accounting for variables that are correlated with the Catholic religion variable), Catholic students are more likely to finish high school and to go to college. In the full sample, 88.4 percent of Catholics graduated from high school as compared with 79.0 percent of non-Catholics. Among students who finished high school, 42.8 percent of Catholics entered college as compared with 33.5 percent of non-Catholics. These differences could lead us to estimate of the effect of a Catholic school education that is large but possibly misleading.

The following simple calculation makes this point clear. With our discrete instrument and assuming a bivariate linear model where the only right-hand-side variable is *CATHOLIC SCHOOL*, we can generate an instrumental variable estimate for the *CATHOLIC SCHOOL* effect through a comparison of means. Using the results in Wald [1940], the instrumental variable estimate is simply the difference in graduation rates for Catholics and non-Catholics, divided by the difference in the probability that Catholics and non-Catholics attend Catholic high schools. In the full sample, 39.1 percent of Catholics and 6.4 percent of non-Catholics go to Catholic schools. Thus, the Wald instrumental variable estimate for the impact of Catholic schools in the dropout model is $(.884 - .790)/(.392 - .064) = .287$. For the sample that has completed high school, 43.1 percent of Catholics and 7.8 percent of non-Catholics are in Catholic high schools, implying a Wald estimate for the college entrance model of $(.428 - .335)/(.431 - .078) = .263$.

These raw numbers suggest that, on average, Catholics are better educated than non-Catholics. This will pose a problem for our estimation if, *after controlling for other observed characteristics*, the Catholic religion instrument is correlated with a student's unobserved propensity to graduate from high school or enter college. The most straightforward way to address this issue is to include *CATHOLIC RELIGION* in the single-equation probits we discussed in Table III. We recognize that this is not a formal test since if the correct specification is a bivariate probit then single-equation models are misspecified, but it does offer a clear sense of the patterns in the data. If we include *CATHOLIC RELIGION* in a single-equation dropout model, its estimated coefficient is positive but statistically insignificant. The estimated marginal effect of the *CATHOLIC RELIGION* variable in that model is very small compared with the effect of going to a Catholic school. Although this is not a direct test of whether our instrument is valid, it does indicate that, as a group, Catholics are no different from non-Catholics.

We performed three further tests in order to explore this issue. First, we have constructed additional sets of instruments that recognize that there is heterogeneity in the demand for Catholic schools among Catholics. These models, for example, allow for the possibility that Catholics who attend church regularly are more likely to send their children to Catholic schools than are Catholics who rarely go to church. Second, following Neal [1994] and Hoxby [1994], we have used a very different instrument: the proportion of the population in the county where a student attends school that is Catholic.[21] They argue that it is probable that there will be more Catholic schools in predominantly Catholic areas and thus students (given their observable characteristics) who live in such areas are more likely to attend a Catholic school.[22] There is no reason, however, to suspect that the probability that a student will finish high school or start college depends on her neighbors' religion. Third, we have formed a final set of instruments by combining the Catholic religion and Catholic population variables. The models, like the models that incorporate church attendance, allow for heterogeneity among Catholics (e.g., Catholics who live in heavily Catholic neighborhoods are more likely to send their children to Catholic schools).

This research strategy is particularly attractive since it leads to several models that are overidentified. In those models, we can use Newey's [1985] method of moments specification tests to look at the internal consistency of the model; i.e., whether the variables we use as instruments can be excluded from the structural equation. In a 2SLS model the test statistic is constructed by regressing the estimated errors from the structural model of interest on all exogenous variables in the system. The number of observations times the uncentered R^2 from this synthetic regression is distributed as χ^2 with degrees of freedom equal to the number of instruments minus the endogenous right-hand-side variables in the structural equation of interest. Here again, we recognize that this is not a proper formal test. Although the Angrist [1991] result allows us to accurately estimate the average treatment effect via 2SLS, it is not clear that the assumptions necessary to perform the tests of overidentifying restrictions are met when both Y and C are discrete. This class of tests, however, is the best available diagnostic.

Table VII summarizes the estimates of models that rely on these alternative instruments. All of the models include the exogenous variables that we included in the basic versions of our probits presented in Table III. In lines (1) and (7) we repeat the estimates of the Catholic school effect from lines (2) and (7) in Table VI. For the *HIGH SCHOOL GRADUATE* models, we first interact Catholic religion with the religious attendance variables. Next, we use *% CATHOLIC IN COUNTY* as an instrument. We next use both *CATHOLIC RELIGION* and *% CATHOLIC IN COUNTY* as instruments, and then add the interaction of these variables to the previous model. Finally, in line (6) we use *% CATHOLIC IN COUNTY* as our instrument and include *CATHOLIC RELIGION* as a covariate in both the Catholic school and dropout equations.

Our estimates of the Catholic school effect from the bivariate probit models in lines (1)–(5) fall between 0.114 and 0.141. The 2SLS estimates are quite similar to the bivariate probit estimates in all cases. We cannot construct a test of overidentifying restrictions for the models in lines (1) and (3) since those models are exactly identified. For the other three models, however, all test statistics are well below their 95 percent critical value. The 2SLS estimate of the Catholic school effect in line (6) is consistent with our other estimates, though this effect is measured imprecisely (the standard

Table VII. System Estimates of *HIGH SCHOOL GRADUATE* and
COLLEGE ENTRANT Models with Alternative Instruments

Instruments	Bivariate probit estimates of average treatment effect, *CATHOLIC SCHOOL*	2SLS estimate of *CATHOLIC SCHOOL*	Test of overidentifying restrictions, (d.o.f.), [95% critical value]
High School Graduate[a]			
(1) Catholic Religion	0.141	0.127	
	(0.014)	(0.024)	
(2) Catholic Religion × Attendance at Religious Services	0.141 (0.013)	0.107 (0.022)	3.29 (2) [5.99]
(3) % Catholic in County	0.114	0.130	
	(0.033)	(0.076)	
(4) Catholic Religion and % Catholic in County	0.139 (0.044)	0.127 (0.024)	0.10 (1) [3.84]
(5) Catholic Religion, % Catholic in County and Catholic Religion × % Catholic in County	0.137 (0.014)	0.127 (0.024)	0.84 (2) [5.99]
(6) % Catholic in County[b]	0.061 (0.038)	0.144 (0.373)	
College Entrant[a]			
(7) Catholic Religion	0.098	0.148	
	(0.028)	(0.030)	
(8) Catholic religion × Attendance at Religious Services	0.122 (0.127)	0.167 (0.027)	6.3 (2) [5.99]
(9) % Catholic in County	0.240 (0.053)	0.656 (0.093)	
(10) Catholic religion and % Catholic in County	0.115 (0.037)	0.161 (0.029)	33.7 (1) [3.84]
(11) Catholic Religion and Catholic Religion × % Catholic in County[c]	0.071 (0.028)	0.104 (0.031)	0.81 (1) [3.84]

Asymptotic standard errors are in parentheses. The number of observations in the *HIGH SCHOOL GRADUATE* and *COLLEGE ENTRANT* models is 13,294 and 10,983, respectively.
a. Other exogenous variables include those listed in Table III.
b. *CATHOlIC RElIGION* is included as an exogenous variable in the model.
c. *% CATHOlIC IN COUNTY* is included as an exogenous variable in the model.

error is more than ten times as large as the standard errors in most of the first five models). The bivariate probit estimate of model (6) is somewhat smaller than the other estimates in the upper panel of Table VII. It thus appears that our graduation results are fairly robust, though the results where we depend on *CATHOLIC RELIGION* as an instrument are estimated more precisely.

The *COLLEGE ENTRANT* models in lines (7) through (10) parallel the graduation models in lines (1) through (4). The *COLLEGE ENTRANT* models are much more sensitive to the choice of instruments than are the *HIGH SCHOOL GRADUATE* models. In particular, versions of the model that use *% CATHOLIC IN COUNTY* as an instrument sometimes lead to results that are substantially different from the results we reported earlier. For example, in line (9) where we use *% CATHOLIC IN COUNTY* as the single instrument, the 2SLS estimate of *CATHOLIC SCHOOL* is implausibly large. The tests of overidentifying restrictions in the college model where we interact *CATHOLIC RELIGION* with the religious attendance variable is slightly larger than the critical value (the *p*-value is approximately 0.043), but the college model in line (10) clearly rejects the null hypothesis of internal consistency.

We suspect that the problem is that Catholics are likely to live in states where large numbers of students go on to college. To test this hypothesis, we used the data files from the 1980–1982 October Current Population Surveys and calculated state-level averages of the percent of 18 to 22 year-olds who are enrolled in college. The raw correlation between these values and the percent of the population in a state that is Catholic is 0.38 (*p*-value of 0.006). Because *% CATHOLIC IN COUNTY* may be capturing some unobserved state characteristics in the college models, in line (11) we included it as an exogenous variable and use *CATHOLIC RELIGION* and the interaction *CATHOLIC RELIGION* and *% CATHOLIC IN COUNTY* as instruments. In that model the estimated average treatment effect is 10.4 percent, and the statistic required for the test of overidentifying restrictions is well below the 95 percent critical value.

C. Heterogeneity in the Catholic School Effect

We have also explored the impact of Catholic schools on different subgroups of our sample, and thus, for example, we have estimated separate models for blacks and whites and Catholics and non-Catholics. When we divide the sample into Catholics and non-Catholics, we clearly cannot use *CATHOLIC RELIGION* as an instrument and thus must rely on *% CATHOLIC IN COUNTY* to identify those bivariate probit models. As we showed in Table VII, *% CATHOLIC IN COUNTY* led to several implausible results in the college models. We therefore focus on high school graduation in this section of the paper.

Table VIII presents estimates of the average treatment effect of a Catholic school education for various subgroups. In the single-equation probits and bivariate probits where we use *CATHOLIC RELIGION* as an instrument, Catholic schools have a larger impact on students who have the

Table VIII. Heterogeneity of the Average Treatment Effect, *HIGH SCHOOL GRADUATE* Models

Sample	Number of obs.	Mean *HIGH SCHOOL GRADUATE*	Single-equation probit	% CATHOLIC IN COUNTY	CATHOLIC RELIGION
				Average treatment effect, CATHOLIC SCHOOL[a]	
				Bivariate probit estimates with instructions:	
White	7831	0.826	0.141	0.086	0.128
			(0.007)	(0.039)	(0.016)
Black	1833	0.803	0.134	0.111	0.146
			(0.019)	(0.101)	(0.044)
Urban[b]	3150	0.774	0.172	0.139	0.184
			(0.016)	(0.069)	(0.037)
Suburban	6696	0.862	0.109	−0.003	0.120
			(0.008)	(0.052)	(0.017)
Sophomore Test, First Quartile	2842	0.658	0.213	0.113	0.242
			(0.025)	(0.145)	(0.051)
Sophomore Test, Second Quartile	2842	0.829	0.105	0.128	0.110
			(0.016)	(0.066)	(0.087)
Sophomore Test, Third Quartile	2854	0.916	0.069	0.176	0.071
			(0.010)	(0.039)	(0.020)
Sophomore Test, Fourth Quartile	2841	0.960	0.030	−0.217	0.012
			(0.007)	(0.188)	(0.031)
Catholic	5104	0.884	0.107	0.328	
			(0.008)	(0.033)	
Non-Catholic	8190	0.790	0.145	0.072	
			(0.013)	(0.098)	

Asymptotic standard errors are in parentheses.
a. Other exogenous variables include those listed in Table III.
b. Schools in the South were deleted from this subsample because there were no urban Catholic schools.

lowest probability of finishing high school: blacks, students in urban areas, and students with low test scores. We still find, however, a large, statistically significant Catholic school effect for white and suburban students. These results are in contrast to Neal [1994], who found that Catholic schools raise the probability that urban black students will graduate but have little impact on other groups of students.

Some of these patterns emerge in bivariate probits where we use % *CATHOLIC IN COUNTY* as an instrument, though in general, these models are estimated less precisely. The effect on black and white students is similar, but the average treatment effect for blacks is not significantly different from zero. The pattern across test score groups is difficult to interpret, and the Catholic school effect for Catholics is implausibly large. In all, these results and the *COLLEGE ENTRANT* results in Table VII lead us to conclude that while the argument in favor of using *% CATHOLIC IN COUNTY* to identify the bivariate probit models is quite plausible, the actual gains from doing so are not as clear as we had first hoped.[23]

V. Summary and Conclusions

Spurred by the work of Coleman et al., academics and policymakers have been involved in a decade-long debate over the relative effectiveness of public and private schools. This debate has been waged largely over a single outcome measure: standardized test scores. But, as Card and Krueger [1992, p. 37] have argued, "success in the labor market is at least as important a yardstick for measuring the performance of the educational system as standardized tests." In this paper we have looked at two measures of education that are clearly linked to virtually every measure of success in the labor market: the decisions to finish high school and go to college. We find that teens enrolled in Catholic schools have a significantly higher probability of completing high school and starting college, that the results appear to be robust, and that we cannot attribute the differences between sectors to sample selection bias. Catholic schools appear to have particularly large effects for urban students. This result has some potentially important policy implications given the concern over the quality of public schools in many inner cities. Most of our conclusions are consistent with other work on this problem including Neal [1994], who uses a different data set but a similar econometric approach, and Sander and Krautmann [1995] (which we learned of only after finishing the research for this paper), who use the same data set, a somewhat different econometric approach, and different instruments.

Our research leaves open a number of questions. First, it is possible that further analysis of the *HS&B* data or other data will make the Catholic school effect go away. For example, perhaps we have missed an important omitted variables problem or possibly a different approach to selectivity bias will yield different conclusions. Second, if Catholic schools are as effective as our results suggest, then we are left with a puzzle: why do not more families (particularly lower income Catholic families) make a fairly modest investment and send their children to a Catholic school? Third, if Catholic schools are more effective than public schools, we need to know more about the source of their effectiveness. Coleman et al. attribute this success to Catholic schools' emphasis on discipline, attendance, and homework. Our research does not address this issue, but it is an obvious next step. Finally, we need to know whether it will ever be possible to apply the lessons we learn from the Catholic schools to nonreligious private schools. In some ways, Catholic schools are like other private schools—they must meet the test of the market. But in other ways they are obviously fundamentally different, and it is not clear that they succeed because of the importance of religion or the discipline of competition.[24]

References

Alexander, Karl L., and Aaron M. Pallas, "School Sector and Cognitive Performance: When Is a Little a Little?" *Sociology of Education*, LVIII (1985), 115–28.

Angrist, Joshua D., "Instrumental Variables Estimation of Average Treatment Effects in Econometrics and Epidemiology," *National Bureau of Economic Research*, Technical Working Paper No. 115, November 1991.

Angrist, Joshua D., and Alan B. Krueger, "Does Compulsory Schooling Affect Schooling and Earnings?" *Quarterly Journal of Economics*, CVI (1991), 979–1014.

Bishop, John, "Achievement, Test Scores, and Relative Wages," in Marvin H. Kosters, ed., *Workers and Their Wages* (Washington, DC: AEI Press, 1991).

Bound, John, and George Johnson, "Changes in the Structure of Wages in the 1980's: An Evaluation of Alternative Explanations," *American Economic Review*, LXXXII (1992), 371–92.

Bryk, Anthony S., Valerie E. Lee, and Peter B. Holland, *Catholic Schools and the Common Good* (Cambridge, MA: Harvard University Press, 1993).

Cameron, Stephen V., and James J. Heckman, "The Nonequivalence of High School Equivalents," *Journal of Labor Economics*, XI (1993), 1–47.

Card, David, and Alan B. Krueger, "Does School Quality Matter? Returns to Education and the Characteristics of Public Schools in the United States," *Journal of Political Economy*, C (1992),1–40.

Card, David, and Alan B. Krueger, "The Economic Returns to School Quality: A Partial Survey," Working Paper No. 334, Industrial Relations Section, Princeton University, 1994.

Chubb, John E., and Terry M. Moe, *Politics, Markets, and America's Schools* (Washington, DC: The Brookings Institution, 1990).

Coleman, James S., and Thomas Hoffer, *Public and Private Schools: The Impact of Communities* (New York, NY: Basic Books, Inc., 1987).

Coleman, James S., Thomas Hoffer, and Sally Kilgore, *High School Achievement: Public, Catholic, and Private Schools Compared* (New York, NY: Basic Books, Inc., 1982).

Cookson, Peter W., Jr., "Assessing Private School Effects: Implications for School Choice," in Edith Hasell and Richard Rothstein, eds., *School Choice: Examining the Evidence* (Washington, DC: Economic Policy Institute, 1993).

Evans, William N., Wallace E. Oates, and Robert M. Schwab, "Measuring Peer Group Effects: A Study of Teenage Behavior," *Journal of Political Economy*, C (1992), 966–91.

Evans, William N., and Robert M. Schwab, "Who Benefits from Private Education: Evidence from Quantile Regressions," Department of Economics Working Paper, University of Maryland, August 1993.

Goldberger, Arthur S., and Glen C. Cain, "The Causal Analysis of Cognitive Outcomes in the Coleman, Hoffer and Kilgore Report," *Sociology of Education*, LV (1982), 103–22.

Griliches, Zvi, Bronwyn H. Hall, and Jerry A. Hausman, "Missing Data and Self-Selection in Large Panels," *Annals de l'INSEE*, XXX (1978), 137–76.

Hanushek, Eric A., Steven G. Rivkin, and Dean T. Jamison, "Improving Educational Outcomes While Controlling Costs," *Carnegie–Rochester Conference Series on Public Policy*, XXXVII (1992), 205–38.

Hashimoto, Masanori, and John Raisian, "Employment, Tenure and Earnings Profiles in Japan and the United States," *American Economic Review*, LXXV (1985), 721–35.

Henig, Jeffrey R., *Rethinking School Choice: Limits of the Market Metaphor* (Princeton, NJ: Princeton University Press, 1994).

Hoxby, Caroline M., "Do Private Schools Provide Competition for Public Schools?" Harvard University Department of Economics Working Paper, 1994.

Jencks, Christopher, and Susan E. Mayer, "The Social Consequences of Growing Up in a Poor Neighborhood," in Laurence E. Lynn, Jr, and Michael G. H. McGeary, eds., *Inner-City Poverty in the United States* (Washington, DC: National Academy Press, 1990).

Kane, Thomas J., and Cecilia Elena Rouse, "Labor Market Returns to Two-Year and Four-Year College: Is a Credit a Credit and Do Degrees Matter?" *National Bureau of Economic Research*, Working Paper No. 4268, January 1993.

Lee, Valerie E., and Anthony S. Bryk, "Curriculum Tracking as Mediating the Social Distribution of High School Achievement," *Sociology of Education*, LXI (1988), 78–94.

Levy, Frank, and Richard J. Murnane, "U. S. Earnings Levels and Earnings Inequality: A Review of Recent Trends and Proposed Explanations," *Journal of Economic Literature*, XXX (1992), 1333–81.

Markey, James P., "The Labor Market Problems of Today's High School Dropouts," *Monthly Labor Review*, CXI (1988), 36–43.

Mayer, Susan E., "How Much Does a High School's Racial and Socioeconomic Mix Affect Graduation and Teenage Fertility Rates," in Christopher Jencks and Paul E. Peterson, eds., *The Urban Underclass* (Washington, DC: The Brookings Institution, 1991) pp. 321–41.

Mensch, Barbara S., and Denise B. Kandel, "Dropping out of High School and Drug Involvement," *Sociology of Education*, LXI (1988), 95–113.

Murname, Richard J., "A Review Essay—Comparisons of Public and Private Schools: Lessons from the Uproar," *Journal of Human Resources*, XIX (1984), 263–77.

Murphy, Kevin M., and Robert H. Topel, "The Evolution of Unemployment in the United States: 1968–1985," in Stanley Fischer, ed., *NBER Macroeconomic Annual: Vol. 2, 1987* (Cambridge, MA: MIT Press, 1987), pp. 11–58.

Neal, Derek, "The Effects of Catholic Secondary Schooling on Educational Attainment," University of Chicago, unpublished paper, July 1994.

Newey, Whitney K., "Generalized Methods of Moments Estimation and Testing," *Journal of Econometrics*, XXIX (1985), 229–56.

Noell, Jay, "Public and Catholic Schools: A Reanalysis of Public and Private Schools," *Sociology of Education*, LV (1982), 123–32.

Quinn, Bernard, et al., *Churches and Church Memberships in the United States, 1980* (Atlanta, GA: Glenmary Research Center, 1982).

Rouse, Cecilia E., "Democratization or Diversion? The Effect of Community Colleges on Educational Attainment," *Journal of Business and Statistics*, XIII (1995), 217–24.

Sander, William, and Anthony C. Krautmann, "Catholic Schools, Dropout Rates and Educational Attainment," *Economic Inquiry*, XXXIII (1995), 217–33.

Taubman Paul J., *Sources of Inequality in Earnings: Personal Skills, Random Events, Preferences Towards Risk and Other Occupational Characteristics* (Amsterdam: North-Holland Publishing Company, 1975).

Thornbery, Terence P., Melanie Moore, and R. L. Christenson, "The Effect of Dropping out of High School on Subsequent Criminal Behavior," *Criminology*, XXIII (1985), 3–18.

Tomes, Nigel, "The Effects of Religion and Denomination on Earnings and the Returns to Human Capital," *Journal of Human Resources*, XIX (1984), 472–88.

Wald, Abraham, "The Fitting of Straight Lines if Both Variables Are Subject to Error," *Annals of Mathematical Statistics*, XI (1940), 284–300.

Weiss, Andrew, "High School Graduation, Performance, and Wages," *Journal of Political Economy*, XCVI (1988), 785–820.

Witte, John F., "Understanding High School Achievement: After a Decade of Research, Do We Have Any Confident Policy Recommendations?" paper presented at the annual meeting of the American Political Science Association, August 1990.

Witte, John F., "Private School versus Public School Achievement: Are There Findings That Should Affect. the Educational Choice Debate?" *Economics of Education Review*, XI (1992), 371–394.

Notes

1. Henig [1994], for example, found that out of 125 questions in *HS&B* dealing with vocabulary, reading, mathematics, science, writing, and civics, public school students improved by 7.16 items (from 67.07 as sophomores to 74.23 as seniors), while Catholic school students improved by 8.98 items. Thus, even before accounting for differences in family characteristics, Coleman's Catholic school effect represents only $8.98 - 7.17 = 1.81$ additional correct answers.

2. For example, Chubb and Moe [1990], in their 318-page analysis of effective schools, use test scores as virtually their sole measure of school performance. Coleman does discuss differences in dropout rates briefly, but the analysis is limited to simple cross tabulations of the data. Neal [1994] and Sander and Krautmann [1995] are similar in some ways to this paper.

3. For a review of the effects of cognitive development on labor market performance, see Hanushek, Rivkin, and Jamison [1992] and Bishop [1991].

4. The test score we report is the sum of the "formula" score on the mathematics, vocabulary, and reading exams. Students received one point for each correct answer and lost a fraction of a point for each incorrect answer (where the fraction depends on the number of possible answers). The maximum possible score on the *10TH GRADE TEST SCORE* is 68.

5. All individual and school variables were constructed from either the composite variables in the *HS&B* data set or were taken from the base-year survey. The summary statistics in Table I are unweighted and thus do not represent an accurate picture of 1980 high school sophomores. We have not used sample weights in our econometric work.

6. The definition of these two outcome measures is not quite as straightforward as one might think. For example, we do not count students earning GED's as high school graduates. This is a reasonable restriction given recent work by Cameron and Heckman [1993], who find that graduates with GED's do not perform as well in the labor market as students with regular high school diplomas. Similarly, we do not count people who went to college long after graduating from high school and people who attended a two-year college as college students. Restricting our attention to students entering a four-year college is arguable given work by Kane and Rouse [1993] who find that credit hours from two- and four-year colleges are rewarded equally in the workforce. Rouse [1995] also finds that, on net, community colleges increase total years of schooling but do not alter the probability of obtaining an undergraduate degree. As we demonstrate later, these assumptions are not critical.

7. There is reason to believe that most of the missing income values are from families with low income. Students were given a breakdown of family

income by thirds and asked in what portion of the income distribution does their family fall. Using sample weights from the second follow-up survey, a total of 29 percent and 27 percent of the students reported being in the top two-thirds of the income distribution, respectively, while only 13 percent said that their family was in the bottom third (the rest did not respond).

8. The test quartiles were calculated for the entire sample using second follow-up sample weights.

9. Bryk, Lee, and Holland [1993] found similar results for Catholic schools in their analysis of the *HS&B* test score data. Using quantile regression techniques, Evans and Schwab [1993] also found that the benefits of a Catholic education on test scores are concentrated among the least able students, students whose parents have little education and students from low-income families.

10. We calculated the marginal effects for the "average" public school student, who we defined as a seventeen-year-old white female, living with both natural parents, in a family where at least one parent has a high school diploma, family income is between $16,000 and $20,000, who attends religious services regularly, and who lives in a suburb in the south.

11. All of the college graduation models we present in this paper are estimated on the subsample of students who graduated from high school. Within the entire sample, 26 percent of the public school students and 53 percent of the Catholic school students entered college. The average treatment effect in the college model presented in Table III using the entire sample is 0.217 with a standard error of 0.020.

12. In our sample, the sophomore test score is missing for 20 percent of the public school students and 11 percent of the Catholic school students. High school completion rates are 84 percent for students with a valid test score, but only 74 percent for students without a score.

13. The marginal effects are calculated for the reference individual defined in Table Ill. In addition, we assume that this student's test score equals the median public school score in our sample. The marginal effects (standard errors) for the *10TH GRADE TEST SCORE* in the high school completion and college entrance models are 0.004 (0.0002) and 0.013 (0.001), respectively. These results suggest that a one-standard-deviation increase in the test score over the median value (about a fifteen-point increase) would increase high school completion and college entrance probabilities by six and twenty percentage points, respectively.

14. See Jencks and Mayer [1990] for a review of the literature on peer effects, and see Mayer [1991] for an estimate of the effects of peer groups on high school completion rates. Both of these studies are concerned with single-equation estimates of the effects of peers on the economic outcomes

of teens. Evans, Oates, and Schwab [1992] argue that because families can choose among schools and neighborhoods, a student's peer group is a potentially endogenous variable. We do not consider the endogeneity of the peer measures in this paper.

15. *HS&B* did collect information at the school level which could be used directly to form peer group measures. As with the test score data, however, these variables are missing for many schools (especially public schools). Although the peer group measures we constructed are based on a sample rather than a census of students from a high school, the large number of observations per school should provide us with a good approximation of the composition of the school. We have tested this argument by using this same procedure to construct a measure of the proportion of the students in a school who are black and comparing this estimate with the figure reported in the school survey. The correlation coefficient for these two series is 0.97.

16. The marginal effects are calculated for a student who has an average public school value of the peer group variables. Because of space limitations, we do not report the parameter estimates for all seven peer group measures in both models. We note that the peer group variables measuring parents' education tended to be more important determinants of high school completion and college entrance than measures of income. In fact, once we included parents' education, the peer measures for income became largely insignificant. The marginal effects (standard errors) for the peer group variables measuring parents' education in the high school completion model are as follows: *% PARENT EDUCATION MISSING* −0.24 (0.06), *% PARENT EDUCATION LESS THAN HIGH SCHOOL* −0.12 (0.05), *% PARENT EDUCATION HIGH SCHOOL GRADUATE* −0.11 (0.04), *% PARENT EDUCATION SOME COLLEGE* −0.20 (0.05). The corresponding values for the college entrance model are −0.63 (0.10), −0.35 (0.07), −0.57 (0.05), −0.50 (0.08). The reference group in both models is the percent of students in the school whose parents are college educated.

17. To calculate the marginal effects for these two models, we assume that the individual owned all four items.

18. We calculated marginal effects for a student who lived in the state with the most observations in our data set.

19. The evidence from the existing literature on the role of student selection in the success of Catholic schools is somewhat mixed. Bryk, Lee, and Holland [1993] argue that, in general, Catholic schools are not highly selective in their admissions. They find that the typical Catholic school accepts 88 percent of the students who apply. They also argue that contrary to widespread belief, very few students are expelled from Catholic schools for either academic or disciplinary grounds. On average, Catholic high schools

dismiss fewer than two students per year. Witte [1990] presents evidence that Catholic schools do in fact screen admissions so that they are able to avoid students who are likely to do poorly. For example, he finds that 55.5 percent of Catholic school principals, as compared with only 8.4 percent of public school principals, indicated that prior academic record was an important factor in admission decisions.

20. These results are available upon request.

21. The Association of Statistics of American Religious Bodies (ASARB) provided us with data on the Catholic population by county. Their data are drawn from a survey of over 200,000 congregations and churches with total membership of nearly 115 million. See Quinn et al. [1982] for a discussion of these data. With the ASARB data and data from the 1980 Census, we then constructed an estimate of the percent Catholic at the county level. County identifiers are not available in the public use *HS&B* data. We have entered into an agreement with the U. S. Department of Education where we created a data set that included the percent Catholics in a county and county FIPS codes. The contractor for the *HS&B* data set then merged the data set we created with student identification numbers. In order to protect the confidentiality of the data, the percent Catholic in the county variable was grouped (0.0–4.9 percent, 5.0–9.9 percent, etc.) and top-coded at 70 percent.

22. This hypothesis is easily validated. In a first-stage probit model where *CATHOLIC SCHOOL* is the dependent variable, the coefficient on *% CATHOLIC IN COUNTY* is .001 with a standard error of 3.1×10^{-4}. To put this result into perspective, moving a student from the twenty-fifth percentile *% CATHOLIC IN THE COUNTY* to the seventy-fifth percentile increases the probability that the student will attend a Catholic school by ten percentage points.

23. Implicitly, we have treated *% CATHOLIC IN COUNTY* as an exogenous variable. It will be correlated with the error term in the outcome equations if, for example, families that care a great deal about education move to counties where many Catholics live in order to take advantage of the availability of Catholic schools or lower tuition as a member of the parish. This argument could explain the problems we have found when we try to use this variable as an instrument.

24. There is substantial disagreement over this issue in the literature. See, for example, Bryk, Lee, and Holland [1993] and Chubb and Moe [1990] for two very different views.

Education and Fertility: Implications for the Roles Women Occupy

Ronald R. Rindfuss, University of North Carolina
Larry Bumpass, University of Wisconsin
Craig St. John, University of North Carolina*

Ronald R. Rindfuss, Larry Bumpass, and Craig St. John, 1980, Education and Fertility: Implications for the Roles Women Occupy, *American Sociological Review* 45 (June), 431–47. © American Sociological Association, reproduced with permission.

* Direct all communication to: Ronald R. Rindfuss; Department of Sociology; Hamilton Hall 050A; University of North Carolina; Chapel Hill, N.C. 27514. This is a revised version of a paper presented at the annual meeting of the American Sociological Association, San Francisco, September, 1978.This analysis was supported in part by grants from the Spencer Foundation to Ronald R. Rindfuss, from the University Research Council of the University of North Carolina to Ronald R. Rindfuss, from a subcontract to Ronald Rindfuss of NICHD grant No. 10751, from the Graduate School of the University of Wisconsin-Madison to Larry Bumpass, by a Center for Population Research grant No. 05798 to the Carolina Population Center from the Center for Population Research of the National Institute of Child Health and Human Development, and by a Center for Population Research grant No. 05876 to the Center for Demography and Ecology of the University of Wisconsin from the Center for Population Research of the National Institute of Child Health and Human Development. The authors gratefully acknowledge the able research assistance of Barb Witt and Cheryl Knobeloch, and the comments of two anonymous reviewers.

Abstract

The interplay between education and fertility has a significant influence on the roles women occupy, when in their life cycle they occupy these roles, and the length of time spent in these roles. The overall inverse relationship between education and fertility is well known; but little is known about the theoretical and empirical basis of this relationship. This paper explores the theoretical linkages between education and fertility and then examines the relationships between the two at three stages in the life cycle. It is found that the reciprocal relationship between education and age at first birth is dominated by the effect from education to age at first birth with only a trivial effect in the other direction. Once the process of childbearing has begun, education has essentially no direct effect on fertility; but it has a large indirect effect through age at first birth.

No factor has a greater impact on the roles women occupy than maternity. Whether a woman becomes a mother[1] the age at which she does so, and the timing and number of her subsequent births set the conditions under which other roles are assumed. Some may deplore this situation and it may be changing, but the dominance of motherhood continues to be a fact for the vast majority of women. While there is clearly variance in this role dominance, the assumption of nonfamilial roles varies markedly with the fact, timing, and extent of maternity.

Education is another prime factor conditioning female roles, Education is expected to impart values, aspirations, and skills which encourage and facilitate nonfamilial roles. It is possible that better educated women may assume less traditional role patterns than less-educated women with identical fertility histories. However, it is also likely that education affects women's roles through differing patterns of fertility. This paper discusses some of the possible linkages between education and fertility and reports analyses bearing on: (1) the relationship between education and age at first birth, (2) the effects of education on the timing of subsequent births, particularly on the experience of short birth intervals, and (3) educational differences in wanted family sizes.

Education–Fertility Linkage

Given the importance of the interplay between education and fertility for the roles women occupy in industrialized societies, there has been surprisingly little attention paid to the causal linkages between the two.[2] In part, this may be because the possible causal connection between fertility and education is exceedingly complex. Some have assumed that education affects fertility (e.g., Westoff and Ryder, 1977; Rindfuss and Sweet, 1977; Cho et al., 1970;

Whelpton et al., 1966), and some have argued that fertility also affects education (Waite and Moore, 1978).

Most of the theory and research concerned with education and fertility conceptualizes both in terms of their end products: completed education and children ever born. In fact, children come one at a time (usually), and education is completed a year at a time, sometimes a course at a time. Children can come close together, or at intervals of 10, 15 or even 20 years. Formal schooling can be completed without interruption; or it can be completed after short or long interruptions (Davis and Bumpass, 1976). Models of education and fertility should reflect the fact that education and fertility are processes which take time to complete and which can intercept each other in complex ways.

The overall relationship between education and fertility has its roots at some unspecified point in adolescence, or perhaps even earlier. At this point aspirations for educational attainment as a goal in itself and for adult roles that have implications for educational attainment first emerge. The desire for education as a measure of status and ability in academic work may encourage women to select occupational goals that require a high level of educational attainment. Conversely, particular occupational or role aspirations may set standards of education that must be achieved. The obverse is true for those with either low educational or occupational goals. Also, occupational and educational aspirations are affected by a number of prior factors, such as mother's education, father's education, family income, intellectual ability, prior educational experiences, race, and number of siblings (for example, see Hout and Morgan, 1975).

Occupational and educational aspirations are also reciprocally related to evolving fertility preferences. These fertility preferences include both number and timing preferences, that is, whether a first birth is wanted ever and, if so, when. The number and timing preferences may be related if, for example, a desire for many children leads to a desire to begin childbearing as soon as possible (Bumpass and Westoff, 1970). Moreover, the preference for postponing a first birth may lead to interests in other areas which may then lead to a decision not to have any children. There is evidence that repeated postponement of the first birth is a typical pattern among those who are voluntarily childless (Veevers, 1973). Such preferences for timing are necessarily vague, but nonetheless important. Some young women may wish to have a baby as soon as possible, perhaps to establish an adult identity separate from their parents, or to fulfill strong nurturing needs, Such aspirations among young women are likely to have a negative effect on evolving role and educational aspirations. Similarly, a young woman who is sure she does not want to have a child any time soon, if at all, may expand her role and educational aspirations accordingly. Influences in the opposite direction operate as the threat of early fertility to educational attainment are recognized and fertility desires are adjusted accordingly.

Both of these preference sets (occupational and educational aspirations as well as fertility preferences) influence actual age and education at first birth through a set of intervening variables[3] that include the standard intermediate variables affecting exposure to intercourse, conception risk and gestation and parturition (Davis and Blake, 1956; Bongaarts, 1978).

Adolescents with higher educational and occupational goals may choose social patterns that are less likely to lead to early marriage, that is, "not wanting to go steady or get serious with boys," because they want to go to college. They may be less willing to engage in intercourse because of the threat of possible pregnancy to their educational or career plans. Sexually active adolescents with high educational aspirations may be more likely to try to control the risk of pregnancy through careful contraceptive use.

Adolescent women who desire early motherhood (and presumably early marriage) are likely to follow social patterns that lead to early intensive emotional involvement; and, when sexually active, this group may have relatively low motivation to avoid pregnancy. Such patterns may lead indirectly to lower educational achievement because of an early age at first birth.

Early marriage may have a direct effect on reducing educational attainment,[4] for example, when a girl leaves school in order to be married. These social patterns also have an indirect effect on education through factors affecting pregnancy and early age at first birth.

It should be noted that in the reciprocal relationship between education and age at first birth, the effects of education on age at first birth can only be the result of the intermediate variables discussed above (also, see Davis and Blake, 1956; Bongaarts, 1978) whereas the effect of age at first birth on education may also include a direct effect.

Both age and education at first birth can affect subsequent role and educational aspirations, and subsequent preferences for the timing and number of children. These subsequent aspirations and preferences are also reciprocally related. After the birth of their first child some women may find that they wish to reduce their fertility goals, increase their occupational goals, and return to school. Others who had planned on continuing their education may decide to have more children, or to quickly become pregnant again, either because of great satisfaction in the mother role, or because of a sense that it is an all-consuming role that precludes other options, or because they are not sure of what else to do.

Education, age at first birth, the possibly revised occupational and educational aspirations, as well as timing and number preferences all affect various aspects of the intermediate variables in a process similar to that elaborated above with respect to the period before the first birth. The period prior to the first birth includes an unmarried and sexually inactive period as well as a married interval for most women. For most (but not all) women, the period following the first birth begins within marriage. Some women will not yet be married and others will have married and separated or divorced by the time of the first birth. At the second birth, a woman may be never married, currently married, widowed, divorced or separated (Rindfuss and Bumpass, 1977). Marital instability is an important social factor in the social patterns category in each segment to the extent that it affects other intermediate variables such as frequency of intercourse, periods of abstinence, and use of contraceptives.

Fecundity is largely exogenous to the processes we are examining, though it has a clear effect on the timing of the first birth and may mediate the effect of age at first birth on subsequent fertility.

While these potential intersections in the relationship between education and fertility warrant more intensive study, that is not our purpose in this paper. The point we are attempting to make in the preceding discussion is that the observed relationship between completed education and completed family size is the cumulative outcome of a complex process that involves attitudes and decisions about both education and fertility that may change as time passes or as the woman moves from one stage to the next, and that it is necessary to examine empirically the various stages in the process.

Data

The data used are from the 1970 National Fertility Study (NFS), a multipurpose study based on a national probability sample of 6,752 ever married women under 45 years of age residing in the continental United States (Westoff and Ryder, 1977). Complete birth and pregnancy histories were obtained, thus permitting analysis of age at first birth and of birth intervals. Unfortunately, a complete educational history was not obtained. Only education at interview and education at marriage were obtained. This means that we have to use education at marriage as a proxy for education at first birth. For most women this is a reasonable proxy, since the correlation between age at first birth and age at first marriage is 0.74. In order to check the reasonableness of using education at marriage, we reran all the analyses using education at interview, and results were unaffected. However, it should be recognized that for younger mothers the first birth is likely to precede the first marriage. Finally; it should be noted that there were no questions asked about educational or occupational aspirations during the adolescent and young adult years.

Although not reported in detail here, wherever possible we have also examined data from the 1973 National Survey of Family Growth (FGS) (NCHS, 1978) (a national probability sample of 9,797 women under age 45 who had ever been married or who were never married mothers in 1973), and essentially comparable results were found in both data sets.

Education and Age at First Birth

In the absence of accurate data on the intermediate variables, the relationship between the fertility and educational processes can be conceptualized as a simple causal process. The aspirations, plans, and decisions (and "apparent" nondecisions) leading to an early first birth may result in lowered educational aspirations and achievement. Women who desire and obtain a high level of education may adjust their fertility preferences accordingly. Both the educational and the first birth process are affected by a set of exogenous factors reflecting background characteristics and characteristics of early adolescence. A model of these relationships is shown in Figure 1. The rationale for this set of exogenous variables, and their effects on education and age at first birth, is considered elsewhere (Rindfuss and St. John, 1979); in the present paper we concentrate only on the relationship between education and age at first birth. Table 1 indicates the measurement of these exogenous variables, and the Appendix reports the zero-order correlations among all the variables in Figure 1.

That the relationship between education and age at first birth should be viewed as potentially reciprocal is often overlooked: one direction of causation is usually emphasized to the exclusion of the other. For example, Jaffe (1977:22) asserts: "Pregnancy is the most common cause of school dropout among adolescent girls in the U.S." Others, however, contend that education determines age at first birth; and, further, that women who get pregnant while still in school do so to have an "acceptable" reason for dropping out of school (Cutright, 1973). Since, there is considerable overlap in the time when women leave school and the time when they have their first child (median age at first birth is currently about 22, and, of recent cohorts, 25% have their first birth by the end of the 19th year), it is important to investigate the extent to which the educational attainment process and age at first birth process are reciprocally related.

The part of the model shown in Figure 1 of direct interest here is the relationship between education and age at first birth. We allow for a reciprocal relationship between these two variables, with each affected by other variables in the model as well. Age at first birth is computed from the date of respondent's birth and date of birth of respondent's first child. Education, as noted, is education at marriage, not education at first birth. In order to

Figure 1. A Model of the Relationship between Educational Attainment and the Beginning of Motherhood.

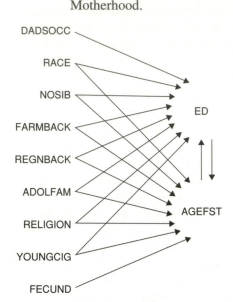

$$\widehat{ED} = b_0 + b_1DADSOCC + b_2RACE + b_3NOSIB + b_4FARMBACK \\ + b_5REGNBACK + b_6ADOLFAM + b_7RELIGION \\ + b_8YOUNGCIG + b_9AGEFST + U$$

$$\widehat{AGEFST} = c_0 + c_1RACE + c_2NOSIB + c_3FARMBACK \\ + c_4REGNBACK + c_5ADOLFAM + c_6RELIGION \\ + c_7YOUNGCIG + c_8FECUND + c_9ED + V$$

estimate the reciprocal relationships between education and age at first birth, instrumental variables are needed for each of the two endogenous variables— that is, variables are needed which directly affect one of the endogenous variables but not the other, which are not causally determined by the endogenous variables, and which are not correlated with the unspecified source of the endogenous variable for which it is not an instrument (Duncan, 1975; Heise, 1975). As can be seen from Figure 1, fecundity is used as the instrument for age at first birth and respondent's father's occupation as the instrument for education. Fecundity is measured by whether or not the respondent had a miscarriage prior to her first birth.[5] A miscarriage before the first birth postpones the first birth in a direct and obvious way: it takes time to conceive again and carry that conception to successful parturition. It also gives the woman a second chance if she wants to contracept. The additional time involved as the result of a miscarriage before the first birth can be substantial since approximately one-fourth of the women who have one miscarriage before their first birth have two or more miscarriages before their first birth.

A miscarriage before the first birth should have no effect on education, except indirectly through age at first birth. This would occur only if the woman dropped out or was expelled from school prior to the miscarriage because of the pregnancy. If this were the case, then the miscarriage would be correlated with the disturbances in the education equation and would be unsuitable as an instrument. However, this is unlikely because the vast majority of miscarriages occur in the early months of a pregnancy, before it is obvious to observers that the woman is pregnant, and often before the woman knows that she is pregnant (see National Center for Health Statistics, 1966). If the woman is unmarried, she is unlikely to notify the school that she is pregnant until it becomes absolutely necessary. It is probably in part for this reason that unmarried women often do not seek prenatal care until very late in pregnancy (National Academy of Sciences, 1973). Furthermore, education should not have any effect on whether or not there is a miscarriage before the first birth. The only exception to this statement would involve a woman obtaining an induced abortion in order to complete her education. However, induced abortions are so grossly underreported in United States fertility surveys that reported miscarriages are essentially spontaneous miscarriages.

That respondent's father's occupation affects respondent's educational attainment is well known (Alexander and Eckland, 1974; Blau and Duncan, 1967; Kerckhoff and Campbell, 1977; Sewell and Hauser, 1977) and does not require further elaboration here. We also argue that father's occupation does not have a direct relationship with age at first birth. Rather, we would argue that the relationship is indirect through education. It can be expected that families of an orientation in which the father has a high status job would be more likely to encourage daughters to postpone the first birth than families of an orientation in which the father has a low status job. However, the most likely explicit and implicit justification for this encouragement would be to allow daughters time to complete their education, and thus the effect on age at first birth would be indirect. However, there may also be an intergenerational transmission of norms regarding age at first birth. (Leonetti [1978] provides a good example of this in the case of Japanese–Americans.) To the extent that socioeconomic status directly affects the intergenerational transmission of norms regarding age at first birth—that is, in addition to the indirect transmission through educational aspirations—then respondent's father's occupation would not be a suitable instrument for education. Recent work by Thornton (forthcoming) suggests that there is no direct transmission

Table 1. Measurement of Variables in the Education and Age at First Birth Model

Variable Label	Variable Name	Units of Measurement	Description	Mean	Standard Deviation
DADSOCC	Respondent's father's occupation	Duncan's SEI scores	Missing data were given the mean value	30.209	21.370
RACE	Race of respondent	Dummy variable	Blacks are coded 1, all others 0	0.099	0.299
NOSIB	Respondent's number of siblings	Actual number is coded		3.889	3.035
FARMBACK	Respondent's farm background	Dummy variable	Coded 1 if respondent grew up on a farm, all others coded 0	0.330	0.470
REGNBACK	Region where respondent grew up	Dummy variable	Coded 1 if respondent grew up in South, all others coded 0	0.357	0.479
ADOLFAM	Household composition when respondent was 14	Dummy variable	Coded 1 if respondent lived with 0 or 1 of her parents at age 14, coded 0 if she lived with both parents at age 14	0.183	0.387
RELIGION	Respondent's religous preferences when growing up	Dummy variable	Coded 1 if Catholic, all others coded 0	0.231	0.421
YOUNGCIG	Whether respondent smoked at a young age	Dummy variable	Coded 1 is respondent smoked before 16, coded 0 otherwise	0.136	0.342
FECUND	Whether respondent had a miscarriage before the first birth	Dummy variable	Coded 1 if respondent had a miscarriage before the first birth; coded 0 otherwise	0.099	0.298
ED	Respondent's education	Years of schooling completed	This is education at first marriage	11.595	2.360
AGEFST	Respondent's age at first birth	Years		22.012	4.079

Note: Throughout, the usual practice of deleting missing data is followed of fertility norms from parental status. Instead, this influence was transmitted through the education of the offspnng.

In order to examine our assumption that parental socioeconomic status does not have a direct effect on the transmission of norms regarding age at first birth, we examined the determinants of ideal age at first birth. The 1970 NFS included the following question:

"Q. 3: What do you think is the ideal age for a woman to have her *first* child?"

Although this question suffers from all the problems of "ideal" questions (Blake, 1966; Bumpass and Westoff, 1970; Rindfuss, 1974; Ryder and Westoff, 1969) as well as some problems specific to this question (Rindfuss and Bumpass, 1978), it does provide the best measure available for norms regarding age at first birth. Using a sample of recently married women in order to minimize the possibility that the responses to the question would be affected by the cumulative maternal experience of the woman, we find that, after other appropriate factors are controlled, father's occupation has no significant direct effect on the ideal age to have a first birth. This further supports Thornton's results and supports the theoretical argument that parental socioeconomic status influences age at first birth only indirectly through its effect on the offspring's educational aspirations, and thus supports the use of father's occupation as an instrument for education in our model.

However, somewhat less consistent support was found in an examination of the 1971 National Survey of Young Women data. Since father's occupation was not available, the relationship between father's education and ideal age at first birth was considered for this sample of teenagers 15–19 years of age. While most of the association is accounted for by educational aspirations, ideal age at first birth is 0.4 years lower among the children of high school graduates than among those of fathers who attended college, net of other factors. While this modest net effect of father's education cautions our theoretical position, we would expect the net effect of father's occupation on ideal age at first birth to be considerably weaker.

Before presenting the results, it is necessary to discuss some of the variables which are not included in Figure 1 and the possible biases their exclusion might introduce. The first is marriage. Although we recognized the role of age at marriage in the earlier discussion in this paper (especially since it is incorporated in sexual experience), it is age at first birth that is emphasized both there and in our analysis here. Clearly, age at first marriage and age at first birth are closely related, normatively and empirically. However, we feel that the first birth has greater consequences for the life style and roles of the woman (Rindfuss, 1979), and that the effects of the first birth are more permanent than those of first marriage. Marini (1978) has recently argued that age at marriage is more important than age at first birth in the transition to adulthood because age at marriage "usually sets a lower limit on the age at which first birth occurs." We disagree for the following reasons: In the first place, motherhood frequently precedes first marriage. (And this is more likely to be the case the younger the age at first birth.) Second, some people may initiate the serious consideration of marriage on the basis of when they want to begin parenthood, as reflected in the phrase "time to settle down and start a family." The high incidence of premarital intercourse argues against the notion that age at first marriage sets a lower bound on exposure to the risk of conception. Third, "becoming a parent" is the modal response of married parents to the question of what marks the transition to adulthood (Hoffman, 1978). Fourth, parenthood is more permanent than marriage, particularly for women since children tend to stay with the mother following a marital disruption. Preston (1975) has estimated that almost half of the current marriages will end in divorce; thus, women often move in and out of the wife role. Finally, and perhaps most importantly, motherhood roles more severely constrain other life options of a woman than do marital roles, especially during the early childbearing years. For these reasons, our emphasis is on age at first birth. Given the high correlation between age at first birth and age at first marriage, and given that both are affected by similar exogenous variables, we have not included both in the analysis. Furthermore, given the assumptions of the model, the exclusion of age at first marriage will not bias our estimates of the relative importance of the processes leading to educational attainment and to the first birth.

In order to allow women sufficient time to get married (and, thus, be eligible to be in the sample) and to have a first birth, the analysis of the education-age at first birth relationship will be limited to women aged 35–44. Most of those who will ever marry before the end of the reproductive period are married by age 35. For example, the proportion of women ever married increases from 0.873 at ages 25–29 to 0.926 at ages 30–34 to 0.941 at ages 35–39. But the proportion of women ever married increases only slightly to 0.946 at ages 40–44 (U.S. Bureau of the Census, 1972). The same holds true for first births. Most of those who will ever give birth do so by age 35. For example, 79.2% of the birth cohort of 1930–1934 had a live birth by ages 25–29, 87.7% did so by ages 30–34, and 90.2% had a live birth by ages 35–39. This percentage increased only slightly to 90.8% by ages 40–44. Less than 3% of the women in this birth cohort who had a live birth had it after age 35 (Heuser, 1976).

Childless women are excluded from this analysis at age at first birth. Only a small proportion (less than 10%) of the married women in these

cohorts remained childless (Heuser, 1976). To the extent that postponement leads to voluntary childlessness (Veevers, 1973), this exclusion could
lead to a weaker estimated effect of age at first birth than actually exists.
However, childlessness in these cohorts was primarily a product of fecundity
impairments.

The model shown in Figure 1 includes background characteristics, aspects of early adolescence, and the reciprocal relationship between education
and age at first birth. Period factors are not included, and this needs to be kept
in mind when interpreting our results. The respondents in this analysis were
aged 35–44 in 1970. Taking 15 as the youngest age at first birth and 35 as
the oldest means that these women were having their first births from 1941
to 1970. During this long period, there were a number of events affecting the
timing of fertility, including World War II, the Korean War, and the Vietnam
War. Those women who postponed their first birth were, of course, exposed
to more of these period factors, which could affect the timing of their first
birth. Since so little is known about the nature of period factors that affect the timing of fertility (Rindfuss et al., 1978), they cannot be explicitly
included in the analysis. Furthermore, the younger women in our sample
experienced the period factors at different ages than the older women in the
sample. To see if this would affect our results, we ran the model separately
for women aged 35–39 and 40–44. The results were virtually identical for
the two groups.

The work of Easterlin (1962; 1966 and 1973) and others suggests that
the financial status of the respondent's family of orientation while the respondent was an adolescent will affect the age at which she has her first
child. Unfortunately, we do not have a direct measure of the respondent's
parents' financial status while the respondent was an adolescent. However, a
number of background variables in the model, such as race, number of siblings, farm background, regional background, and family composition when
respondent was 14, indirectly control for the respondent's family's financial
situation.

Further, the model shown in Figure 1 also does not include the labor
force experiences of women. As noted earlier, labor force experiences and
aspirations are likely to affect, and be affected by, childbearing and childbearing preferences. In fact, there is a long literature on this relationship
(see Waite and Stolzenberg, 1976; and Smith-Lovin and Tickamyer, 1978,
for recent summaries of this literature). Unfortunately, adequate labor force
participation information is not available.

Estimation of the effects shown in Figure 1 was accomplished by using two-stage least squares regression analysis (Goldberger, 1964; Johnston,
1972). The estimates were made using ordinary least squares in two steps,

Table 2. Metric and Standardized Coefficients Measuring the Reciprocal Relationship between Education and Age at First Birth, 1970 NFS[a].

Independent Variable	Dependent Variable	Metric Coefficient	Standardized Coefficient
Education	Age at First Birth	0.741*	0.429*
Age at First Birth	Education	0.075	0.130

Correlation of Disturbances (U and V): −0.255

[a] N = 1,766.
* Significant at 0.05.

making the appropriate corrections as outlined by Hout (1977). The results are shown in Table 2.

This table shows only the results for the endogenous variables; results for the complete model are reported and discussed elsewhere (Rindfuss and St. John, 1979).

The effect of education on age at first birth is significant—both statistically and substantively. Each additional year of schooling results in the delay of the first birth by approximately three-quarters of a year. However, the effect of age at first birth on education is not statistically significant; and even if it were, the effect would be trivial substantively.

The results shown in Table 2 are based on the assumption of linear effects. It might be argued that the effect of age at first birth on education is not linear. The inclination to have a birth at a very young age may have more serious effects on educational plans than the preference to have a child at a later age. The potential conflict between school and motherhood is greatest at the younger ages at first birth. This suggests that a nonlinear age at first birth effect on education should be specified. Such a specification should force a difference of a year at the younger ages at first birth to be larger than a difference of a year at the older ages at first birth. We used three different transformations of age at first birth (AGEFST) to explore this possibility: (1) LN (AGEFST), (2) 1/AGEFST, and (3) $1/(AGEFST)^2$. The model shown in Figure 1 was reestimated for each of these three transformations. In each case the results are the same as the linear model: age at first birth does not have a significant effect on educational attainment.

Furthermore, there is some evidence to suggest that the family building process may be different for whites and blacks. For example, blacks have higher illegitimacy rates than whites (NCHS, 1977), and blacks appear to rely more heavily on relatives to take temporary, but primary, care of children born to young mothers than whites (Rindfuss, 1977). In order to check for a potential interaction with race, we reran the analysis separately for whites and blacks.[6] The important point for the present analysis is that, for both blacks and whites, education has a strong and significant effect on age at first birth, but age at first birth has an insignificant effect on education. Thus, our results are unaffected by any racial interaction.

In the relationship between education and age at first birth, the principal direction of causality is from education to age at first birth. Those who have recently examined the relationship between education and age at first marriage have found corroborating results (Marini, 1978; Alexander and Eckland, 1978), namely, that education has a much stronger effect on age at first marriage than age at first marriage has on education. Given the sheer amount of time the mother role requires in contrast to the wife role, the timing of the first birth has greater consequences for the roles women occupy. Yet, it is interesting to note that (ignoring the differences between the samples used here and those used by Marini [1978] and Alexander and Eckland [1978]), age at first marriage appears to have a somewhat greater effect on education than age at first birth. Even though age at first birth has a greater effect on the roles occupied by women, age at first marriage could have a stronger effect on educational attainment because first marriage schedules are younger and more compact than first birth schedules. Thus, more marriages take place during the years in which women are in school.

The finding that age at first birth has only a very small effect on educational attainment may seem paradoxical, given the social policy concern with the pregnant girls who have to drop out of school and face reduced social opportunities as a consequence. Such a fate is unquestionably experienced by some women, particularly those among the 3% to 6% of the American cohort that have had a first birth before age 17. But the fact is that the vast majority of women do not get pregnant while they are enrolled in school. Even among those who do become mothers at ages at which society expects one to be in school, the direction of causality might run from education to fertility. Zelnik and Kantner (1978) and Ross (1978) suggest that a significant minority of premarital pregnancies were intentional. To further explore this issue, we compared the age at leaving school[7] with age at first birth for women who become mothers at age 17 or younger. If leaving school and the first birth occur in the same year, it is ambiguous which process dominates. But for those who left school more than a year before their first birth, one can assume that the educational process is affecting the fertility process. Surprisingly, more than 40% of the women who had a first birth at age 17 or less dropped out of school at least a year prior to becoming a mother—which suggests that even at the very young ages at motherhood, the fertility process is being affected by the educational process.[8] Further, there is longitudinal evidence showing a negative relation between educational aspirations and age at first birth (Marshall and Cosby, 1977; Card and Wise, 1978, Table 3), which suggests that many of those who have a first birth while they are of school age do so after deciding not to continue in school—and, perhaps, do so to justify dropping out of school. Finally, Haggstrom and Morrison (1979) find that among teenagers who do not drop out of high school, the effects of adolescent parenthood on subsequent educational aspirations are extremely small when other appropriate factors are controlled. All of this does not mean that fertility never truncates education, but only that it does so rarely. In the vast majority of the cases, education and educational aspirations determine age at first birth.

It is important that scientific discourse clarify the difference between a social policy concern that requires amelioration and the characterization of the overall process in which that concern is embedded.

Education and the Lengths of Birth Intervals

As discussed in the first section of this paper, we would expect to find a variety of reasons why women with more education would want to avoid very short birth intervals and we would expect them to be more effective at implementing their preferences. In this section we examine the relationship between education and the probability of having a short interbirth interval. Unlike the previous section, here, we assume that the direction of causality runs from education to the length of birth intervals.[9]

The birth history information contained in the 1970 NFS allows us to compute the length of each birth interval. Given the well-known difficulties involved in the analysis of birth intervals (see Bumpass et al., 1977, for a fuller discussion), we initially constructed life tables for each birth interval. These preliminary life tables were constructed for intervals begun in the period 1959–1968. By restricting the analysis to intervals begun in this period, we avoid a young-age-at-initiation bias (see Rindfuss and Bumpass, 1979).

The preliminary life table analyses showed the expected positive relationship between education and length of intervals. However, this conclusion is based on a bivariate analysis, and there are numerous other factors affecting the length of birth intervals (e.g., Bumpass et al., 1978), and the effects of these factors should be controlled. Unfortunately, the sample size of the 1970 NFS (or the 1973 FGS) is far too small to permit the simultaneous control of all these factors by using conventional life table techniques. Consequently, we used regression analysis to examine the probability of giving birth within a relatively short time interval—specifically, the probability of giving birth within 18 months of the previous birth. Because the life table

Table 3. Differentials in the Proportion Experiencing Birth Intervals of
18 Months or Less, for All Second, Third and Fourth Birth Intervals Begun
1959–1968, by Education, Gross and Net[a] Percentages: 1970 NFS

Education at Marriage	Second Birth Interval			Third Birth Interval			Fourth Birth Interval		
	N	Gross	Net	N	Gross	Net	N	Gross	Net
Total	2612	25		2236	19		1551	17	
1–8	155	33	31	168	32	29	164	29	25
9–11	657	28	26	592	21	18	433	20	17
12	1218	24	23	1016	16	17	670	15	16
13–15	388	22	24	312	17	20	202	11	14
16+	194	20	29	149	16	21	82	13	18

[a] Adjusted through a dummy variable regression analysis for the effects of race, religion, region, age at first birth, marital status at first birth, contraceptive use before first birth, planning status of first birth and smoking before age 16.

results suggested that the differences in interbirth interval length are greater between adjoining categories at the lower educational categories than at the higher educational categories, we used a variant of multiple regression analysis, Multiple Classification Analysis (Andrews et al., 1973), to see if this pattern continued when other factors were controlled. The results are summarized in Table 3.

Controlling for other factors that affect the length of interbirth intervals eliminates much of the relationship between education and the probability of having a short birth interval. Compare the gross and net columns for the second, third and fourth birth intervals.[10] The difference which remains after controlling for other variables is primarily between those with a grade school education and all others. Given that those with only a grade school education are a small proportion of the population, and since the proportion with only a grade school education is declining, the principal result to emerge from Table 3 is that, when the effects of other factors are controlled, the respondent's education at first marriage has essentially no effect on the probability of having a short second, third or fourth birth interval.[11]

Education and Fertility Preferences

As discussed earlier in this paper, educational preferences and fertility preferences affect each other; and, since neither is fixed, their interrelationship develops over time. To examine adequately this complex set of interrelationships would require longitudinal data of the kind not currently available. However, in the absence of the appropriate longitudinal data, it is still possible to examine part of the process by looking at the effect of education at marriage on fertility preferences at time of interview. Framed this way, the causal direction is essentially unambiguous.

Education at marriage can affect fertility preferences in two ways. First, education at marriage can have a direct effect on fertility preferences. Insofar as increased education makes a larger variety of roles available to women, we could expect education to have a direct and negative effect on fertility preferences. In addition, specific topics covered while in school might have a direct negative effect on fertility preferences. Second, education at marriage can have an indirect effect on fertility preferences through its effect on age at first birth. As shown earlier, higher levels of educational attainment result in older ages at first birth. An older age at first birth, in turn, leads to longer intervals between births (Bumpass et al., 1978). Thus, education leads to older ages at any given parity; and older ages at any given parity have a negative effect on the probability of wanting another child (Rindfuss and Bumpass, 1978).

The measure of fertility preferences used here, FERTPREF, is the sum of the number of "wanted" children the woman had had by the time of the interview plus the additional number of children she intended to have. For each live birth, the woman was asked a series of questions to determine whether or not, before that child was conceived, she wanted to have a birth of that order at some time during her reproductive life (see Westoff and Ryder, 1977, for a more detailed description). Such a series of questions minimizes the possibility of post factum rationalization of unwanted births (Rindfuss, 1974). The additional number intended is obtained from a question asking the respondent how many additional children she intended. This fertility preference measure is coded in numbers of children and has a mean of 2.9, and a standard deviation of 1.5.[12]

Because one of our interests is in the mediating effect of age at first birth, the sample being analyzed is limited to mothers, that is, women who have had at least one live birth. As in the previous two sections, in order to allow women sufficient time to get married and have a first birth, younger women are excluded from the analysis. The analysis in this section, like the age at first birth analysis, will be restricted to respondents aged 35–44 at the time of the interview. Because the full set of questions used in constructing our fertility preference measure was not asked of postmarried women (i.e., those widowed, divorced or separated at the time of the interview), the analysis will be limited to currently married women. Finally, for ease of presentation, the set of exogenous variables to be used here, in addition to education at first marriage, is exactly the same as those shown in Figure 1 and described in Table 1. We have experimented with other sets of exogenous variables and with other definitions of the sample, and the results are similar in all cases.

Figure 2. A Model[a] of the Relationship between Education and Marriage and Fertility Preferences (Standardized Coefficients)[b]

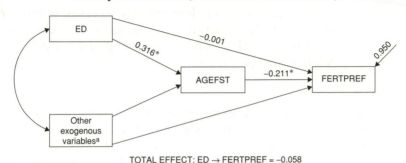

TOTAL EFFECT: ED → FERTPREF = –0.058

[a] The other exogenous variables in the model are: DADSOCC, RACE, NOSIB, FARMBACK, REGNBACK, ADOLFAM, RELIGION, YOUNGCIG, AND FECUND. See Table 1 for a description of the measurement of these variables.
[b] N = 1,551.
[c] Significant at 0.01.

The results are summarized in Figure 2. In order to focus on the education-fertility preference relationship, only the direct and indirect effects of education are shown. It can be seen that the direct effect of education on fertility preferences is trivial and insignificant. Virtually all of the effects of education at marriage on fertility preferences operates through age at first birth. Furthermore, the importance of age at first birth in influencing fertility preferences at time of interview should be underscored. Although it is not shown in Figure 2, age at first birth has a stronger direct effect on fertility preferences measured at time of interview than any of the listed exogenous variables. Thus, it appears that education affects fertility preferences by sorting women into various ages at first birth.

For approximately four-fifths of these women, education at first marriage is the same as education at interview; but one-fifth of these women have attended school, since their first marriage (Davis and Bumpass, 1976). For many women, this school attendance takes place a considerable time after the first marriage. For example, for women first married between 1951 and 1955 who returned to school after marriage, 62% last attended school 10 or more years after the first marriage. This additional schooling could affect fertility preferences, or could be affected by fertility preferences. We do not have the appropriate data to sort out these possibly reciprocal influences. But we did rerun the analysis in Figure 2 using education at interview instead of education at marriage, and the results are suggestive. The finding, as before, is that most of the relationships between education and fertility preferences operate through age at first birth. However, the direct relationship between education and fertility preferences is somewhat larger when education at interview is used than when education at first marriage is used. Without being able to sort out the potential reciprocal effects, we can only speculate that education after marriage operates to provide options that would not otherwise be available, or is itself a response to (or simultaneous with) a decision to terminate childbearing earlier than planned. This issue is something that warrants further examination.

Conclusion

To summarize, the reciprocal relationship between education and age at first birth is dominated by the effect from education to age at birth, with only a trivial effect in the other direction.

Once the process of childbearing has begun, education has essentially no direct effect on that process. Education has little direct effect on either the length of interbirth interval or on fertility preferences. Work by Vaughn and her colleagues (1977) shows that education has no direct effect on contraceptive efficacy. However, education has a significant indirect effect on these various components of fertility because it is the major determinant of age at the beginning of childbearing; in fact, education has a substantially greater influence on age at first birth than any other variable (Rindfuss and St. John, 1979). Thus, it is the postponing of motherhood that produces the oft-observed negative bivariate relationship between education and children ever born.

The powerful mediating effect of age at first birth is of interest in its own right. Older ages at first birth lead to longer interbirth intervals (Bumpass et al., 1978), more effective contraceptive use (Vaughn et al., 1977), and preferences for fewer children (as shown in the previous section of this paper).

These results, particularly if they are supported by future research on more recent cohorts, raise a set of interesting policy issues about which we can only speculate at present. Because the postponement of something is always more amenable to policy initiatives than its prevention, policies aimed at influencing age at first birth would be more likely to succeed than policies aimed at directly influencing children ever born. Furthermore, how adolescents spend their time has been accepted (although not universally) as something governments can legitimately influence—the military draft system is the most obvious example.

We began with the observation that a major way education might affect the roles women occupy is through altering the structure of childbearing experience, given the dominance of mother roles. We conclude that such educational effects as we can identify are explicable more in terms of education's effect on age at first motherhood than in terms of other values or aspirations that might derive from advanced schooling.

Appendix.

Zero Order Correlations Among Background Factors, Early Adolescent Characteristics, Education and age at First Birth: 1970 NFS, Women Aged 35–40 With 1 + Children

	DADS OCC	RACE	NOSIB	FARM BACK	REGN BACK	ADOL FAM	RELIGION	YOUNG CIG	FECUND	ED	AGEFST
DADSOCC	1.000										
RACE	-.144	1.000									
NOSIB	-.244	.156	1.000								
FARMBACK	-.323	.088	.274	1.000							
REGNBACK	-.129	.315	.150	.218	1.000						
ADOLFAM	-.056	.150	-.039	-.030	.071	1.000					
RELIGION	.053	-.152	.014	-.149	-.292	-.052	1.000				
YOUNGCIG	-.043	.030	.028	-.060	-.011	.067	-.010	1.000			
FECUND	.037	.035	.002	-.032	-.027	.018	-.002	.009	1.000		
ED	.370	-.222	-.328	-.185	-.211	-.157	-.012	-.171	.038	1.000	
AGEFST	.186	-.189	-.115	-.118	-.177	.111	.098	-.122	.216	.380	1.000

Note: Definitions, means and standard deviations for these variables are found in Table 1.

References

Alexander, K. L. and B. K. Eckland. 1974. "Differences in the educational attainment process." *American Sociological Review* 39:668–82.

——— 1978. "Family formation and educational attainment: alternative models." Manuscript.

Andrews, F. M., J. N. Morgan, J. A. Sonquist, and L. Klem. 1973. *Multiple Classification Analysis*. Ann Arbor: Institute for Social Research.

Blake, J. 1966. "Ideal family size among white Americans: a quarter of a century's evidence." *Demography* 3:154–73.

Blau, P. M. and O. D. Duncan. 1967. *The American Occupational Structure*. New York: Wiley.

Bongaarts, J. 1978. "A framework for analyzing the proximate determinants of fertility." *Population and Development Review* 4:105–32.

Bumpass, L. L., R. R. Rindfuss, and R. B. Janosik. 1978. "Age and marital status at first birth and the pace of subsequent fertility." *Demography* 15:75–86.

Bumpass, L. L. and C. F. Westoff. 1970. *The Later Years of Childbearing*. Princeton: Princeton University Press.

Card, J. J. and L. L. Wise. 1978. "Teenage mothers and teenage fathers: the impact of early childbearing on the parents' personal and professional lives." *Family Planning Perspectives* 10:199–205.

Cho, L. J., W. H. Grabill, and D. J. Bogue. 1970. *Differential Current Fertility in the United States*. Chicago: Community and Family Study Center.

Cutright, P. 1973. "Timing the first birth: Does it matter?" *Journal of Marriage and the Family* 35:585.

Davis, K. and J. Blake. 1956. "Social structure and fertility: an analytic framework." *Economic Development and Cultural Change* 4:211–35.

Davis, N. J. and L. L. Bumpass. 1976. "The continuation of education after marriage among women in the United States: 1970." *Demography* 13:161–74.

Duncan, O. D. 1975. *Introduction to Structural Equation Models*. New York: Academic Press.

Easterlin, R. A. 1962. "The American baby boom in historical perspective." National Bureau of Economic Research, Occasional Paper 79. New York.

——— 1966. "On the relation of economic factors to recent and projected fertility changes." *Demography* 3:131–53.

——— 1973. "Relative economic status and the American fertility swing." Pp. 170–223 in Eleanor Bernert Sheldon (ed.), *Family Economic Behavior*. Philadelphia: J. B. Lippincott.

Funderburk, S. J., D. Guthrie and D. Meldrum. 1976. "Suboptimal pregnancy outcome among women with prior abortions and premature births." *American Journal of Obstetrics and Gynecology* 126:55–60.

Goldberger, A. S. 1964. *Econometric Theory*. New York: Wiley.

Haggstrom, G. W. and P. A. Morrison. 1979. "Consequences of parenthood in late adolescence: findings from the national longitudinal study of high school seniors." Paper presented at the annual meeting of the Population Association of America, Philadelphia, April.

Hout, M. 1977. "A cautionary note on the use of two-stage least squares." *Sociological Methods and Research* 5:335–46.

Hout, M. and W. R. Morgan. 1975. "Race and sex variations in the causes of the expected attainments of high school seniors." *American Journal of Sociology* 81:364–94.

Heise, D. R. 1975. *Causal Analysis*. New York: Wiley.

Heuser, R. L. 1976. *Fertility Tables for Birth Cohorts by Color*. Rockville, Md.: National Center for Health Statistics.

Hoffman, L. W. 1978. "Effects of the first child on the woman's role." Pp. 340–67 in W. B. Miller and L. F. Newman (eds.), *The First Child and Family Formation*. Chapel Hill: Carolina Population Center.

Holsinger, D. B. and J. D. Kasarda. 1976. "Education and human fertility: sociological perspectives." Pp. 154–81 in R. G. Ridker (ed.), *Population and Development*. Baltimore: Johns Hopkins University Press.

Jaffe, F. S. 1977. "View from the United States." Pp. 19–29 in D. Bogue (ed.), *Adolescent Pregnancy*. Chicago: Community and Family Study Center.

Johnston, J. 1972. *Econometric Methods*. 2nd ed. New York: McGraw-Hill.

Kantner, J. F. and M. Zelnik. 1972. "Sexual experiences of young unmarried women in the United States." *Family Planning Perspectives* 4:9–17.

Kerckhoff, A. C. and R. T. Campbell. 1977. "Black-white differences in the educational attainment process." *Sociology of Education* 50:15–27.

Leonetti, D. C. 1978. "The biocultural pattern of Japanese-American fertility." *Social Biology* 25:38–51.

Marini, M. M. 1978. "Transition to adulthood." *American Sociological Review* 43:483–507.

Marshall, K. P. and A. G. Cosby. 1977. "Antecedents of early marital and fertility behavior." *Youth and Society* 9:191–212.

National Academy of Sciences. 1973. *Infant Death: An Analysis by Maternal Risk and Health Care*. Washington, D. C.: National Academy of Science.

National Center for Health Statistics. 1966. *Infant, Fetal and Maternal Mortality, U. S., 1963*. Series 20, No. 3. Washington, D. C.: U. S. Government Printing Office.

———— 1977. *Vital Statistics of the United States, 1973, Volume 1—Natality*. Washington, D. C.: U. S. Government Printing Office.

———— 1978. "National survey of family growth, cycle I: sample design, estimation procedures, and variance estimation." *Vital and Health Statistics Series* 2, No. 76.

Preston, S. 1975. "Estimating the proportion of American marriages that end in divorce." *Sociological Methods and Research* 3:435–60.

Rindfuss, R. R. 1974. *Measurement of Personal Fertility Preferences*. Ph.D. dissertation, Department of Sociology, Princeton University.

—— 1977. "Methodological difficulties encountered in using own-children data: illustrations fromthe United States." *East-West Population Institute Paper Series*. Honolulu: East-West Population Institute.

—— 1979. "Changes in the timing of fertility: implications for industrialized societies." Unpublished manuscript.

Rindfuss, R. R. and L. L. Bumpass. 1977. "Fertility during marital disruption." *Journal of Marriage and the Family* 39:517–28.

—— 1978. "Age and the sociology of fertility: How old is too old?" Karl Taeuber, Larry Bumpass and James Sweet (eds.), *Social Demography*. New York: Academic Press.

—— 1979. "The analysis of childspacing: illustrations from Korea and the Philippines." Presented at the annual meeting of the Population Association of America, Philadelphia, April.

Rindfuss, R. R., J. S. Reed and C. St. John. 1978. "A fertility reaction to a historical event: southern white birthrates and the 1954 desegregation ruling." *Science* 201:178–80.

Rindfuss, R. R. and C. St. John. 1979. "Social determinants of age at first birth." Manuscript.

Rindfuss, R. R. and J. A. Sweet. 1977. *Postwar Fertility Trends and Differentials in the United States*. New York: Academic Press.

Ross, S. 1978. *The Youth Values Project*. Washington: The Population Institute.

Ryder, N. B. and C. F. Westoff. 1969. "Relationships among intended, expected, desired and ideal family size: United States, 1965." *Center for Population Research Working Paper* (March). Washington, D.C.

Sewell, W. H. and R. M. Hauser. 1977. *Education, Occupation and Earnings*. New York: Academic Press.

Shapiro, S., H. S. Levine and M. Abramowicz. 1971. "Factors associated with early and late fetal loss." *Advances in Planned Parenthood* 6:45–58.

Smith-Lovin, L. and A. R. Tickamyer. 1978. "Nonrecursive models of labor force participation, fertility behavior and sex role attitudes." *American Sociological Review* 43:541–57.

Thornton, A. Forthcoming. "The difference of first generation fertility and economic status on second generation fertility." *Journal of Population: Behavioral, Social, and Environmental Issues*.

U.S. Bureau of the Census. 1972. *Census of the Population: 1970 Marital Status*. PC(2)–4C. Washington: U. S. Government Printing Office.

Vaughn, B., J. Trussell, J. Menken, and L. Jones. 1977. "Contraceptive failure among married women in the U.S., 1970–1973." *Family Planning Perspectives* 9:251–7.

Veevers, J. E. 1973. "Voluntary childless wives: an exploratory study." *Sociology and Social Research* 57:356–65.

Voss, P. R. 1977. "Social determinants of age at first marriage in the United States." Paper presented at the annual meeting of the Population Association of America, St. Louis, April.

Waite, L. J. and K. A. Moore. 1978. "The impact of an early first birth on young women's educational attainment." *Social Forces* 56:845–65.

Waite, L. J. and R. M. Stolzenberg. 1976. "Intended childbearing and labor force participation of young women: insights from nonrecursive models." *American Sociological Review* 41:235–52.

Westoff, C. F. and N. B. Ryder. 1977. *The Contraceptive Revolution*. Princeton: Princeton University Press.

Whelpton, P. K., A. A. Campbell, and J. E. Patterson. 1966. *Fertility and Family Planning in the United States*. Princeton: Princeton University Press.

Zelnik, Melvin and J. F. Kantner. 1978. "First pregnancies to women aged 15–19: 1976 and 1971." *Family Planning Perspectives* 10:11–20.

Notes

1. Here, and throughout the paper, we use the term "mother" in its social rather than biological sense. The biological mother is the female who gives birth to the child. The social mother need not be the biological mother; but, typically, the two are the same. It is the social mother that has primary responsibility for the care and nurture of the child. This role need not be occupied by a female, but, typically, it is. Also, the word "children" throughout this paper is used in its social, rather than biological, sense.

2. The work of Holsinger and Kasarda (1976) for developing countries is an exception.

3. In actual practice, we know of no case where all the intermediate variables are adequately measured. Models are evaluated as if there were direct effects, with researchers unable to specify the precise nature of the social and economic effects on fertility as they operate through the intermediate variables.

4. Note, however, that Voss (1977) finds a negative effect of age at first marriage on educational attainment. Marini (1978) argues, and we agree, that this finding of Voss is the result of the lack of an adequate instrument for age at first marriage.

5. There is some evidence that a history of miscarriage greatly increases the chance that subsequent conceptions will be terminated by a miscarriage (Funderburk et al., 1976; Shapiro et al., 1971). Given the unreliability with which fetal losses are reported in pregnancy histories (Bumpass and Westoff, 1970) and given the fact that very early miscarriages are often unnoticed by the woman, we experimented with alternative and more complex measures of fecundity which incorporated information from the woman's history subsequent to the first birth. However, the simple measure of whether or not the

woman had a miscarriage prior to the first birth proved to be the strongest predictor of age at first birth, and this is the measure that has been used in the final models.

6. Other nonwhites were not included.

7. Age at leaving school was computed by assuming a normal starting age, and assuming that education is obtained one year at a time,

8. To further explore this issue, and to explore whether a gating mechanism existed, we reran the two-stage least squares analysis for women who became mothers at a young age. Although caution is necessary in interpreting such an analysis because the variance of the endogenous variables has been reduced, age at first birth does not have a significant effect on education.

9. It should be noted, however, that it is possible that, for some women, short interbirth intervals prevent the return to school. Virtually nothing is known about returning to school after becoming a mother, although there has been some research on education after marriage. Approximately one in five women attend school after marriage; but the average addition to their educational attainment is relatively small: 1.0 years (Davis and Bumpass, 1976). Whether this schooling takes place before or after the start of childbearing is unknown. In order to minimize the possibility of education after the first birth being affected by the pace of fertility, we have primarily used education at marriage (rather than education at interview) for this analysis.

10. We follow the standard convention of indexing birth intervals by the order of the fertile pregnancy terminating the interval. Thus, the second birth interval is the interval terminated by the second fertile pregnancy.

11. The results in Table 3 are based on all birth intervals. Thus, both wanted or intended intervals and unwanted or unintended intervals are included. To make sure that the relationships shown in Table 3 were not the result of differences in fertility intentions, we calculated a set of life tables for "intended" intervals, excluding the following two types of intervals: (a) closed intervals that were closed by an unwanted birth, and (b) open intervals where the respondent indicates she does not intend to have another child. These results (not shown) are virtually identical to those shown in Table 3. Also, in order to see if the finding was sensitive to the particular measure of education used, we reran the analysis using respondent's education at interview, and then we reran it again using respondent's husband's education at respondent's first marriage. These alternative analyses lead to the same conclusions.

12. It should be noted that there is little variance in fertility preferences. Three-fourths of the sample gave a preference of 2, 3 or 4. This, of course, reduces the possibility of any variable significantly affecting fertility preferences.

Institutional Arrangements and the Creation of Social Capital: The Effects of Public School Choice

Mark Schneider, State University of New York at Stony Brook
Paul Teske, State University of New York at Stony Brook
Melissa Marschall, State University of New York at Stony Brook
Michael Mintrom, Michigan State University
Christine Roch, State University of New York at Stony Brook*

Mark Schneider, Paul Teske, Melissa Marschall, Michael Mintrom, and Christine Roch, 1997, Institutional Arrangements and the Creation of Social Capital: The Effects of Public School Choice, *American Political Science Review* 91 (1): 82–93. © American Political Science Association, published by Cambridge University Press, reproduced with permission.

* Mark Schneider is Professor of Political Science; Paul Teske is Associate Professor of Political Science; Melissa Marschall is a Ph.D. candidate, Department of Political Science; and Christine Roch is a graduate student, Department of Political Science, State University of New York at Stony Brook, Stony Brook, NY 11794-4392. Michael Mintrom is Assistant Professor, Department of Political Science, Michigan State University, East Lansing, MI 48824-1032. The research reported in this paper was supported by a grant from the National Science Foundation, Number SBR9408970. Paul Teske thanks the National Academy of Education's Spencer Foundation Post-Doctoral Fellowship for support on this project.

Abstract

While the possible decline in the level of social capital in the United States has received considerable attention by scholars such as Putnam and Fukuyama, less attention has been paid to the local activities of citizens that help define a nation's stock of social capital. Scholars have paid even less attention to how institutional arrangements affect levels of social capital. We argue that giving parents greater choice over the public schools their children attend creates incentives for parents as "citizen/consumers" to engage in activities that build social capital. Our empirical analysis employs a quasi-experimental approach comparing parental behavior in two pairs of demographically similar school districts that vary on the degree of parental choice over the schools their children attend. Our data show that, controlling for many other factors, parents who choose when given the opportunity are higher on all the indicators of social capital analyzed. Fukuyama has argued that it is easier for governments to decrease social capital than to increase it. We argue, however, that the design of government institutions can create incentives for individuals to engage in activities that increase social capital.

The delivery of services by local governments involves a complex relationship between the institutions that supply them and the citizens who use them. To improve the delivery of public services, many reformers argue that governments should imitate private markets by increasing the number of suppliers and by "empowering" citizens to shop across this expanded choice set. In this model, "citizen/consumers" become better consumers of public services by becoming more informed about their options and by more carefully selecting services that meet their preferences.

We suggest that the benefits of such market-like reforms can extend beyond the consumer behavior that has been the focus of previous analysis. Specifically, we argue that by expanding the options people have over public services, citizen/consumers can also become better *citizens*, and by so doing, increase the nation's stock of social capital. We test this hypothesis in the context of public school choice—a set of reforms that increases the control parents have over the selection of schools their children attend. These reforms are of long standing in some communities and are emerging in many others. In this research, we show that the design of public institutions charged with delivering education can affect the formation of social capital.

Social Capital and Local Citizenship

An intense scholarly debate recently has emerged concerning the role of social capital in economic and political development (e.g., Brehm and Rahn forthcoming; Fukuyama 1995; Granato, Inglehartand Leblang 1996a, 1996b; Inglehart 1990; Jackman and Miller 1996a, 1996b; Lipset 1995; Putnam 1993, 1995a, 1995b; Swank 1996; Tarrow 1996).[1] One theme in this debate is that social capital may be important to strong democracies for the same reasons that it is important for the functioning of strong economies: High levels of social capital engender norms of cooperation and trust, reduce transaction costs, and mitigate the intensity of conflicts.

While political scientists have only recently adopted the concept of social capital, the term has been used by sociologists for some time (see, e.g., Bourdieu 1980, Loury 1977). Coleman (1988, 1990) brought the term into wider circulation and argued (1988, S101) that social capital is generated as a byproduct of individuals engaging in forms of behavior that require sociability. In his study of 20 subnational governments in Italy, Putnam (1993) argued that the quality of governance is determined by the level of social capital within a region. Fukuyama concurs (1995, 356):

"The ability to cooperate socially is dependent on prior habits, traditions, and norms, which themselves serve to structure the market. Hence it is more likely that a successful market economy, rather than being the cause of stable democracy, is codetermined by the prior factor of social capital. If the latter is abundant, then both markets and democratic politics will thrive, and the market can in fact play a role as a school of sociability that reinforces democratic institutions."

While comparisons across nations and the identification of trends over time are obviously important, less scholarly work has focused on how government policies affect the stock of social capital. This is especially true for the analysis of the formation of social capital at the local level, where a small but growing body of work has developed addressing the link between government policies and social capital. Stone and his colleagues have been examining the role of "civic capacity," a concept similar to social capital, in local economic development and the politics of education (see, e.g., Stone 1996). Berry, Portney, and Thomson (1993) examined the importance of local community activity in the formation of social capital. And, in the context of education, Astone and McLanahan (1991), Coleman and Schneider (1993), and Lee (1993) have examined social capital as a function of the interactions among administrators, teachers, parents, and children.

We follow the approach of Berry, Portney, and Thomson, who emphasize the importance of communities where neighbors talk to each other about politics. In these face-to-face meetings, these authors argue that "democracy moves politics away from its adversarial norm, where interest groups square off in conflict and lobbyists speak for their constituents. Instead, the bonds of friendship and community are forged as neighbors look for common solutions to their problems" (1993, 3). (Also see Mansbridge 1980 on "unitary democracy" and Barber 1984 on "strong democracy.") Berry, Portney, and Thomson's emphasis on "face-to-face" interactions parallels Fukuyama's (1995) focus on "spontaneous sociability" and Putnam's (1993) emphasis on the role of networks and membership in voluntary and social organizations as supports for representative democracy (see also the review by Diamond 1992).

In this article, we go beyond documenting levels of social capital by identifying the effects of institutional arrangements governing the delivery of education, the most important public good local governments provide, on the formation of social capital. Whereas scholars have recognized the importance of schools in creating social capital for the next generation (see, e.g., Henig 1994, 201–3), for us, schools are also arenas in which social capital can be generated among today's parents.

We explore the relationship between schools and social capital by considering how school choice can influence parental behavior. Specifically, we examine how school choice may increase levels of voluntary parental involvement in the schools, face-to-face discussions between parents, and levels of parental trust in teachers—behaviors that have all been identified as components of social capital. We test these relationships empirically using a quasi-experimental design that allows us to isolate the link between school choice and citizen behavior. Fukuyama has argued that "social capital is like a ratchet that is more easily turned in one direction than another; it can be dissipated by the actions of governments much more readily than those governments can build it up again" (1995, 62). We show that institutional arrangements that increase parental control over the schools their children attend may be able to reverse that ratchet.

Some scholars are skeptical that government policies expanding choice can increase social capital For example, Anderson argues that expanded citizen choice, at best, will cultivate only a "passive understanding" of the demands of democratic participation and that this "consumer's skill" is not a sufficient basis for "competent citizenship" (1990, 197–8). Carnoy (1993, 187) and Henig (1994, 222) both argue that school choice will increase the social stratification between parents who are more involved and interested in their children's education and those who are not, fundamentally reducing the ability of communities to address collective problems. And Handler (1996, 185) notes that while choice plans require parents to choose, they cannot force parents to become actively engaged in school activities.

In contrast, other scholars argue that choice and related reforms will foster social capital. As Ravitch (1994, 9) notes: "The act of choosing seems to make parents feel more responsible and become more involved." And Berry, Portney, and Thomson (1993,294) cite the shift to parental control over local schools in Chicago in the late 1980s as a rare example of a successful attempt to get low-income parents more involved in local public affairs (also see Handler 1996).

In the analysis that follows, we show that reforms introducing choice can affect the level of social capital within communities. While our findings are limited to one particular aspect of local communities—schools—they provide important evidence that government or community-initiated policies can indeed ratchet up the preexisting levels of social capital and enhance the social fabric necessary for building and maintaining effective democracy. And, we demonstrate that this can be done both in suburban communities, where most Americans now live, and in inner-city neighborhoods, where the stock of social capital may be most depleted and where its absence may have the most deleterious effects (e.g., Berry, Portney, and Thomson 1993; Wilson 1987).

School Choice

School choice is perhaps the most widely discussed approach to addressing persistent problems in primary and secondary education in the United States. School choice advocates, liberals and conservatives alike, contend that changing the institutions governing school organization will improve student performance by changing the incentives faced by educators and by changing the behavior of students and parents (see Handler 1996, 9).[2]

It is possible to define school choice in such a way that it is already the norm. Many families already use residential location to choose the public schools their children attend. Even after the residential decision is made, many private alternatives to public education are available and about 10% of parents nationwide choose that option. School choice, however, is typically construed to involve policies that reduce the constraints that traditional public schooling arrangements place on schools and students. (For a discussion of distinctions among choice approaches, see Witte and Rigdon 1993.) Most important, school choice policies are designed to break the one-to-one relationship between residential location and the schools students attend.[3]

Responding to intense policy debates and the growing recognition of the problems of American schools, over the past two decades a growing number of local school districts have changed the institutional frameworks governing the provision of local education giving parents expanded choice over the schools their children attend. We take advantage of this diffusion of the innovation in school choice policy, employing a quasi-experimental approach comparing parental behavior in two pairs of school districts that are demographically similar but vary on institutional arrangements. We analyze the effects of choice on the formation of social capital in a matched pair of inner-city school districts, one with a long history of extensive choice and one without much choice. We then replicate this analysis in two suburban school districts. In each matched pair, the populations are similar demographically, but the institutional arrangements allowing parental choice over the schools their children attend differ.

Our analysis is based on interviews of approximately 300 parents of children in public school grades K–8 across four districts. (Appendix A describes the sample design.) Two of these are inner-city districts in New York City: District 1, which has only recently introduced limited choice, and District 4, which has offered programs of choice for 20 years. The other two are suburban communities in New Jersey: Morristown, which strictly maintains assignment to neighborhood schools, and Montclair, which has had a program of choice since the 1970s.

We begin with a discussion of the two New York school districts, describing in detail the evolution of choice in District 4. We then present an empirical analysis of effects of choice on social capital in the New York setting. Finally, we replicate the analysis using our New Jersey sample.

District 4: A School Choice Innovator

District 4 is located in East or "Spanish" Harlem, one of the poorest communities in New York City. The district serves roughly 12,000 students from pre-kindergarten through the ninth grade. In the early 1970s, the district's performance was ranked the lowest of 32 city public school districts in math and reading scores. Choice was part of a response to this poor performance.

Fliegel (1990) described the evolution of school choice in District 4 as resulting from "creative noncompliance" with New York City rules and regulations. The factors shaping the emergence of the District 4 can be traced back to the late 1960s when the administration of New York City's public school system was decentralized to allow for greater community control. Thirty-two separate community school districts were established, each of which was governed by an elected community school board and by the central Board of Education. High schools remained under the authority of the Board of Education. Decentralization was supposed to promote greater parental participation, but it has also led to problems with corruption, over-politicization, and poor performance (Cookson 1994, 50–1).

District 4 took full advantage of decentralization, in large part due to the entrepreneurial efforts of Anthony Alvarado, district superintendent from 1972 until 1982. As Boyer (1992, 41–2) notes, Alvarado bent rules, attracted outside grants, and won support from powerful teacher and principal unions. When Alvarado took over as superintendent, District 4 ran 22 schools in 22 buildings. In 1974, the first alternative school, Central Park East Elementary, was developed, followed by an alternative program for seventh and eighth graders with serious emotional and behavioral problems and by the East Harlem Performing Arts School, a program for fourth through ninth graders. These schools were open to parental choice and, as minischools, they were located within existing buildings where space was available. These schools were given greater flexibility over staffing, use of resources, organization of time, and forms of assessment.

The differences between the administration of these alternative schools and the traditional schools led to complaints of favoritism from some teachers and principals in the traditional schools. In response, new opportunities were offered to develop alternate schools using funding from the Magnet Schools Assistance Act (Wells 1993, 56). The district also exceeded its annual budget for many years as these alternative schools were being developed (Henig 1994, 164).

The focus on educational goals was shaped by Seymour Fliegel, appointed District 4's first director of alternative schools in 1976, who developed small schools designed to provide students, parents, and professional staff with flexibility and a sense of school "ownership" (Fliegel 1990, 209). Fliegel also used choice to encourage this sense of ownership. During the late 1970s and the 1980s more than 20 alternative schools were developed, many with distinctive curricular themes. As the number of schools increased, the differences between schools became more apparent. With many new schools and the potential for parents and students to make meaningful choices, Smith and Meier (1995, 94) suggest that it "became hopeless" to tell parents or teachers that their assignments would be determined bureaucratically. Thus, in 1982, the district decided to provide all parents with choice. Sixteen neighborhood elementary schools remained intact, with space reserved first for those living in the designated zones. While the emphasis was placed on providing choice at the junior high school level, the district also created a considerable number of alternative elementary schools, many of them bilingual (Smith and Meier 1995, 94).

In District 4, all students must make an explicit choice about the junior high school they will attend. Each sixth-grader receives a copy of a booklet describing the alternative junior high schools. Parents and students attend orientation sessions led by the directors of various alternative schools and are encouraged to visit the schools (Wells 1993, 55). Students and their parents rank and discuss their six choices of junior high schools. Sixty percent of the students in the district are accepted into their first-choice school, 30% into their second-choice school, and 5% into their third-choice school. The remaining 5% are placed in schools thought to be most appropriate for them (Boyer 1992, 52–3). To ensure that all students have viable choices, District 4 administrators monitor the popularity of the various alternative schools, closing or restructuring less popular schools (Wells 1993, 55).

District 1: Limited Choice

Our other New York City research site is District 1 on Manhattan's Lower East Side. Largely Hispanic and poor, the residents of District 1 share many characteristics with those of District 4. District 1 was created out of the Two Bridges School District, one of most active districts in New York City's fights over school decentralization in the 1960s. Despite this high initial level of community activism, the schools have foundered over the years. Following the success of District 4, District 1 began experimenting with school choice, and in 1992 created a small number of alternative schools.[4]

As a result of entrepreneurial efforts to develop choice, District 4 has developed a reputation in the city and in the nation as an innovative, successful district. A sense of mission is evident among parents, teachers, and administrators. While there is some dispute about how much of the success can be attributed to choice per se (see Henig 1994, 124–44), there is no question that performance in District 4 improved from its original low level as choice was implemented. In contrast, despite the high level of community activities during the push for decentralization, District 1 has faced considerable administrative turnover and turmoil for the last few years.

We report some comparative data on the districts in Table 1. Both districts are geographically compact, have large numbers of students from very poor families (more than eight of ten students are eligible for free lunches), and have a majority Hispanic student population.

The Survey Respondents

We contracted Polimetrics Laboratory for Political and Social Research, a survey research facility at Ohio State University, to interview 400 residents in each district in spring 1995, sampling parents (or the person in a household who "makes the decisions about the education of children"). To focus on the

Table 1. District 4 and District 1 Population and Sample Demographics

	District 4		District 1	
	Population	Sample	Population	Sample
Number of students	13,806	333	12,519	295
Number of schools	50	46	24	24
Hispanics	63%	68%	63%	71%
Blacks	33%	26%	12%	11%
Whites	2%	2%	10%	10%
Asian	1%	1%	13%	2%
Percentage in poverty	54%	NA	49%	NA
Income < $20,000 per year	NA	67%	NA	66%
Employed	35%	38%	48%	43%
High school degree or more	48%	65%	63%	65%
Single parent	NA	61%	NA	46%
Female	50%	90%	55%	87%

Source: For district information: *School District Data Book Profiles*. 1989–90.
NA: Since both districts are administrative units for the New York City school system rather than, e.g., census designated units, some demographic data are not available.

schools controlled by the districts, the sample frame was limited to parents with children in grades K–8.[5] To randomize, respondents were asked to answer school-specific questions based on the experience of their child in grades K–8 whose birthday came next in the calendar year.

As Table 1 illustrates, the sample of public school parents in each district is fairly representative of the student population on many key demographic variables. (We chose to interview parents of children who live in the districts but attend private schools as these parents are exercising a form of choice. However, they are not included in the analyses presented below. In District 1, 26% of the respondents sent their child to private school, compared to 17% in District 4.)[6]

Overwhelmingly, we sampled females, both because there are many single mothers in these districts and because we asked to speak with the person in the family who makes the decisions about school. More than 60% of the households were headed by a single parent in District 4, compared to 46% in District 1, and in both districts, more than 85% of the respondents were female.

Constructing the Models

With this background in place we now turn to our major goal: to assess the degree to which giving parents more control over the schools their children attend increases their level of social capital. In our analysis we use four measures of social capital, three of which are directly derived from Putnam (1993) and Fukuyama (1995) and the fourth a logical extension.

The first measure is whether the parent is a member of the PTA. Putnam uses declining participation in PTAs as one of his indicators of the erosion of social capital.[7] Second, we analyze a slightly broader measure of parental involvement in the schools, asking parents if in the past year they had engaged in any volunteer activities for their child's school. The third measure we investigate is the number of other parents our respondent talked with about school matters. We use this measure to reflect the "spontaneous sociability" Fukuyama emphasizes as underlying social capital and the importance of "face-to-face democracy" emphasized by Berry, Portney, and Thomson (1993). Our final measure reflects the level of trust parents have in their child's teacher to do the "right thing" for their child.[8] For Fukuyama the general level of trust in society is the critical dimension of social capital, since it lubricates economic, political and social transactions. In this research, we concentrate on a single domain-specific dimension of trust (trust in teachers). These activities not only are central to building social capital, they are also critical to building good schools (see, e.g., Anson et al. 1991).

In our selection of independent variables, we measure elements of motivation, resources, time constraints, and school policies that Kerbow and Bernhardt (1993, 116) argue are critical features of parental involvement in the schools. Thus we employ variables related to individual demographic characteristics as well as those related to the schools children are attending.

Three different types of institutional arrangements exist in the two central city districts in our study. The oldest and most traditional form of school organization is the neighborhood model, in which children are assigned to schools based on residential location. The second is universal choice, which characterizes the intermediate school system (grades 6–8) in District 4. Under this type of arrangement all parents must choose a school for their children (i.e., there is no "default" school). Finally, an "option demand" system of choice (see Elmore 1991), which exists in both districts but is much more developed in District 4, allows parents to select a school other than their neighborhood school. We refer to those parents who have decided to exercise choice as "active choosers." About 20% of our sample fall into the universal choice category (all in District 4), while about 9% of all of the sampled parents in New York are active choosers.

Active choosers present us with the same fundamental problem faced by any research on the behavior of parents in school choice settings—parents choosing alternative schools may not be a random selection of all parents in a school district. And, if parents who self-select alternative schools are also high on social capital then our results will be biased. While other studies have acknowledged this problem and made various efforts to control for selection bias (Chubb and Moe 1990; Coleman and Hoffer 1987; Coleman, Hoffer and Kilgore, 1982; Smith and Meier 1995), we correct for it by constructing a two-stage nonrandom assignment model, in which the first equation models the assignment process and the second equation the "outcome." The method, described in Appendix B and based on the work of Heckman (1978), Heckman, Hotz, and Dabos (1987) and Lord (1967, 1969), corrects for both the nonrandom selection process and other econometric problems associated with the use of dichotomous dependent variables (see Achen (1986) and Alvarez and Brehm (1994) for discussions of the applicability of this method in political science).[9]

By limiting the possibility that parents likely to make active choices are also likely to engage in other activities that we refer to as part of social capital, the use of this methodology is critical to our argument that making an active choice influences parental behavior.

As noted in detail in Appendix B, we begin with an explicit assignment equation:

$$Active\ choosers\ =\ a\ +\ B[\text{Demographics}]\ +\ B[\text{Values}]$$
$$+\ B[\text{Diversity}]\ +\ error, \tag{1}$$

where *Active choosers* is a dichotomous variable indicating whether a parent has elected an alternative school or program for their child ($1 = $ yes, $0 = $ no); *Demographics* is a vector consisting of a set of dummy variables for self-identified racial group membership (black, Hispanic, Asian—white is the excluded category), a continuous variable measuring years of schooling of the parent, a continuous variable reflecting the length of residence in the school district, and a 7-point scale measuring frequency of church attendance ($1 = $ never, $7 = $ once per week). We also include two dichotomous variables reflecting the gender of the respondent ($1 = $ female) and whether or not the respondent is employed ($1 = $ yes). The racial, gender, and employment variables reflect the resources and demographic factors that may influence activities related to social capital. Parental education level may be particularly important—Putnam (1995b, 667) reports that it "is by far the strongest correlate ... of civic engagement in all its forms." The length of residence variable reflects the argument advanced by Brehm and Rahn (forthcoming) and by Putnam, who both argue that mobility decreases social capital. In addition, Teske et al. (1993) found that length of residence affected knowledge of school policies. Church attendance is a control variable representing an alternative form of interaction and involvement with the local community.

The *Values* and *Diversity* variables indicate whether a parent thought either particular values or diversity as school attributes were important in their choice of schools. In our survey parents were asked to name up to four attributes they thought were most important in a school. Two attributes in particular, the values espoused by the school and the diversity of the student body, were considered important by parents of children in alternative public schools but not by parents of children in neighborhood public schools.[10] We therefore include these variables in the assignment equation for theoretical reasons, as they are important predictors of active school choosers. We have no theoretical reason, however to expect these variables to affect social capital and, indeed, they are not empirically related to the activities we have measured. These are used as exclusions in our outcome equation and provide the necessary leverage for estimating the system of equations.[11]

Thus, as described in greater detail in Appendix B, we estimate this assignment equation and the predicted value of the active chooser variable is used in estimating the following outcome equation:

$$\text{Social capital} = a + B[\text{"Predicted" active choosers}]$$
$$+ B[\text{School factors}] + B[\text{Demographics}] + error, \quad (2)$$

where *Demographics* are as noted in equation (1) and *Values* and *Diversity* are excluded. *"Predicted" active choosers* is the estimated values from equation (1), transformed into a linear functional form following Goldberger (1964; also see Achen 1986, Heckman 1978). *School factors* measure other aspects of the school environment. These factors include a variable measuring the enrollment in the school the child attends, as smaller schools are often considered to be better arenas for building social capital (Harrington and Cookson 1992); a dummy variable (= 1) when the respondent had made a universal choice at the junior high level in District 4; and a measure of parental dissatisfaction with her child's school.[12] Previous research (e.g., Witte 1991) has demonstrated that parental dissatisfaction is negatively correlated with levels of parental involvement and participation in school activities.

When the dependent variable in the outcome equation is continuous, as in our analysis of the number of parents with whom a respondent has talked about schools, the two-stage estimation technique is fairly straightforward. When the dependent variable is a dichotomous variable, however, another round of corrections is necessary because the disturbances are heteroskedastic (see Appendix B; also see Achen 1986, 40–7). In our analysis of the other three measures of social capital we report these generalized two-stage least squares (G2SLS) results. Note that since the results are generalized linear probability estimates, the coefficients have a straightforward interpretation: They represent the change in the probability of finding an event given a unit change in the independent variable.

The Effects of Choice in the Central City

With these corrections in place, we are now able to estimate the effects of school choice on the behavior of parents controlling for the nonrandom "assignment" across alternative schools.[13] We present the results in Table 2. Turning first to PTA membership, reported in the first column, we find strong evidence that school choice affects this widely used measure of social capital: Ceteris paribus, participation in the PTA among active choosers is 13% higher than among nonchoosers ($p < .05$), the largest effect in our model, apart from gender.

The effects of some other variables are worth noting. First, note that as the length of residence increases, so does participation in the PTA ($p < .05$, using a one-tail test). Similarly, frequency of church attendance increases participation in the PTA. These findings confirm empirically the arguments presented by Putnam and Fukuyama, as well as findings by education researchers (Kerbow and Bernhardt 1993, Muller and Kerbow 1993). Note too that participation in the PTA increases with the level of parental education—individual human capital and social capital flow together.

In the second column of Table 2, we turn to more general patterns of participation in voluntary events. Here we find that active choosers are over 12% more likely to engage in such activities than are nonchoosers. Paralleling the results reported for PTA membership, church attendance and longer residence are associated with volunteering, as is more years of parental education.

We have shown that active participation in school choice increases levels of involvement with voluntary organizations. We turn next to a measure of "spontaneous sociability"—how many other parents do our respondents engage in discussions about schools? The same cluster of variables emerges as important: Ceteris paribus, active choosers talked with four more parents than nonchoosers (see the third column of Table 2). Again, longer term residents, more educated respondents, and frequent churchgoers talk with more parents than do other respondents.

Finally, we examine trust in teachers. As shown in the final column of Table 2, school factors dominate this model. Active choosers are almost 10% more likely to trust teachers all or most of the time and universal choosers are 9% more likely to do so. In contrast, parents who are dissatisfied with their

Table 2. The Effects of Choice on the Formation of Social Capital in Two
New York Districts

	PTA Member	Voluntary Activities	Parents Talked to	Trust Teacher
Active chooser	.128*	.123*	4.053*	.095*
	(.064)	(.064)	(2.295)	(.049)
Universal choice	−.035	.025	−.613	.096*
	(.066)	(.062)	(.651)	(.056)
Dissatisfaction	−.042	−.003	.234	−.239***
	(.041)	(.040)	(.404)	(.039)
School size	−.000	−.000	−.000	.000
	(.000)	(.001)	(.001)	(.000)
Black	.092	.048	−.401	−.057
	(.072)	(.068)	(1.30)	(.044)
Hispanic	−.068	−.021	.419	−.066
	(.066)	(.062)	(1.22)	(.036)
Asian	.041	.149	1.61	.059
	(.187)	(.157)	(2.47)	(.097)
Length of residence	.005*	.005*	.085**	−.002
	(.003)	(.003)	(.030)	(.002)
Education	.015**	.020**	.148*	−.009*
	(.005)	(.006)	(.063)	(.004)
Employed	−.046	.031	.038	.033
	(.044)	(.042)	(.427)	(.029)
Female	.277***	.110	.370	−.052
	(.056)	(.067)	(.708)	(.036)
Attend church	.041***	.023***	.242*	.010
	(.009)	(.009)	(.108)	(.006)
Constant	.336**	.327**	.739	1.05
	(.129)	(.135)	(2.34)	(.090)
	$N = 580$,	$N = 580$,	$N = 568$,	$N = 578$,
	$F = 66$	$F = 107$	$F = 4.4$	$F = 4.3$

Note: Numbers in parentheses are adjusted standard errors. We do not report
R-squared statistics because in the adjustment process necessary to correct for
the nonrandom assignment problem, this statistic becomes inappropriate (see
Aldrich and Nelson 1994, 14–5).
* $p \le .05$; ** $p \le .01$; *** $p \le .001$

child's school and have considered moving the child to a different school are 24% less likely to trust their child's teachers. Of the demographic factors, only education is related to trust—but this relationship is negative.

Note also that while choosing significantly increases social capital on all four dimensions we measure, school size is not related to any of these measures. Harrington and Cookson (1992) have argued that the introduction of smaller schools in District 4 was the most important innovation accounting for the improvements found in the district. Our results differ—it is choice and not school size that matters.

Taking Advantage of the Quasi-Experimental Design: Replicating the New York Findings

Replication is one of the most powerful tools available for validating social scientific findings. In the next stage of our analysis, we take advantage of our quasi-experimental design to replicate the results of our New York study in another pair of school districts. This replication allows us to explore the robustness of our findings by testing their sensitivity to changes in the context of choice. In our next comparison, we explore the effects of community composition on our findings. In our first analysis, we demonstrated that school choice fosters behavior that builds social capital among parents in low-income central city school districts. Given the multitude of problems facing central cities, this is obviously an important finding. The next question is obvious: Does this relationship hold among suburban parents who now make up a larger share of the American population than do those in the central city?

Second, and more important for us, the institutional factors that define the extent of school choice varies across our two sets of communities. In our next "experiment," we compare patterns of activities in a traditional neighborhood school district (where no one can choose a school except by changing their residential location or by opting out of the public sector altogether) with those in a universal choice district (where there are no neighborhood schools). These institutional arrangements represent more extreme points on the policy continuum than do those in District 1 and District 4. Are the results we found in New York replicated under these different community and institutional conditions? Are the magnitude of the effects similar?

School Choice and Social Capital in Suburban Communities

To answer these questions we turn to our second paired set of communities, Montclair and Morristown, New Jersey, two suburban communities within commuting distance of New York City. Given the institutional arrangements governing the schools in these two districts, we can test the effects of universal choice directly, since everyone in Montclair's public schools chooses and no one in Morristown's can.

Montclair and Morristown, New Jersey

In both communities, court-ordered desegregation decisions in the 1970s led to fundamental changes in the school assignment mechanisms; however, very different responses were developed to achieve racial balance. Montclair adopted school choice, with parents given the right to choose schools from kindergarten through the eighth grade (there is only one high school), with choice constrained by racial balancing. In Morristown, residential zones were created for neighborhood schools. These zones are frequently adjusted so that each school in each zone has the same racial balance, but once set the zones are strictly enforced.

School choice has been operating in Montclair for about as long as in District 4. In 1969, the New Jersey Commissioner of Education ordered Montclair to desegregate or lose state funding. A forced busing plan was implemented in 1972, which caused conflict and considerable white flight. A limited choice program was implemented in 1975 to try to encourage voluntary racial balancing by establishing magnet schools. Several changes were made to the choice plan in Montclair, and in 1984 choice was introduced to the whole district by the symbolic act of turning all schools into magnets.

While choice was initially a solution to racial balancing, parents, teachers, and administrators used it to promote competition and better schools (Boyer 1992, 33). Parents in Montclair are provided with considerable information about the schools. In choosing schools, parents request two options and students are placed in their first choice if it matches the racial balancing goals. The schools are nearly uniformly good and about 95% of parents receive their first choice (Strobert 1991, 56–7). Between 60 and 80% of students are bused to their schools, but now such busing is voluntary.

Table 3 shows the demographics of the public school parents in these two New Jersey districts, overall and for our surveyed sample of 400 parents in each community.

Under the universal system of choice in Montclair, all parents are required to choose a school for their child. Therefore, it is not necessary to specify the selection process as we did for the analyses of our New York City parents—that is, no assignment equation is needed and the extensive corrections noted in Appendix B are not necessary. Thus, the results reported in Table 4 are the results of straightforward multivariate analyses. For comparability with the linear probabilities reported in our analysis of New York, we report the percentage point change for a unit change in the independent

Table 3. Montclair and Morristown Population and Sample Demographics

	Montclair		Morristown	
	Population	Sample	Population	Sample
Number of students	5850	356	5080	286
Number of schools	10	10	9	9
Hispanics	4%	3%	9%	7%
Blacks	36%	34%	17%	16%
Whites	56%	57%	70%	70%
Asian	3%	1%	4%	5%
Percentage in poverty	7%	NA	6%	NA
Income < $20,000 per year	16%	8%	21%	14%
Employed	59%	80%	58%	71%
High school degree or more	88%	98%	86%	94%
Single parent	11%	23%	23%	22%
Female	54%	78%	53%	76%

Source: For district information, *School District Data Book Profiles, 1989–90.*

variable (for the dummy variable, this is the effect of having the characteristic [1] versus not having it [0]). Since all Montclair parents *must* choose their children's school and no one in Morristown public schools can choose (except by moving), the coefficient of the dummy variable for Montclair represents the effects of universal choice, ceteris paribus.

The results in Table 4 show patterns consistent with those in our New York analysis. Choosers are significantly more likely to engage all measures of social capital—PTA membership, volunteering for a school activity, talking to people about schools, and trusting teachers—controlling for other important factors.[14]

School Choice Can Help Build Social Capital

At the heart of calls for the introduction of market-like reforms into the public sector lies the belief that giving people choices over public goods will increase efficiency. Research into the effects of reforming the "supply side" of the provision of public goods has established that such competitive mechanisms can in fact pressure the producers of public goods to be more efficient and more responsive (for local public goods, see, e.g., Ostrom 1972, Schneider 1989, Schneider and Teske 1995, Tiebout 1956). Recently, scholars have begun to study the effects of reforms on the demand-side of the market, leading to debates about the level of information held by citizens and the levels necessary for markets for public goods to work (e.g., Lowery,

Table 4. The Effects of Choice on the Formation of Social Capital in Two New Jersey Districts

	PTA Member (standard error)	% Change	Voluntary Activity (standard error)	% Change	Parents Talked To (standard error)	% Change	Trust Teacher (standard error)	% Change
Universal choice	0.35** (.11)	13%	0.21* (.13)	6%	1.24** (.38)	13%	0.28* (.14)	6%
Black	−0.55** (.13)	−21%	−0.48** (.14)	−14%	−3.38** (.44)	−30%	−0.41** (.15)	−9%
Hispanic	−1.24** (.29)	−45%	−0.96** (.26)	−34%	−2.86** (.91)	−12%	0.34 (.38)	6%
Asian	−0.57 (.33)	−22%	0.15 (.39)	4%	−3.49** (1.17)	−11%	0.49 (.55)	8%
Length of residence	−0.01 (.01)	−0.07%	0.02** (.01)	0.6%	0.07*** (.03)	9%	0.01 (.01)	0.02%
Education	0.09** (.02)	3%	0.06** (.02)	2%	0.31*** (.08)	16%	0.03 (.03)	0.5%
Employed	−0.07 (.14)	−3%	−0.06 (.16)	−1%	−0.78* (.47)	−6%	−0.27 (.18)	−5%
Female	0.40** (.13)	15%	0.52** (.14)	16%	1.22*** (.44)	10%	−0.02 (.16)	−0.5%
Attend church	0.09** (.03)	4%	0.06* (.03)	2%	0.24*** (.08)	11%	−0.01 (.03)	−0.01%
Dissatisfaction	−1.76** (.42)	−8%	−0.01 (.14)	−0.1%	0.51 (.41)	6%	−0.73** (.14)	−18%
Constant	−0.92 (.41)		−0.45 (.44)		1.71 (1.4)		1.04 (.49)	
	$N = 629$		$N = 629$		$N = 626$		$N = 622$	
	$\chi^2 = 91$		$\chi^2 = 61$		$F = 14$		$\chi^2 = 43$	
	(.00)		(.00)		(.00)		(.00)	

Note: In the three probit equations the percentage point change figures indicate the effect of a change from 0 to 1 for the dummy variables and represent the effect of a unit change for the nondummy variables. For the regression equation (parents talked to) the percentage changes are calculated from the normalized beta coefficients.

* $p \leq .05$; ** $p \leq .01$; *** $p \leq .001$

Lyons, and DeHoog 1995, Lyons, Lowry, and DeHoog 1992, Teske et al. 1993, 1995). This debate has focused on only a limited aspect of the behavior of the "citizen/consumer" in the market for public goods, revolving around the question of whether competition can enhance the behavior of citizens as consumers. We broaden the question by asking if government policies that enhance choice over public goods can increase the capacity of the citizen/consumer to act as a responsible, involved citizen. Our results show that in the domain we study, local public education, the answer is yes.

According to Putnam, societies can evolve two different equilibria as they solve collective action problems.   One equilibrium is built on a "virtuous circle" that nurtures healthy norms of reciprocity, cooperation, and mutual trust. The other relies on coercion and creates an environment in which only kin can be trusted. Civic engagement is at the core of Putnam's concept of social capital because it breeds cooperation and facilitates coordination in governing. Public schools constitute a domain in which the virtuous circle is essential for improving the quality of education. Hillary Rodham Clinton (1996) has argued that "it takes a village" to raise a child. It may also take a "village" to educate a child: High quality education is dependent on parental involvement supported by high levels of community involvement. In turn, higher quality education is associated with activities that build social capital—a virtuous circle is created.

Our research shows that the design of the institutions delivering local public goods can influence levels of social capital. No present statistical method can fully correct for problems in estimation introduced by the complex causal linkages that motivate our study. Our two-stage modeling, however, clearly addresses the biases introduced by the nonrandom "assignment" of parents as active choosers in New York. Our research shows that in both an urban and a suburban setting and under different institutional settings of choice, the act of school choice seems to stimulate parents to become more involved in a wide range of school-related activities that build social capital. Our results support arguments linking participation and urban democracy and, within the domain of schools that we studied, are directly congruent with Berry, Portney, and Thomson's (1993, 254) claim that "increased participation does lead to greater sense of community, increased governmental legitimacy, and enhanced status of governmental institutions."

Clearly, many factors affecting the formation of social capital are individual-level characteristics effectively beyond the control of government (e.g., social capital increases with church attendance and with length of residence in a community). This fundamentally limits the role that government can play in nurturing the formation of social capital. Despite this, we believe that governmental policies can and do affect the level of social capital. The careful design of governmental institutions may be able to reverse the ratchet that Fukuyama believes has only driven social capital down.

Appendix A: Survey Methodology

We contracted the Polimetrics Research and Survey Laboratory at Ohio State University to carry out the survey. To start, Polimetrics identified the zip codes in each of the four school districts. All listed telephone numbers for each zip code were identified. From this, a list was developed using random generation of the last two digits of the appropriate telephone exchanges, so that unlisted numbers were included as well. All known business telephone numbers were removed as they were not eligible to be interviewed. Then, a random sample was taken of the remaining numbers.

To be eligible to be interviewed, respondents needed to live within the school district, have children between grades K–8, be the adult responsible for decisions affecting that child's education, and identify the school their child attended (which could be either a private school or a district public school).

The actual interviews were conducted from March through June 1995. The interviewers were given extensive training and some interviews were conducted in Spanish. Interviews were monitored randomly and, to ensure validity, 15% of all completed interviews were verified with respondents by the supervisors.

The goal was to obtain 400 completed interviews in each of the four districts. The following table shows the call dispositions in each district.

Table A-1. Disposition of Survey Telephone Calls

	District 4	District 1	Montclair	Morristown
Completed	400	401	408	395
Refusals	113	522	109	174
No final disposition	225	1,642	281	343
Nonhousehold	5,237	17,883	5,268	12,913
Ineligible	5,722	13,469	3,935	5,918

Appendix B: Correcting for Nonrandom Assignment

As Achen (1986) demonstrates, ordinary regression fails to produce unbiased estimates of treatment effects in quasi-experiments when the "assignment" to different conditions is not random (see LaLonde and Maynard 1987; Lord 1967, 1969; Heckman 1978; Heckman, Hotz, and Dabos 1987). Consequently, in addition to specifying the behavioral outcome, we must explicitly model the assignment process. To deal with the dichotomous nature of three of our dependent variables, we apply Achen's generalized two-stage least squares estimator (G2SLS). The steps for this estimation procedure, as well as the standard 2SLS we employ to estimate our continuous outcome equation, are summarized below.

Table B-1. Assignment (First-Stage) Equation: Active Public
School Choosers in New York

	Coefficient	Standard Error
Diversity	.090*	.038
Values	.115**	.037
Length of residence	.005*	.002
Years of schooling	.006	.004
Black	−.252***	.048
Hispanic	−.216***	.045
Asian	−.318**	.116
Employed	.079**	.028
Female	−.003	.043
Attend church	−.003	.006
Constant	.127	.114

$^*p \leq .05$; $^{**}p \leq .01$; $^{***}p \leq .001$
$N = 584$; $F(10, 573) = 10.36$; $p = .000$

The first stage consists of estimating the assignment equation. This can be done in a straightforward manner by applying the linear probability model. Goldberger's (1964) two-step weighted estimator can be employed to correct for the problems of ordinary least squares (OLS) regression with a dichotomous dependent variable. Before calculating the weights, the predicted values outside the 0–1 interval from the OLS regression should be reset to the bounds. It should also be noted that in order for the system of equations to be estimated, at least one variable in the assignment equation must be excluded from the outcome equation. This variable provides the necessary statistical leverage to estimate the system, so its coefficient in the assignment equation must be nonzero. See Table B-1 for the results of the assignment equation.

For the second stage, the forecast values of the treatment variable (the dependent variable from the assignment equation) are inserted into the outcome equation. When the dependent variable in this equation is continuous (as in the case of our "spontaneous sociability" model) ordinary regression can be applied. The resulting coefficients are 2SLS estimates. The only remaining step in the continuous variable case consists of correcting the standard errors of the coefficients. To accomplish this we first denote the variance of the residuals from our OLS regression ω^2. Next we generate a new forecast value for the dependent variable by using the second-stage coefficients and the original variables. We then compute the variance of the new set of residuals, σ^2, by taking the difference between the two equations. The standard errors of the 2SLS coefficients are corrected by multiplying each standard error by the square root of σ^2/ω^2.

If the dependent variable in the outcome equation is dichotomous, as in our three other models, additional steps are necessary. Once again we insert the forecast values of the treatment variable into the outcome equation. After applying OLS to the outcome equation we compute a new forecast value for the dependent variable using the regression coefficients and the original variables. Once again, predicted values outside the 0–1 interval are reset to the bounds. Next we apply Goldberger's two-step weighted estimator to the outcome equation. The coefficients of the final estimation are the 2GSLS estimates, but again, the reported standard errors are wrong. To correct them we first denote the variance of the residuals from the final stage regression as ω^2. We then multiply each standard error by the square root of $1/\omega^2$. We report these corrected coefficients and standard errors in our tables. Note too that once these corrections are implemented the R^2 statistic is no longer meaningful and is not reported for any of our New York models.

References

Achen, Chris. 1986. *The Statistical Analysis of Quasi-Experiments*. Chicago: University of Chicago Press.

Aldrich, John, and Forrest Nelson. 1994. *Linear Probability, Logit and Probit Models*. Newbury Park, CA: Sage Publications.

Almond, Gabriel, and Sidney Verba. 1963. *The Civic Culture*. Princeton: Princeton University Press.

Alvarez, R. Michael, and John Brehm. 1994. "Two-Stage Estimation of Non-Recursive Choice Models." California Institute of Technology, Social Science Working Paper 905. Typescript.

Anderson, Charles. 1990. *Pragmatic Liberalism*. Chicago: University of Chicago Press.

Anson, Amy, Thomas Cook, Farah Habib, Michael Grady, Norris Haynes, and James Comer. 1991. "The Comer School Development Program: A Theoretical Analysis." *Urban Education* 26 (April):56–82.

Astone, Nan Marie, and Sara McLanahan. 1991. "Family Structure, Parental Practices, and High School Completion." *American Sociological Review* 56 (June):309–20.

Barber, Benjamin. 1984. *Strong Democracy: Participatory Politics for a New Age*. Berkeley: University of California Press.

Berry, Jeffrey, Kent Portney, and Ken Thomson. 1993. *The Rebirth of Urban Democracy*. Washington, DC: Brookings Institution.

Blank, Rolf K. 1990. "Educational Effects of Magnet High Schools." In *Choice and Control in American Education.* Vol. 2: *The Practice of Choice, Decentralization and School Restructuring*, ed. William H. Clune and John F. Witte. London: Falmer Press.

Bourdieu, Pierre. 1980. *Questions de Sociologie.* Paris: Minuit.

Boyer, Ernest. 1992. *School Choice.* Princeton, NJ: Carnegie Foundation.

Brehm, John, and Wendy Rahn. N.d. "Individual Level Evidence for the Causes and Consequences of Social Capital." *American Journal of Political Science.* Forthcoming.

Carnoy, Martin. 1993. "School Improvement: Is Privatization the Answer?" In *Decentralization and School Improvement*, ed. Jane Hannaway and Martin Carnoy. San Francisco: Jossey-Bass.

Chubb, John, and Terry Moe. 1990. *Politics, Markets and America's Schools.* Washington, DC: Brookings Institution.

Clinton, Hillary Rodham. 1996. *It Takes a Village: And Other Lessons Children Teach Us.* New York: Simon and Schuster.

Clune, William H., and John F. Witte, eds. 1990. *Choice and Control in American Education.* Vol. 2: *The Practice of Choice, Decentralization and School Restructuring.* London: Falmer Press.

Coleman, James. 1990. *Foundations of Social Theory.* Cambridge, MA: Harvard University Press.

Coleman, James, and Thomas Hoffer. 1987. *Public and Private High Schools Compared.* New York: Basic Books.

Coleman, James, Thomas Hoffer, and Sally Kilgore. 1982. *High School Achievement: Public, Catholic, and Private Schools Compared.* New York: Basic Books.

Coleman, James. 1988. "Social Capital in the Creation of Human Capital." *American Journal of Sociology* 94 (Supplement):S95–120.

Cookson, Peter W., Jr. 1994. *School Choice: The Struggle for the Soul of American Education.* New Haven, CT: Yale University Press.

Coons, John E., and Stephen D. Sugarman. 1978. *Education by Choice: The Case for Family Control.* Berkeley: University of California Press.

Diamond, Larry. 1992. "Economic Development and Democracy Reconsidered." In *Reexamining Democracy*, ed. Gary Marks and Larry Diamond. New York: Sage.

Elmore, Richard F. 1991. "Public School Choice as a Policy Issue." In *Privatization and Its Alternatives*, ed. William T. Gormley, Jr. Madison: University of Wisconsin Press.

Fantini, Mario D. 1973. *Public Schools of Choice.* New York: Simon and Schuster.

Fliegel, Seymour. 1990. "Creative Non-Compliance." In *Choice and Control in American Education,* Vol. 2: *The Practice of Choice, Decentralization and*

School Restructuring, ed. William H. Clune and John F. Witte. New York: Falmer Press.

Friedman, Milton. 1955. "The role of government in education." In *Economics and the Public Interest*, ed. R. A. Solo. New Brunswick, NJ: Rutgers University Press.

Friedman, Milton. 1962. *Capitalism and Freedom*. Chicago: The University of Chicago.

Fukuyama, Francis. 1995. *Trust: Social Virtues and the Creation of Prosperity*. New York: Free Press.

Goldberger, Arnold. 1964. *Econometric Theory*. New York: John Wiley.

Granato, Jim, Ronald Inglehart, and David Leblang. 1996a. "The Effect of Cultural Values on Economic Development: Theory, Hypotheses, and Some Empirical Tests." *American Journal of Political Science* 40 (August):607–31.

Granato, Jim, Ronald Inglehart, and David Leblang. 1996b. "Cultural Values, Stable Democracy, and Economic Development: A Reply." *American Journal of Political Science* 40 (August):680–96.

Handler, Joel. 1996. *Down From Bureaucracy: The Ambiguity of Privatization and Empowerment*. Princeton: Princeton University Press.

Harrington, Diane, and Peter Cookson, Jr. 1992. "School Reform in East Harlem: Alternative Schools versus Schools of Choice." In *Empowering Teachers and Parents*, ed. G. Alfred Hess. Westport, CT: Bergin and Garvey.

Heckman, James. 1978. "Dummy Endogenous Variables in a Simultaneous Equation System." *Econometrica* 46 (July):931–59.

Heckman, James J., V. Joseph Hotz, and Marcelo Dabos. 1987. "Do We Need Experimental Data to Evaluate the Impact of Manpower Training on Earnings?" *Evaluation Review* 11 (August):395–427.

Henig, Jeffrey. 1994. *Rethinking School Choice: Limits of the Market Metaphor*. Princeton: Princeton University Press.

Inglehart, Ronald. 1990. *Culture Shift in Advanced Industrial Society*. Princeton: Princeton University Press.

Jackman, Robert, and Ross Miller. 1996a. "A Renaissance of Political Culture?" *American Journal of Political Science* 40 (August):632–57.

Jackman, Robert, and Ross Miller. 1996b. "The Poverty of Political Culture?" *American Journal of Political Science* 40 (August):697–716.

Jencks, Christopher. 1966. "Is the Public School Obsolete?" *The Public Interest* 2 (Winter):18–27.

Kerbow, David, and Annette Bernhardt. 1993. "Parental Intervention in the School: The Context of Minority Involvement." In *Parents, Their Children and School*, ed. Barbara Schneider and James S. Coleman. Boulder, CO: Westview Press.

LaLonde, Robert, and Rebecca Maynard. 1987. "How Precise are Evaluations of Employment and Training Programs: Evidence from a Field Experiment." *Evaluation Review* 11 (August):428–51.

Lee, Dwight R. 1991. "Vouchers—The Key to Meaningful Reform." In *Privatization and Its Alternatives*, ed. William T. Gormley, Jr. Madison: University of Wisconsin Press.

Lee, Seh-Ahn. 1993. "Family Structure Effects on Student Outcomes." In *Parents, Their Children and Schools*, ed. Barbara Schneider and James S. Coleman. Boulder, CO: Westview Press.

Lipset, Seymour Martin. 1995. "Malaise and Resiliency in America." *Journal of Democracy* (January):2–16.

Lord, Frederic M. 1967. "A Paradox in the Interpretation of Group Comparisons." *Psychological Bulletin* 68 (November):304–5.

Lord, Frederic M. 1969. "Statistical Adjustments When Comparing Preexisting Groups." *Psychological Bulletin* 72 (November):336–7.

Loury, Glenn. 1977. "A Dynamic Theory of Racial Income Differences." In *Women, Minorities, and Employment Discrimination*, ed. P.A. Wallace and A. LeMund. Lexington, MA: Lexington Books.

Lyons, William, David Lowery, and Ruth Hoogland DeHoog. 1992. *The Politics of Dissatisfaction*. Armonk, NY: Sharpe.

Lowery, David, William Lyons, and Ruth Hoogland DeHoog. 1995. "The Empirical Evidence for Citizen Information and a Local Market for Public Goods." *American Political Science Review* 89 (September):705–9.

Mansbridge, Jane. 1980. *Beyond Adversary Democracy*. New York: Basic Books.

Muller, Chandra, and David Kerbow. 1993. "Parental Involvement in the Home, School and Community." In *Parents, Their Children and Schools*, ed. Barbara Schneider and James S. Coleman. Boulder, CO: Westview Press.

Ostrom, Elinor. 1972. "Metropolitan Reform: Propositions Derived from Two Traditions." *Social Science Quarterly* 53 (December): 474–93.

Putnam, Robert. 1993. *Making Democracy Work: Civic Traditions in Modern Italy*. Princeton: Princeton University Press.

Putnam, Robert. 1995a. "Bowling Alonc: America's Declining Social Capital." *Journal of Democracy* 6 (January):65–78.

Putnam, Robert. 1995b. "Tuning in, Tuning out: The Strange Disappearance of Social Capital in America." *PS: Political Science and Politics*, 28 (December): 664–83.

Ravitch, Diane. 1994. "Somebody's Children: Expanding Educational Opportunities for All America's Children." *Brookings Review* (Fall)4–9.

Schneider, Barbara, and James Coleman, eds. 1993. *Parents, Their Children and Schools*. Boulder, CO: Westview Press.

Schneider, Mark. 1989. *The Competitive City*. Pittsburgh: University of Pittsburgh Press.

Schneider, Mark, and Paul Teske. 1995. *Public Entrepreneurs: Agents for Change in American Government*. Princeton: Princeton University Press.

School District Data Book v. 1.0 [CD-ROM]. 1995. U.S. Department of Education, National Center for Education Statistics and The MESA Group.

Smith, Kevin, and Kenneth Meier. 1995. *The Case against School Choice: Politics, Markets, and Fools*. Armonk, NY: M. E. Sharpe.

Stone, Clarence. 1996. "The Politics of Urban School Reform: Civic Capacity, Social Capital, and Intergroup Context." Presented at the annual meeting of the American Political Science Association, San Francisco, August 29–September 1.

Strobert, Barbara. 1991. *Factors Influencing Parental Choice in Selection of a Magnet School in the Montclair, New Jersey Public Schools*. Ed.D. diss. Columbia University Teachers College.

Swank, Duane. 1996. "Culture, Institutions, and Economic Growth: Theory, Recent Evidence, and the Role of Communitarian Polities." *American Journal of Political Science* 40 (August):660–79.

Tarrow, Sidney. 1996. "Making Social Science Work across Space and Time: A Critical Reflection on Robert Putnam's Making Democracy Work." *American Political Science Review* 90 (June):389–97.

Teske, Paul, Mark Schneider, Michael Mintrom, and Samuel Best. 1993. "Establishing the Micro Foundations of a Macro Theory: Information, Movers, and the Competitive Local Market for Public Goods." *American Political Science Review* 87 (September):702–13.

Teske, Paul, Mark Schneider, Michael Mintrom, and Samuel Best. 1995. "The Empirical Evidence for Citizen Information and a Local Market for Public Goods." *American Political Science Review* 89 (September):705–9.

Tiebout, Charles. 1956. "A Pure Theory of Local Expenditure." *Journal of Political Economy* 64 (October):416–24.

Verba, Sidney, Kay Lehman Schlozman, and Henry Brady. 1995. *Voice and Equality: Civic Voluntarism in American Politics*. Cambridge: Harvard University Press.

Wells, Amy Stuart. 1993. *Time to Choose: America at the Crossroads of School Choice Policy*. New York: Hill and Wang.

Wilson, William Julius. 1987. *The Truly Disadvantaged: The Inner City, the Underclass, and Public Policy*. Chicago: University of Chicago Press.

Witte, John F. 1991. "The Milwaukee Parental Choice Program." In *School Choice: Examining the Evidence*, ed. Edith Rasell and Richard Rothstein. Washington, DC: Economic Policy Institute.

Witte, John F., and Mark E. Rigdon. 1993. "Education Choice Reforms: Will They Change American Schools?" *Publius: The Journal of Federalism* 23 (Summer):95–114.

Wohlstetter, Priscilla, Richard Wenning, and Kerri L. Briggs. 1995. "Charter Schools in the United States: The Question of Autonomy." Working paper, University of Southern California's Center on Educational Governance. Typescript.

Notes

1. The current debate in political science is focused on somewhat different issues than we address here. However, our research *is* directly relevant to one central theme of that debate—the role of government in creating social capital. In critiquing what he sees as a critical omission by Putnam (1993), Tarrow (1996, 395) asks: "Can we be satisfied interpreting civic capacity as a home-grown product in which the state has no role?" Similarly, Jackman and Miller (1996a, 655) argue that a political institutional approach that endogenizes civic culture can help explain differential political and economic development.

2. Classic theoretical treatments include: Chubb and Moe 1990; Coons and Sugarman 1978; Fantini 1973; Friedman 1955, 1962; Jencks 1966. For reviews of school choice in practice, see Cookson 1994; Clune and Witte 1990; and Wells 1993.

3. These policies include publicly provided vouchers that can be used in a variety of schools, both public and private (see, e.g., Lee 1991), the introduction of magnet schools (see, e.g., Blank 1990), the introduction of charter schools (see, e.g., Wohlstetter, Wenning, and Briggs 1995), and public school choice plans such as those we analyze here.

4. In 1993, the New York City Board of Education established a new policy of *interdistrict* choice. If space is available (usually it is not), students can go to schools outside of their district. The Board did not mandate choice programs within districts.

5. Recall that high schools in New York are run by the central Board of Education.

6. The table in Appendix A shows that telephone interviewers had greater difficulty completing interviews in District 1 than District 4; however, as evident in Table 1 our samples of public school parents are nonetheless representative of the population of the districts as a whole.

7. We recognize a limitation inherent in the cross-sectional nature of our research design. Ideally, research on changes in social capital would employ a longitudinal, interrupted time-series analysis, involving panel responses. In this ideal research design, data would be collected prior to institutional changes and, by interviewing the same subjects over time, researchers could isolate the specific effect of institutional changes. Unfortunately, few researchers had the foresight or the resources to conduct such a study; trade-offs must inevitably be made. For example, Putnam (1993) used aggregate level and (some would say) problematic measures of social capital (see, e.g., Jackman and Miller 1996a) and went beyond his data to explore historical differences in the development of Italian regions. The trade-off in our case is that while we can not gather detailed individual-level data on parents in these

districts before they chose a school, we do have detailed individual measures today that our cross-sectional design allows us to test while controlling for individual-level demographic and socioeconomic factors. With replication across four different institutional settings, our quasi-experimental design provides a strong cross-sectional test of the causal relationships postulated in the existing social capital literature.

8. Participation in the PTA and in voluntary activities is a dichotomous variable, with 1 indicating membership in the PTA (52% report membership) or voluntary activity (66% report such activity). As Verba, Schlozman, and Brady (1995, 74–9) note and our data confirm, levels of voluntary activity in social organizations are considerably higher in America than is participation in electoral activities. The number of parents a respondent reported talking with is a continuous variable based on the midpoints of categories presented (mean = 4.5; s.d. = 4.6). Trust in teachers is operationalized as a dichotomous variable (1 = trusts teachers most of the time or always [77% report this level of trust]; 0 = never or only sometimes).

9. While it is also plausible that there could be a two-way or reciprocal relationship between social capital and school choice, the timing of our research design makes this unlikely: Parents made their school choice in spring 1994. They were not interviewed until spring 1995, during which time they answered questions about activities during the previous school year. Thus, they chose first and engaged in the activities we measured later.

10. Smith and Meier find that religion and race help explain why some parents choose private schools for their children (1995, 71–2). Our values and diversity variables for the public schools are closely related to these concepts. Alternative schools in New York tend to emphasize themes and pedagogical approaches that are based on particular social, educational, or civic values. Diversity has a somewhat different meaning in districts where two-thirds of the children are Hispanic.

11. To estimate two stage models there must be at least one exclusion in the assignment equation. In other words, we must find at least one variable that significantly influences assignment but not the outcome (Achen 1986, 38). We use these two variables, diversity and values, as exclusions.

12. Our specific measure, indicating whether or not the parent has often thought about moving her child to another school, is a dummy variable coded 1 = yes, the parent has thought about moving her child to a different school. We expect a negative relationship between this measure and our measures of involvement in the schools.

13. While the two-stage results are the technically correct ones, we should also note that these findings are robust with a simpler methodology. Using a one-stage model, the results are essentially the same.

14. We should also note that, for both urban and suburban districts, parents who chose to send their children to private schools are significantly more likely to engage in all of these social capital building activities than public school parents and more so than even active public choosers, with the exception of PTA involvement. This result is not surprising, and has been documented in the literature on private schools.

Index